MATHEMATICAL
METHODS in PHYSICS
and ENGINEERING with
MATHEMATICA

CHAPMAN & HALL/CRC APPLIED MATHEMATICS
AND NONLINEAR SCIENCE SERIES
Series Editors   *Goong Chen and Thomas J. Bridges*

Published Titles

Mathematical Methods in Physics and Engineering with Mathematica,
   Ferdinand F. Cap

Forthcoming Titles

An Introduction to Partial Differential Equations with MATLAB,
   Matthew P. Coleman
Mathematical Theory of Quantum Computation, Goong Chen and Zijian Diao
Optimal Estimation of Dynamic Systems, John L. Crassidis and John L. Junkins

CHAPMAN & HALL/CRC APPLIED MATHEMATICS
AND NONLINEAR SCIENCE SERIES

MATHEMATICAL METHODS in PHYSICS and ENGINEERING with *MATHEMATICA*

Ferdinand F. Cap

CRC Press
Taylor & Francis Group
Boca Raton London New York

CRC Press is an imprint of the
Taylor & Francis Group, an **informa** business

A CHAPMAN & HALL BOOK

CRC Press
Taylor & Francis Group
6000 Broken Sound Parkway NW, Suite 300
Boca Raton, FL 33487-2742

First issued in paperback 2019

ISBN-13: 978-1-58488-402-6 (hbk)
ISBN-13: 978-0-367-39518-6 (pbk)
Library of Congress Card Number 2003043974

Library of Congress Cataloging-in-Publication Data

Cap, Ferdinand
    Mathematical methods in physics and engineering with Mathematica / Ferdinand F. Cap.
        p. cm.
    Includes bibliographical references and index.
    ISBN 1-58488-402-9 (alk. paper)
    1. Mathematical physics--Data processing. 2. Engineering--Data processing. 3.
    Mathematica (Computer program language) 4. Mathematica (Computer file) I. Title.
QC20.6.C36 2003
530.15'0285—dc21   2003043974

Visit the Taylor & Francis Web site at
http://www.taylorandfrancis.com

and the CRC Press Web site at
http://www.crcpress.com

Preface

Nowadays, not only tackling the phenomena of physics, but also many issues in technology, engineering and production, oftentimes require complicated mathematical processing. Recently, a CEO of an industrial plant revealed that, through application of a boundary value problem of elastic stress, he was able to save 30 percent in material. It seems that close cooperation of science with technology and production is more relevant today than ever. Many processes of production and technology can be simulated through differential equations. Knowledge of these mathematical procedures and their applications is therefore not only essential in theory for the student, but also vital for the practitioner, the engineer, physicist or technologist.

Grounded on 40 years of teaching experience in classical and applied physics, the author directs this book to the practitioner, who finds him/herself confronted with boundary value problems. The book does not contain mathematical theorems or numerical methods such as boundary elements or finite elements. Many of the processes contained in this book originate from engineering, where they have a history of successful application. Extensive references and a comprehensive bibliography are given.

The book will help to solve ordinary and partial differential equations using the program package Mathematica 4.1. It is not a textbook for this cutting-edge procedure of resolving mathematical issues; however, it is also written for the reader who has no prior knowledge of Mathematica and will support him/her in "learning by doing." All Mathematica commands will be explained in detail at their first occurrence and in the appendix. The book contains many step-by-step recipes for different classical and brand-new applications. It has not only computer programs (Fortran and Mathematica), but also countless problems to solve, including useful hints and the solutions. It also demonstrates how Mathematica can help in solving ordinary and partial differential equations and the related boundary value problems.

Which problems are discussed? What is new in this book? The reader will learn in many examples that in areas where one would not expect it, these mathematical procedures are used. For example, which boundary problem describes the diffusion of a perfume? Which predicts the ripening time and optimal taste for certain cheeses? Has the boundary problem of the movements of sperm cells relevance for the determination of the sex of an embryo? Which methods does one apply for the cooling of radioactive wastes in deposits or to determine the critical mass of a nuclear bomb? How does an electrostatic parametric high-voltage generator work and can the electromagnetic pulse EMP within the antimissile program be estimated? How can mathematics be a valuable tool in the prediction of the spread of infectious diseases? How much power has the irradiation of cellular phones?

The book discusses traditional knowledge in the field as well as brand-new methods and procedures, such as using the LIE series method to solve differential equations. A new method to calculate zeros of Bessel functions is introduced; also, a collocation method that allows the solving of elliptic differential equations with two different boundaries as well as problems of inseparable type and arbitrary shape of the domain, also including corners along the boundary. The reader is invited to view the index and contents sections of this book for further details and an overview.

I thank my colleague Firneis from the Austrian Academy of Sciences for critical review of the manuscript and for many useful suggestions. I also thank my wife Theresia for providing the typed version of this book ready for print. With endless patience, interest and engagement she brought the countless different versions of the often poorly handwritten manuscripts into professional format using the computer program LaTeX. Thanks also go to CRC Press, especially Bob Stern, Helena Redshaw, Sylvia Wood and Debbie Vescio for valuable advice and the fast completion of the book.

Innsbruck, Austria, February 2003.

Contents

1

Introduction

1.1 What is a boundary problem?

An equation containing two variables at most, and derivatives of the first or higher order of one of the variables with respect to the other is called a *differential equation*. The *order n* of such an ordinary differential equation is the order of the highest derivative that appears. To solve a differential equation of n-th order, n integrations are necessary. Each integration delivers an integration constant. These arbitrary constants can be used to adapt the general solution to the particular solution of the problem in question.

Many differential equations represent mathematical models describing a physical or technical problem. The free fall of a parachutist is described by the equation of motion

$$m\frac{\mathrm{d}^2 x(t)}{\mathrm{d}t^2} = -mg + a\left(\frac{\mathrm{d}x}{\mathrm{d}t}\right)^2. \tag{1.1.1}$$

Here m is the mass of the parachutist, $-mg$ is the gravitational attraction and the last term on the right-hand side describes aerodynamic drag. By defining the velocity v of fall

$$v(t) = \frac{\mathrm{d}x(t)}{\mathrm{d}t} \equiv \dot{x}, \tag{1.1.2}$$

the equation (1.1.1), which is of second order, can be reduced to a differential equation of first order

$$\frac{\mathrm{d}v(t)}{\mathrm{d}t} = -g + \frac{a}{m}v^2(t). \tag{1.1.3}$$

This equation is of the form

$$P(t)\mathrm{d}t + Q(v)\mathrm{d}v = 0 \tag{1.1.4}$$

and is called differential equation with variables separable (*separable equation*) since (1.1.3) can be written in the form

$$\mathrm{d}t = \frac{\mathrm{d}v}{-g + av^2/m}. \tag{1.1.5}$$

Integration yields

$$t = -\frac{\sqrt{m}}{2\sqrt{ag}} \ln C \frac{av/m - \sqrt{ag/m}}{av/m + \sqrt{ag/m}}. \tag{1.1.6}$$

This integration has been executed using the *Mathematica* command
Integrate[1/(a*v^2/m-g),v]
and observe $\ln((1 + x)/(1 - x))/2 = \operatorname{arctanh}(x)$.

   If we now assume the *initial condition* $v(0) = 0 \equiv v_0$, then we obtain the integration constant $C = -1$ from (1.1.6). Thus, our solution of (1.1.5) for $v(t)$ reads

$$\frac{av/m + \sqrt{ag/m}}{av/m - \sqrt{ag/m}} = \exp\left(-\frac{2\sqrt{ag}}{\sqrt{m}}t\right). \tag{1.1.7}$$

For $t \to \infty$ this gives the final *falling speed*

$$v(\infty) = -\sqrt{gm/a} \equiv v_\infty. \tag{1.1.8}$$

Inserting (1.1.2) into (1.1.7) yields a differential equation for $x(t)$ and a second integration constant after integration. Instead of solving (1.1.1) (1.1.7) for $x(t)$, we may derive a differential equation for $x(v)$. Substituting dt from (1.1.5) into (1.1.2) we have

$$dx = -v\frac{dv}{g - av^2/m} \tag{1.1.9}$$

and after integration (using the substitution $u = v^2$) one has

$$x + C = \frac{1}{2}\frac{m}{a} \ln\left(v_\infty^2 - v^2\right). \tag{1.1.10}$$

In order to determine the second integration constant C, let us consider that the parachutist leaves the airplane at time $t = 0$ with zero velocity v_0 at a certain height $x_0 \neq 0$ above the earth's surface $x = 0$. Thus, we have

$$x_0 + C = \frac{1}{2}\frac{m}{a} \ln v_\infty^2 \tag{1.1.11}$$

as the second initial condition and the solution of (1.1.9) reads

$$x = x_0 + \frac{m}{2a} \ln\left(1 - \frac{v^2}{v_\infty^2}\right). \tag{1.1.12}$$

In this example (1.1.1) of a differential equation of second order, we used two initial conditions, i.e., two conditions imposed on the solution at the *same* location $x = x_0$. We could, however, impose two conditions at different locations. We then speak of a *two-point condition* or a *boundary condition*.

First, we keep the initial condition $v(0) = 0$, so that (1.1.10) is still valid. Then we can assume that the parachutist should contact the earth's surface $x = 0$ with a velocity v of fall equivalent to one tenth of v_∞. We then have the final condition (*boundary condition*) that for $x = 0$, $v \to v_\infty/10$ or from (1.1.10)

$$C = \frac{1}{2}\frac{m}{a}\ln\frac{99}{100}v_\infty^2. \tag{1.1.13}$$

Then the *general solution* (1.1.10) becomes the *particular solution*

$$x = \frac{1}{2}\frac{m}{a}\ln\left(\frac{100}{99} - \frac{v^2}{v_\infty^2}\right). \tag{1.1.14}$$

If we solve (1.1.1) using the command

`DSolve[m*x''[t]+m*g-a*(x'[t])^2==0, x[t],t]`

we obtain

$$x(t) = C_2 - \frac{m}{a}\ln\left[\cosh\left(\frac{\sqrt{g}(-at + mC_1)}{\sqrt{am}}\right)\right]. \tag{1.1.15}$$

Here C_1 and C_2 are integration constants. Using the two initial conditions $v(0) = 0$ and $x(0) = x_0$, one obtains the particular solution

$$x(t) = x_0 - \frac{m}{a}\ln\left[\cosh\frac{t\sqrt{ga}}{\sqrt{m}}\right]. \tag{1.1.16}$$

For $m = 70$ kg, $a = 0.08$, $x_0 = 5000$ m and $g = 9.81$ m/s^2, this function is shown in Figure 1.1. This figure has been produced with the help of the commands

```
m=70; a=0.08; x0=5000; g=9.81;
x[t_]=x0-m*Log[Cosh[t*Sqrt[g*a]/Sqrt[m]]]/a;
Plot[x[t],{t,0,60}]
```

In this expression we have used a semicolon. This allows more than one command to be written in one line. The command to plot a function needs a function that can be evaluated at any arbitrary value of the independent variable t. To obtain such a variable we replaced t by $t\_$.

We now treat an example to find the force producing a given trajectory. Let us consider the relativistic one-dimensional equation of motion of an astronaut traveling into space along a straight line and returning to earth. Neglecting gravitation, such a motion is described by three equations [1.1]*

relativistic equation of motion

$$\frac{d}{dt}\frac{m_{01}v_1}{\sqrt{1 - v_1^2/c^2}} = \frac{dm_{02}}{dt}\frac{v_2}{\sqrt{1 - v_2^2/c^2}}, \tag{1.1.17}$$

---

*Numbers in brackets designate bibliographical references; see the list at the end of the book.

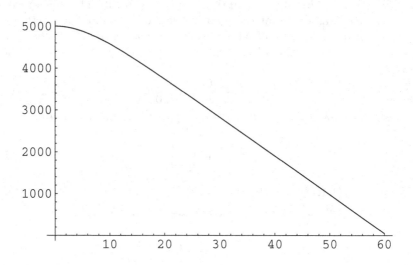

Figure 1.1
Trajectory of a parachutist

conservation of energy

$$\frac{\mathrm{d}}{\mathrm{d}t}\frac{m_{01}c^2}{\sqrt{1-v_1^2/c^2}} + \frac{\mathrm{d}m_{02}}{\mathrm{d}t}\frac{c^2}{\sqrt{1-v_2^2/c^2}} = 0, \tag{1.1.18}$$

addition theorem of velocities

$$v_2 = \frac{w - v_1}{1 - v_1 w/c^2}. \tag{1.1.19}$$

In these equations we used the following designations: m_{01} and m_{02} are the rest masses of the space rocket and the exhausted gases, respectively, v_1 and v_2 are the respective velocities and w is the exhaust velocity relative to the space ship. The repulsive force, which we shall later designate by F, is represented by the rhs term of (1.1.17). The quantities m_{01}, m_{02}, v_1 and v_2 are functions of time t. c is the constant velocity of light.

We now turn to the integration of the equation of motion (1.1.17). Differentiation delivers

$$A_1 \equiv \mathrm{d}m_{01}\frac{v_1}{\sqrt{1-v_1^2/c^2}} + m_{01}\frac{\mathrm{d}v_1}{\sqrt{1-v_1^2/c^2}} + m_{01}v_1\mathrm{d}\left(\frac{1}{\sqrt{1-v_1^2/c^2}}\right) =$$

$$= \mathrm{d}m_{02}\frac{v_2}{\sqrt{1-v_2^2/c^2}}, \tag{1.1.20}$$

whereas the energy theorem (1.1.18) results in

$$A_2 \equiv \mathrm{d}m_{01} \frac{c^2}{\sqrt{1 - v_1^2/c^2}} + m_{01}c^2 \mathrm{d}\left(\frac{1}{\sqrt{1 - v_1^2/c^2}}\right)$$

$$= -\mathrm{d}m_{02} \frac{c^2}{\sqrt{1 - v_2^2/c^2}}. \tag{1.1.21}$$

Here A_1 and A_2 are abbreviations for the l.h.s. terms. Using now the addition theorem (1.1.19), we build the expressions

$$B_1 \equiv \frac{1}{\sqrt{1 - v_2^2/c^2}} = \frac{1 - v_1 w/c^2}{\sqrt{(1 - w^2/c^2}\sqrt{1 - v_1^2/c^2}} \tag{1.1.22}$$

and

$$B_2 \equiv \frac{v_2}{\sqrt{1 - v_2^2/c^2}} = \frac{w - v_1}{\sqrt{(1 - w^2/c^2}\sqrt{1 - v_1^2/c^2}}. \tag{1.1.23}$$

These abbreviations allow the rewriting of the equations (1.1.20) and (1.1.21) in the simple forms

$$A_1 = \mathrm{d}m_{02} \cdot B_2 \quad \text{and} \quad A_2 = -\mathrm{d}m_{02} \cdot c^2 B_1. \tag{1.1.24}$$

This allows the elimination of $\mathrm{d}m_{02}$. The resulting equation is $A_2 B_2 = -c^2 B_1 A_1$. The four abbreviations only contain the time-dependent variables $v_1(t)$ and $m_{01}(t)$. After some straightforward algebra the resulting equation takes the form

$$\frac{\mathrm{d}m_{01} \cdot w}{\sqrt{1 - v_1^2/c^2}}(c^2 - v_1^2) + \frac{m_{01}\mathrm{d}v_1}{\sqrt{1 - v_1^2/c^2}}(c^2 - v_1^2)$$

$$+ m_{01}w(c^2 - v_1^2) \cdot \mathrm{d}\left(\frac{1}{\sqrt{1 - v_1^2/c^2}}\right) = 0 \tag{1.1.25}$$

or simplified

$$m_{01}(t)\frac{\mathrm{d}v_1(t)}{\mathrm{d}t} = -\frac{\mathrm{d}m_{01}(t)}{\mathrm{d}t} \cdot w(t) \cdot \left(1 - \frac{v_1^2(t)}{c^2}\right) \equiv F. \tag{1.1.26}$$

When treating the motion of the parachutist, we assumed a given force and we derived his motion. Now we assume a given trajectory $x_1(t)$ of the space ship and we are determining from (1.1.26) the repulsive force F producing this trajectory. During space travel, the space ship leaving earth will first accelerate, then slow down to zero velocity, make a 180-degree turn and accelerate again to return to earth. Slowing down is again necessary prior to landing. Since always $\mathrm{d}m_{02}/\mathrm{d}t < 0$, the slowing down force F must be realized by a change of sign of the exhaust speed w. Therefore, w will change its sign and must be a function of time. Since we want to calculate the driving force F from a given trajectory $x_1(t)$, we make no setup for $w(t)$ or $m_{01}(t)$, but only for $x_1(t)$. Let τ be the total duration of the trip into space, then we can assume

$$x_1(t) = x_0 \sin \frac{\pi t}{\tau}, \tag{1.1.27}$$

$$v_1(t) = x_0 \frac{\pi}{\tau} \cos \frac{\pi t}{\tau} = v_0 \cos \frac{\pi t}{\tau}. \tag{1.1.28}$$

Here x_0 is the largest distance from earth ($x_1 = 0$). The trajectory satisfies the following conditions: $x_1(0) = 0$ (surface of the earth), $x_1(\tau/2) = x_0$ (maximum distance travelled), $v_1(0) = v_0$ (initial speed) and $v_1(\tau) = -v_0$ (landing speed). Inserting (1.1.28) into (1.1.26) we obtain for $v_0 = fc$, $f < 1$

$$-\frac{1}{m_{01}(t)} \frac{dm_{01}(t)}{dt} w(t) = \frac{-cf \dfrac{\pi}{\tau} \sin \dfrac{\pi t}{\tau}}{1 - f^2 \cos^2 \dfrac{\pi t}{\tau}} \equiv G(t). \tag{1.1.29}$$

There seem to be several possibilities for the choice of $m_{01}(t)$ and $w(t)$. Whereas $w(t)$ is quite arbitrary, the function $m_{01}(t)$ has to satisfy the two boundary conditions $m_{01}(0) = M_s$, $m_{01}(\tau) = M_f$, where M_s and M_f are the start rest mass of the space rocket and M_f the landing rest mass, respectively.

Equations involving more than one independent variable and partial derivatives with respect to these variables are called *partial differential equations*. The order of a partial differential equation is the order of the partial derivative of highest order that occurs in the equation. The problem of finding a solution to a given partial differential equation that will meet certain specified requirements for a given set of values of the independent variables (*boundary conditions*) constitutes now a *boundary value problem*. The given set of values is then given on a (two-dimensional) curve or on a (three-dimensional) surface.

If the values given by the boundary condition are all zero, the *boundary condition* is called *homogeneous*. If the values do not vanish, the *boundary condition* is called *inhomogeneous*.

For instance, the vibrations of a *membrane* are described in cartesian coordinates by the so-called (two-dimensional) *wave equation*

$$\Delta u(x, y, t) = \frac{1}{c^2} \frac{\partial^2 u}{\partial t^2}, \tag{1.1.30}$$

where the Laplacian is given by

$$\Delta u = \frac{\partial^2 u}{\partial x^2} + \frac{\partial^2 u}{\partial y^2} \equiv u_{xx} + u_{yy} \tag{1.1.31}$$

and c is a constant speed describing the material properties of the membrane. Trying to separate the independent variable t, we use the setup

$$u(x, y, t) = v(x, y) \cdot T(t). \tag{1.1.32}$$

Inserting into (1.1.31), we obtain after division by u

$$\frac{\Delta v}{v} = \frac{1}{c^2} \frac{1}{T} \frac{\mathrm{d}^2 T}{\mathrm{d}t^2} \equiv \omega^2/c^2. \tag{1.1.33}$$

Since both sides of this equation depend on different independent variables, the left-hand side and the right-hand side must be equal to the same constant that we called $-\omega^2/c^2$. We thus have to solve the ordinary differential equation

$$\frac{\mathrm{d}^2 T}{\mathrm{d}t^2} + \omega^2 T = 0 \tag{1.1.34}$$

(together with some initial or two point condition) and the partial differential equation

$$v_{xx} + v_{yy} + k^2 v = 0, \tag{1.1.35}$$

where we used $k^2 = \omega^2/c^2$. The dimension of k is given by $[\mathrm{m}^{-1}]$. Equation (1.1.35) is usually called the HELMHOLTZ *equation*.

One may look for the *general solution* of (1.1.35) or one may be interested in the vibrations of a rectangular membrane clamped along its boundary. Considering a rectangle as shown in Figure 1.2, the *homogeneous boundary conditions* for a membrane clamped at the boundaries read

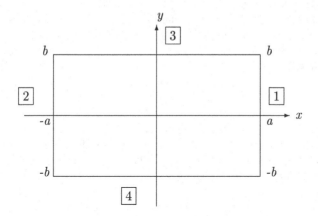

Figure 1.2
Rectangular boundary

$$
\begin{array}{llll}
\boxed{1} & v(a, y) = 0, & \text{for} & -b \le y \le +b, \\
\boxed{2} & v(-a, y) = 0, & \text{for} & -b \le y \le +b, \\
\boxed{3} & v(x, b) = 0, & \text{for} & -a \le x \le +a, \\
\boxed{4} & v(x, -b) = 0, & \text{for} & -a \le x \le +a.
\end{array} \tag{1.1.36}
$$

To find a solution to (1.1.35) we make the ansatz

$$v(x, y) = X(x) \cdot Y(y). \tag{1.1.37}$$

Inserting into (1.1.35) we obtain

$$\frac{X''}{X} = -\frac{Y''}{Y} - k^2. \tag{1.1.38}$$

Since both second derivatives divided by X and Y, respectively, must be constants like $-\alpha^2$ or $-\beta^2$, we now have the following choice:

$$X'' + \alpha^2 X = 0, \tag{1.1.39}$$

$$Y'' + \beta^2 Y = 0, \tag{1.1.40}$$

$$-\alpha^2 - \beta^2 + k^2 = 0, \tag{1.1.41}$$

either a)
$$Y'' + \beta^2 Y = 0,$$

$$X'' + (k^2 - \beta^2)X = 0, \tag{1.1.42}$$

or b)
$$Y'' - \beta^2 Y = 0,$$

$$X'' + (k^2 + \beta^2)X = 0, \tag{1.1.43}$$

which we will discuss later in detail. The constants α, β are usually called *separation constants*. They are dependent on the boundary conditions.

From Figure 1.2 one can see that a solution satisfying the boundary conditions (1.1.36) will be symmetric in both independent variables x and y. We hence take the solutions

$$X(x) = A \cos \alpha x, \tag{1.1.44}$$

$$Y(y) = B \cos \beta y, \tag{1.1.45}$$

where A and B are integration constants (*partial amplitudes*).

If the solutions (1.1.44), (1.1.45) are used, the boundary conditions (1.1.36) determine the separation constants α, β. This is the usual method. From (1.1.36) and (1.1.44), (1.1.45) we have

$$v(x = \pm a, y) = A \cos \alpha a \cdot B \cos \beta y = 0,$$

$$v(x, y = \pm b) = A \cos \alpha x \cdot B \cos \beta b = 0. \tag{1.1.46}$$

These boundary conditions are satisfied by

$$\alpha a = \frac{\pi}{2}(2m - 1), \quad \beta b = \frac{\pi}{2}(2n - 1), \quad m, n = 1, 2 \dots. \tag{1.1.47}$$

Since there appear several (even an infinite number of) solutions for the separation constants, we should write

$$\alpha_m = \frac{\pi}{2a}(2m - 1), \quad \beta_n = \frac{\pi}{2b}(2n - 1). \tag{1.1.48}$$

The separation constants are *here* determined by the dimensions a, b of the membrane. From (1.1.41) we then have

$$k_{mn}^2 = \alpha_m^2 + \beta_n^2. \tag{1.1.49}$$

Apparently our boundary value problem (1.1.35) with (1.1.36) has solutions only for special discrete values of k. If a differential equation like (1.1.35) or (1.1.39) has solutions only for special values of a parameter, the problem is called an *eigenvalue problem*, k is the *eigenvalue* and the solutions $X_m(x) = A_m \cos(\alpha_m x)$ are called *eigenfunctions*. In contrast to this, solutions of (1.1.42), (1.1.43) will be called *modes* or *particular solutions*.

Let us now consider (1.1.34). One possible solution is

$$T(t) = \cos \omega t. \tag{1.1.50}$$

Due to (1.1.49) and the definition used for k in (1.1.35) the constant ω (*angular frequency*) has the dimension $[\sec^{-1}]$ and assumes the discrete values

$$\omega_{mn}^2 = c^2 \frac{\pi^2}{4} \left[\frac{(2m-1)^2}{a^2} + \frac{(2n-1)^2}{b^2} \right]. \tag{1.1.51}$$

The ω_{mn} are called *eigenfrequencies*.

A *partial differential equation* is called *linear*, if it is of the first degree in the dependent variable $u(x, y), u(x, y, t)$ and their partial derivatives u_x, u_{xx}, i.e., if each term either consists of the product of a known function of the independent variables, and the dependent variable or one of its partial derivatives. The equation may also contain a known function of the independent variables only (*inhomogeneous equation*). If this term does not appear, the *equation* is called *homogeneous*. It is a general property of all linear differential equations that two particular solutions can be superposed, which means that the sum of two solutions is again a (new) solution (*superposition principle*). So the general solution is a superposition of all particular solutions. Hence the general symmetric solution of (1.1.30) may be written as

$$u(x, y, t) = \sum_{m,n=0}^{\infty} A_{mn} \cos\left(\frac{\pi x}{2a}[2m-1]\right) \cos\left(\frac{\pi y}{2b}[2n-1]\right) \cos \omega_{mn} t, \tag{1.1.52}$$

where the ω_{mn} are given by (1.1.51) and the partial amplitudes A_{mn} may be determined from an *initial condition* like

$$u(x, y, 0) = f(x, y), \tag{1.1.53}$$

where $f(x, y)$ is a given function, i.e., the deflection of the membrane at the time $t = 0$.

Using now the solutions (1.1.42), we have

$$v(x, y) = A \cos\left(\sqrt{k^2 - \beta^2} x\right) \cdot \cos \beta y. \tag{1.1.54}$$

It might be that the "*separation constants*" β will have discrete values. Then the general solution reads

$$v(x,y) = \sum_{n=1}^{\infty} A_n \cos\left(\sqrt{k^2 - \beta_n^2}\, x\right) \cdot \cos \beta_n y. \qquad (1.1.55)$$

In order to satisfy boundary conditions, the unknowns A_n, β_n and k have to be determined. We will discuss this problem later. (1.1.55) is the starting point for a boundary point collocation method.

Problems

1. In *Mathematica* a function $f(x)$ is represented by **f1[x]**. We now define **f1[x]=1/(1-a*x^2)** and integrate it by the command

   F1[x]=Integrate[f1[x],x]   which results in $\dfrac{\text{ArcTanh}[\sqrt{a}\, x]}{\sqrt{a}}$.

   Now we verify by differentiation **D[F1[x],x]** which gives **f1[x]**.

   Be careful, the letter l looks very much like the number 1.

2. If we want to define an expression like
   R[x]=a+b*x+c*x^2; f2[x]=x/Sqrt[R[x]]; we may end the expression or the line with a semicolon. Then *Mathematica* allows to put several commands on the same line. If there is a last semicolon in the line as above, no outprint will be given.

   Integration **F2[x]=Integrate[f2[x],x]** results in

   $$\frac{\sqrt{a+bx+cx^2}}{c} - \frac{b\, \text{Log}\left(\dfrac{b+2ax}{\sqrt{c}} + 2\sqrt{a+bx+cx^2}\right)}{2c^{3/2}}.$$

   To verify again one may give the command **D[F2[x],x]** yielding

   $$\frac{b+2cx}{2c\sqrt{a+bx+cx^2}} - \frac{b\left(2\sqrt{c} + \dfrac{b+2cx}{\sqrt{a+bx+cx^2}}\right)}{2c^{3/2}\left(\dfrac{b+2cx}{\sqrt{c}} + 2\sqrt{a+bx+cx^2}\right)}.$$

   Apparently this can be simplified. We give the command **Simplify[%]**

   to obtain $\dfrac{x}{\sqrt{a+bx+cx^2}}$.

   Here the symbol % means: use the last result generated and %% means the next-to-last result and %%...% (k-times) indicates the k-th previous result.

3. In order to include a comment in *Mathematica* we write
   (*Some more problems*) Integrate[1/(1-x),x] resulting in
   $-\mathrm{Log}[-1+x]$

4. *Mathematica* knows how to solve differential equations
   DSolve[x''[t]+a^2*x[t]==0, x[t],t]
   This solves $x''(t) + a^2 x(t) = 0$ with respect to the independent variable
   t and gives the solution $x(t)$:
   $x[t] \to C[1]\ Cos[a\ t] + C[2]\ Sin[a\ t]$

   As usual, *Mathematica* drops the sign * for the multiplication. Do not
   use quotation marks " for the second derivative, use two apostrophes
   ". Specifying the integration constants, we can plot the result in the
   interval $0 \le x \le 2\pi$ by the command **Plot[2 Sin[x],{x,0,2 Pi}]**

5. **DSolve[y''[x]-3*y'[x]+2*y[x]==Exp[5*x],y[x],x]**

   $\{\{y[x] \to \dfrac{e^{5\,x}}{12} + e^x\ C[1] + e^{2\,x}\ C[2]\}\}$

6. **DSolve[y''[x]+y'[x]-6*y[x]==8*Exp[3*x],y[x],x]**

   $\{\{y[x] \to \dfrac{4\,e^{3\,x}}{3} + e^{-3\,x}\ C[1] + e^{2\,x}\ C[2]\}\}$

7. **Integrate[(x^3+2)^2*3*x^2, x]** $\qquad 3\left(\dfrac{4\,x^3}{3} + \dfrac{2\,x^6}{3} + \dfrac{x^9}{9}\right)$

   Simplify[%] $\qquad \dfrac{1}{3}\,x^3(12 + 6\,x^3 + x^6)$

8. **Integrate[(x+2)/(x+1),x]** $\qquad x + \mathrm{Log}[1+x]$

9. **Integrate[(x+3)/Sqrt[5-4*x-x^2],x]**

   $-\sqrt{5 - 4\,x - x^2} - \mathrm{ArcSin}\left[\dfrac{1}{3}\,(-2 - x)\right]$

10. **D[ArcTan[b*Tan[x]/a]/(a*b),x]** $\qquad \dfrac{Sec[x]^2}{a^2\left(1 + \dfrac{b^2\ Tan\ [x]^2}{a^2}\right)}$

    Simplify[%] $\qquad \dfrac{Sec[x]^2}{a^2 + b^2\ Tan[x]^2}$

11. **D[Cosh[x^2-3*x+1],x]** $\qquad (-3 + 2\,x)\ Sinh[1 - 3\,x + x^2]$

12. **DSolve[y'[x]-2*x*y[x]-3*x^2-1==0, y[x],x]**

    $\{\{y[x] \to e^{x^2} C[1] + e^{x^2}\left(-\dfrac{3}{2}\,e^{-x^2}\,x + \dfrac{5}{4}\,\sqrt{\pi}\ \mathrm{Erf}[x]\right)\}\}$

13. **DSolve[x''[t]-m g-a*x'[t]==0, x[t],t]**

    $\{\{x[t] \to \dfrac{g\,m\,t}{a} + \dfrac{e^{a\,t}\ C[1]}{a} + C[2]\}\}$

1.2 Classification of partial differential equations

Ordinary differential equations containing only one independent variable (and *two point boundary conditions*) are of minor importance for problems of physics and engineering. Hence, we mainly concentrate on partial differential equations.

Partial differential equations possess a large manifold of solutions. Instead of integration constants, arbitrary functions appear in the solution. As an example, we consider (1.1.15) in a specialization of two independent variables

$$c^2 u_{xx}(x,t) = u_{tt}(x,t). \tag{1.2.1}$$

By inserting into (1.2.1) it is easy to prove that

$$u(x,t) = f(x+ct) + g(x-ct) \tag{1.2.2}$$

is a solution of (1.2.1). Here f and g are arbitrary functions. Such a solution of a partial differential equation of order n with n arbitrary functions is called a *general solution*. If a partial differential equation contains p independent variables $x, y, \ldots$, one can find a *complete solution*, which contains p integration constants. If a function satisfies the partial differential equation and the accompanying boundary conditions and has no arbitrary function or constants, it is called a *particular solution*. If a solution is not obtainable by assigning particular values to the parameters in the general (complete) solution, it is called a *singular solution*. It describes an envelope of the family of curves represented by the general solution. But the expression is also used for a solution containing a *singular point*, a *singularity*. We will later consider such solutions containing singular points where the solution tends to infinity.

There are other essential properties characterizing various types of partial differential equations. We give some examples below:

$$u_{xx}(x,y) + u_{yy}(x,y) = 0. \tag{1.2.3}$$

This LAPLACE *equation* is linear, homogeneous and has constant coefficients. The equation

$$u_t(x,t) - x^2 u_x(x,t) = 0 \tag{1.2.4}$$

is of the first order, linear, homogeneous and has variable coefficients. (For a first-order partial differential equation a boundary problem cannot be formulated.) An example of a nonlinear homogeneous partial differential equation is given by

$$u_x^2(x,y) + u_{yy}(x,y) = 0 \tag{1.2.5}$$

and the equation

$$u_t(x,t) + u(x,t)u_x(x,t) = 0 \tag{1.2.6}$$

is called *quasilinear*. (The derivatives are linear). An inhomogeneous example is given by the linear POISSON *equation*

$$u_{xx}(x,y) + u_{yy}(x,y) = \rho(x,y), \tag{1.2.7}$$

where ρ is a given function. The most *general linear partial differential equation* of two independent variables has the form

$$a(x,y)u_{xx}(x,y) + 2b(x,y)u_{xy}(x,y) + c(x,y)u_{yy}(x,y) + d(x,y)u_x(x,y)$$
$$+ e(x,y)u_y(x,y) + g(x,y)u(x,y) = h(x,y). \tag{1.2.8}$$

Now equations satisfying

$$b^2(x,y) > a(x,y)c(x,y) \tag{1.2.9}$$

in a certain domain in the x,y plane are called *hyperbolic equations*,

$$b^2(x,y) < a(x,y)c(x,y) \tag{1.2.10}$$

characterizes *elliptic equations* and if

$$b^2(x,y) = a(x,y)c(x,y) \tag{1.2.11}$$

the equation is called *parabolic*. Thus, an equation is hyperbolic, elliptic or parabolic within a certain domain in the x,y plane. This is an important distinctive mark determining the *solvability* of a *boundary problem*.

Boundary curves or surfaces may be *open* or *closed*. A closed boundary surface is one that surrounds the domain everywhere, confining it to a finite surface or volume. A simple closed smooth curve is called JORDAN *curve*. An open surface is one that does not completely enclose the domain but lets it extend to infinity in at least one direction. Then the DIRICHLET *boundary conditions* (*first boundary value problem*) fix the value $u(x,y)$ on the boundary, NEUMANN *boundary conditions* (*second boundary value problem*) fix the value of the *normal derivative* $\partial u/\partial \vec{n}$ on the boundary and a CAUCHY *condition* fixes both value and normal derivative at the *same* place. The CAUCHY condition actually represents an *initial condition*. The normal derivative is the directional derivative of a function $u(x,y)$ in the direction of the normal at the point of the boundary where the derivative is taken. A generalized NEUMANN *boundary condition* (*third boundary value problem*) fixes

$$k(x,y)\frac{\partial u(x,y)}{\partial \vec{n}} + l(x,y)u(x,y) = m(x,y) \tag{1.2.12}$$

on the boundary. This condition is of importance in heat flow and fluid mechanics. It is possible to prove [1.2] the *solvability* of *boundary problems*, see Table 1.1.

Table 1.1 Solvability of boundary problem

Boundary condition	Equation hyperbolic	elliptic	parabolic
CAUCHY			
open boundary	*solvable*	indeterminate	overdeterminate
one closed boundary	overdeterminate	overdeterminate	overdeterminate
DIRICHLET			
open boundary	indeterminate	indeterminate	*solvable*
one closed boundary	indeterminate	*solvable*	overdeterminate
NEUMANN			
open boundary	indeterminate	indeterminate	*solvable*
one closed boundary	indeterminate	*solvable*	overdeterminate

In this connection the term solvable means solvable by an analytic solution. If an elliptic boundary problem has *two closed boundaries*, an analytic solution is no longer possible and singularities have to be accepted.

If the value $u(x, y)$ or its derivatives (or $m(x, y)$ in (1.2.12)) vanish on the boundary, the *boundary condition* is said to be *homogeneous*. If the values on the boundary do not vanish, the *boundary condition* is called *inhomogeneous*. A *boundary problem* is called *homogeneous* if the differential equation *and* the boundary condition are both homogeneous. If the differential equation or the boundary condition or both are inhomogeneous, then the *boundary problem* is said to be *inhomogeneous*. Inhomogeneous boundary conditions of a homogeneous equation can be transformed into homogeneous conditions of an inhomogeneous equation. This is made possible by the following fact. The general solution $u(x, y)$ of a linear inhomogeneous differential equation consists of the superposition of the general solution $w(x, y)$ of the matching homogeneous equation and a particular solution $v(x, y)$ of the inhomogeneous equation. We consider an example. Let

$$\Delta u(x, y) = \rho(x, y) \tag{1.2.13}$$

be an inhomogeneous equation with the homogeneous condition $u(\text{boundary}) = 0$. If we insert the ansatz

$$u(x, y) = w(x, y) + v(x, y) \tag{1.2.14}$$

into (1.2.13) we obtain

$$\Delta w(x, y) + \Delta v(x, y) = \rho(x, y). \tag{1.2.15}$$

Putting

$$\Delta v(x, y) = \rho(x, y) \tag{1.2.16}$$

we obtain an inhomogeneous equation for v and a homogeneous equation for w, which reads

$$\Delta w(x, y) = 0. \tag{1.2.17}$$

Due to the assumption of a homogeneous boundary $u(\text{boundary}) = 0$ we have

$$v(\text{boundary}) = -w(\text{boundary}). \tag{1.2.18}$$

We now see that the inhomogeneous condition (1.2.18) that belongs to the homogeneous equation (1.2.17) has been converted into a homogeneous condition $u(\text{boundary}) = 0$ and an inhomogeneous equation (1.2.13). We thus have to construct a function $v(x,y)$ satisfying the inhomogeneous condition (1.2.18) for w and producing the term $\rho(x,y)$ by application of Δ on v as in (1.2.16).

We now give an example of the conversion of an inhomogeneous boundary condition for a homogeneous equation into a homogeneous condition matching an inhomogeneous equation. As the homogeneous equation we choose the Laplacian (1.2.3) and write it now in the form (1.2.17)

$$w_{xx}(x,y) + w_{yy}(x,y) = 0. \tag{1.2.19}$$

We consider again the rectangle of Figure 1.2, but instead of the homogeneous boundary conditions (1.1.36) we now use inhomogeneous conditions

$$\begin{aligned}
w(a, y) &= 0 && \text{for} && -b \leq y \leq +b, \\
w(-a, y) &= 0 && \text{for} && -b \leq y \leq +b, \\
w(x, b) &= f(x) && \text{for} && -a \leq x \leq +a, \\
w(x, -b) &= f(x) && \text{for} && -a \leq x \leq +a.
\end{aligned} \tag{1.2.20}$$

Then the corresponding inhomogeneous equation is given by (1.2.13) and its homogeneous boundary conditions are (1.1.36) written for $u(x,y)$. To find a solution of (1.2.19), (1.2.20) we have thus to solve

$$u_{xx}(x,y) + u_{yy}(x,y) = \rho(x,y) \tag{1.2.21}$$

together with the homogeneous boundary conditions

$$u(\pm a, y) = 0 \qquad \text{for} \qquad -b \leq y \leq +b, \tag{1.2.22}$$

$$u(x, \pm b) = 0 \qquad \text{for} \qquad -a \leq x \leq +a. \tag{1.2.23}$$

The solution of the inhomogeneous equation (1.2.21) is now a superposition of a general solution $u_1(x,y)$ of the homogeneous equation and a particular solution $v(x,y) = u_2(x,y)$. A setup $u_1(x,y) = X(x) \cdot Y(y)$ together with the symmetry expressed by (1.2.22) and (1.2.23) delivers

$$u_1(x,y) = \sum_m A_m \cos(\alpha_m x) \cosh(\alpha_m y). \tag{1.2.24}$$

To make the problem a little easier we assume $\rho(x,y) = p = const.$ This choice will influence $f(x)$ in (1.2.20).

We now derive the particular solution $u_2(x, y)$. Since its Laplacian derivative must give the constant p we use the setup

$$u_2(x, y) = c_1 x^2 + c_2 xy + c_3 y^2 + c_4 x + c_5 y + c_6 \qquad (1.2.25)$$

which seems to be a general possibility where the c_i are constants. We obtain from (1.2.21)

$$\Delta u_2 = 2c_1 + 2c_3 = \rho = p. \qquad (1.2.26)$$

Since the particular solution u_2 has to satisfy separately the boundary conditions we have from (1.2.22)

$$c_1 a^2 \pm c_2 ay + c_3 y^2 \pm c_4 a + c_5 y + c_6 = 0 \qquad (1.2.27)$$

for $-b \le y \le +b$, hence also for $y = 0$, this gives $c_2 = c_3 = c_5 = 0$ and $c_1 = p/2$ from (1.2.26). From (1.2.23) we get

$$u_2(x, \pm b) = \frac{p}{2} x^2 + c_4 x + c_6 = 0 \qquad (1.2.28)$$

or with (1.2.22)

$$u_2(x, \pm b) = \frac{p}{2}(x^2 - a^2) = 0. \qquad (1.2.29)$$

If α_m is given by (1.1.48), the homogeneous solution satisfies (1.2.22) too. From (1.2.24) we have

$$\frac{p}{2}(x^2 - a^2) = -\sum_{m=0}^{\infty} A_m \cos\left(\frac{\pi(2m-1)}{2a} x\right) \cosh\left(\frac{\pi(2m-1)}{2a} b\right). \qquad (1.2.30)$$

This indicates that we have to expand the left-hand side for $-a \le x \le a$ into a cos-FOURIER *series* to satisfy the boundary condition (1.2.23). This procedure gives the A_m. So the solution $u(x, y) = u_1(x, y) + u_2(x, y)$ reads

$$u(x, y) = \sum_{m=0}^{\infty} A_m \cos\left(\frac{\pi(2m-1)}{2a} x\right) \cosh\left(\frac{\pi(2m-1)}{2a} y\right) + \frac{p}{2}(x^2 - a^2). \qquad (1.2.31)$$

It satisfies the inhomogeneous equation (1.2.21) for $\rho(x, y) = p$ and also the associated boundary conditions (1.2.22) and (1.2.23).

The particular solution $u_2 = v(x, y)$ has now to satisfy (1.2.16) or

$$v_{xx}(x, y) + v_{yy}(x, y) = p. \qquad (1.2.32)$$

For $v(x, y)$ we have

$$u_2(x, y) = v(x, y) = \frac{p}{2}(x^2 - a^2). \qquad (1.2.33)$$

Then (1.2.32) gives an identity. Due to (1.2.18) the solution (1.2.33) satisfies the boundary conditions (1.2.20) and $f(x) = \frac{p}{2}(x^2 - a^2)$.

We now have finally solved the problem of equations (1.2.19) and (1.2.20). The inhomogeneous boundary conditions (1.2.20) have been homogenized and according to (1.2.14) the solution of (1.2.19) reads

$$w(x, y) = u(x, y) - v(x, y)$$

$$= \sum_{m=0}^{\infty} A_m \cos\left(\frac{\pi(2m-1)}{2a}x\right) \cosh\left(\frac{\pi(2m-1)}{2a}y\right) \quad (1.2.34)$$

which satisfies (1.2.20) due to the relation (1.2.30). It is clear that the method described can be used for $\rho(x, y) \neq const$, too.

Problems

1. Determine if the following partial differential equations are hyperbolic (h), elliptic (e), parabolic (p) or of a mixed type (m), which means the type depends on the domain in the x, y plane.

$u_{xx} + u_{yy} = 0$	(e);	$u_{xx} - u_{yy} = 0$	(h);
$u_x^2 + u_{yy} = 0$	(p);	$u_{xx} \cdot x^2 + u_{yy} \cdot y^2 = 0$	(m);
$u_{xx} + x u_{yy} = 0$	(m);	$u_{xx} = u_y$	(p);
$u_{xx} + 2u_{xy} + u_{yy} + x = 0$	(p);	$x^2 u_{xx} - u_{yy} + u \cdot u_x = 0$	(h).

Using *Mathematica* calculate:

2. $\int (x^2 + 2)^2 3x^2 \, dx = 3(4x^3/3 + 4x^5 + x^7/7)$

3. $\int \dfrac{x+2}{x+1} dx = x + \log(x+1)$

4. $\int \dfrac{x+3}{\sqrt{5-4x-x^2}} dx = -\sqrt{5 - 4x - x^2} - \arcsin \dfrac{-x-2}{3}$

5. $\dfrac{d}{dx} \dfrac{1}{ab} \arctan\left(\dfrac{a}{b} \tan x\right) = \dfrac{\sec^2(x)}{b^2 + a^2 \tan^2 x}$

6. $\dfrac{d}{dx} \cosh(x^2 - 3x + 1) = \sinh(x^2 - 3x + 1)(2x - 3)$

7. What happens if (1.2.25) reads
$u_2(x) = c_1 x^3 + c_2 y^3 + c_3 x^2$
and if $\rho(x, y) = ax + by + c$?

This is very simple: $u_{2xx} = 6c_1x + 2c_3, u_{2yy} = 6c_2y$. Insertion into (1.2.21) results in $6c_1x + 2c_3 + 6c_2y = \rho = ax + by + c$, so that $6c_1 = a$, $6c_2 = b$, $2c_3 = c$.

8. ```
f1[x]=(x^2+2)^2*3*x^2
F1[x]=Integrate[f1[x],x]
D[F1[x],x];   Simplify[%]
```
Explanations of these commands will be given later.

## 1.3   Types of boundary conditions and the collocation method

If a boundary curve or surface can be described by coordinate lines or surfaces *and if* the partial differential equation in question is separable into ordinary differential equations in this coordinate system by a setup like (1.1.32), the boundary problem can be solved quite easily (compare the calculations (1.1.30) through (1.1.51)). To express boundary conditions in a simple way, one must have coordinate surfaces that fit the physical boundary of the problem. Very often, however, the situation is more complicated even for partial differential equations that are separable in only a few coordinate systems, if the boundary cannot be described in the corresponding coordinate system. On the other hand, there are problems that belong to partial differential equations that cannot be separated at all into ordinary differential equations. As an example we mention equation

$$u_{xx}(x,y) + f(x,y)u_{yy}(x,y) = 0. \qquad (1.3.1)$$

If the coefficient function $f(x,y)$ *cannot* be represented by a product $f(x,y) = g(x) \cdot h(y)$, then a separation of (1.3.1) into ordinary differential equations is rarely possible. However, in a problem of plasma physics (5.1.44) or

$$u_{zz}(r,z) + u_{rr}(r,z) + \frac{1}{r}u_r(r,z) - \frac{1}{r^2}u(r,z)$$
$$+ \gamma^2 u(r,z) + (a + br + cr^2 + cz^2)u = 0 \qquad (1.3.2)$$

is an example demonstrating the opposite but has to be solved [1.3]. In this case the ansatz $u(r,z) = R(r) \cdot Z(z)$ leads to a separation into two ordinary differential equations

$$R'' + \frac{1}{r}R' - \frac{1}{r^2}R + (\gamma^2 + a + br + cr^2 - k^2)R = 0,$$

$$Z'' + (cz^2 + k^2)Z = 0, \qquad (1.3.3)$$
```

where k is the separation constant. It is thus not possible to predict separability in general terms. On the other hand, it is well known that the HELMHOLTZ *equation* $\Delta u + k^2 u = 0$ is separable in 11 coordinate systems only. These coordinate systems are [1.4]

rectangular coordinates	x, y, z	circular cylinder coord.	r, z, φ
elliptic cylinder coord.	η, ψ, z	parabolic cylinder coord.	μ, ν, z
spherical coordinates	r, ϑ, ψ	prolate spheroidal coord.	η, ϑ, ψ
oblate spheroidal coord.	η, ϑ, λ	parabolic coordinates	μ, ν, ψ
conical coordinates	r, ϑ, λ	ellipsoidal coordinates	η, ϑ, λ
paraboloid coordinates	μ, ν, λ		

These 11 coordinate systems are formed from first- and second-degree surfaces. There are also systems built from fourth-degree surfaces [1.5] that may have practical applications.

However, the theory of separability of partial differential equations [1.4] is of no great interest to us since we will be discussing methods to solve nonseparable problems in this book. In principle, each boundary problem has a solution [1.2]. The problem is, though, how to find the solution. In this book we will discuss a method to solve boundary problems of various kinds, see Table 1.2. In this table the term *boundary fitted* means that the boundary can be described by coordinate lines of the coordinate system in which the partial differential equation is separable.

Table 1.2 Various boundary problems

diff. equ. separable	boundary fitted	example	solution
yes	yes	rectangular membrane	classical
yes	2 × yes	circular ring membrane	singularity
yes	no	circ. membrane cartes. coordinates	possible
no	no	CASSINI curve membrane	numerical
yes	yes	2 boundaries from 2 coord.systems	sing. nonuniform
no	no	toroidal problems	singularity
no	corners	non-JORDAN curve	special solution

We will discuss some examples of boundary problems that are mentioned in Table 1.2. A membrane described by the HELMHOLTZ *equation* and bound by a CASSINI *curve* is such an example. Membranes with holes and exhibiting two closed boundaries and toroidal problems or boundaries with corners will be treated. If there are two boundaries, it may happen that they belong to two different coordinate systems. This type of problem can be called a *nonuniform boundary problem*. It can be solved by special methods [1.6].

In principle, each reasonable boundary problem can be solved using the fact that the general solution of a partial differential equation contains one or more arbitrary functions. These arbitrary functions may be used to adapt the general solution to a special boundary condition. According to (1.2.31) any reasonable function can be expanded into an infinite series of partial solutions. Thus, an infinite set of constants like partial amplitudes A_m is equivalent to an arbitrary function.

In the case of different boundary problems one has two possibilities: one can choose an expression that satisfies *either* the differential equation *or* the boundary conditions exactly. Coefficients contained in the expression can then be used to satisfy the boundary condition or the differential equation, respectively. We give an example [1.7]. We consider the problem

$$u_{xx} + u_{yy} = -1 \tag{1.3.4}$$

with the boundary condition on the square $|x| \le 1$, $|y| \le 1$

$$\frac{\partial u}{\partial x} = -u, \qquad x = \pm 1, \qquad -1 \le y \le +1, \tag{1.3.5}$$

$$\frac{\partial u}{\partial y} = -u, \qquad y = \pm 1, \qquad -1 \le x \le 1. \tag{1.3.6}$$

Instead of using the method that we used to solve (1.2.13), we first write down another expression satisfying the differential equation (1.3.4). For the general solution of (1.3.4) we require a particular solution of it together with a solution of the homogeneous equation. A particular solution u_2 is apparently given by

$$u_2 = -\frac{1}{4}(x^2 + y^2), \tag{1.3.7}$$

compare (1.2.25).

As the solution u_1 of the homogeneous equation satisfying the given symmetry conditions, we could use (1.2.24) or the real parts of $(x+iy)^{4n}$, $n = 1, 2, \ldots$, which deliver the so-called *harmonic polynomials*. Multiplying them by coefficients we have

$$u = u_1 + u_2 = -\frac{1}{4}(x^2 + y^2) + a_0 + a_1(x^4 - 6x^2y^2 + y^4)$$
$$+ a_2(x^8 - 28x^6y^2 + 70x^4y^4 - 28x^2y^6 + y^8) + \ldots . \tag{1.3.8}$$

This is a solution of (1.3.4) for $-1 \to 0$. We now must determine the coefficients in such a way that the solution (1.3.8) satisfies the boundary conditions. To do this we use a *collocation method*. This means that we choose a set of n so-called *collocation points* x_i, y_i, $i = 1 \ldots n$ on the boundary curve on which the boundary conditions must be satisfied. The coefficients a_n will then be calculated from the boundary conditions (1.3.5), (1.3.6). Due to the double symmetry, we need only consider the part of the boundary along $x = 1$, where

$$\left(\frac{\partial u}{\partial x} + u\right)_{x=1} = 0. \tag{1.3.9}$$

Inserting u from (1.3.8) we have

$$-\frac{3}{4} - \frac{y^2}{4} + a_0 + a_1(5 - 18y^2 + y^4)$$
$$+ a_2(9 - 196y^2 + 350y^4 - 84y^6 + y^8) = 0. \qquad (1.3.10)$$

In order to determine the three unknown coefficients a_0, a_1, a_2, we need three equations. This means that we have to choose three collocation points y_1, y_2, y_3 along the boundary line $x = 1$, $0 \le y \le 1$. We choose $y_1 = 0$, $y_2 = 0.33$, $y_3 = 0.66$ and obtain from (1.3.10) the three equations, $i = 1, 2, 3$

$$-\frac{3}{4} - \frac{y_i^2}{4} + a_0 + a_1(5 - 18y_i^2 + y_i^4)$$
$$+ a_2(9 - 196y_i^2 + 350y_i^4 - 84y_i^6 + y_i^8) = 0. \qquad (1.3.11)$$

The method to determine the unknown coefficients a_0, a_1 and a_2 from the boundary conditions is called *boundary collocation*.

On the other hand, we can use another ansatz satisfying the boundary conditions from the start. Then the coefficients have to be determined in such a way that the differential equation is satisfied in the whole domain (*interior collocation*). Since we now need more collocation points to cover the whole square, this method is more expensive.

In order first to satisfy the boundary conditions we make the ansatz

$$u = a_1[a_{11} + a_{12}(x^2 + y^2) + a_{22}x^2y^2] + a_2[a_{21} + a_{22}(x^2 + y^2)$$
$$+ a_{23}(x^4 + y^4) + a_{24}(x^4y^2 + x^2y^4)] + \dots \qquad (1.3.12)$$

Each term $[\dots]$ has to satisfy the boundary conditions separately. To obtain this we calculate the a_{11}, a_{12} and so on from (1.3.9), whereas the coefficients a_1, a_2 have to be determined in such a way that the differential equation is satisfied. Inserting (1.3.12) into (1.3.9) we have for the first term

$$2a_{12} + 2a_{22}y^2 + a_{11} + a_{12} + a_{12}y^2 + a_{22}y^2 = 0 \qquad (1.3.13)$$

which is solved by $a_{22} = 1$, $a_{12} = -3$, $a_{11} = 9$. The second term yields

$$2a_{22} + 4a_{23} + 4a_{24}y^2 + 2a_{24}y^4 + a_{21} + a_{22} + a_{22}y^2$$
$$+ a_{23} + a_{23}y^4 + a_{24}y^2 + a_{24}y^4 = 0, \qquad (1.3.14)$$

which delivers $a_{24} = 1$, $a_{23} = -3$, $a_{22} = -5$, $a_{21} = 30$. We now have the ansatz

$$u = a_1[9 - 3(x^2 + y^2) + x^2y^2] + a_2[30 - 5(x^2 + y^2)$$
$$- 3(x^4 + y^4) + x^4y^2 + x^2y^4] + \dots (1.3.15)$$

The coefficients a_1, a_2 etc., have now to be determined in such a way that (1.3.15) satisfies the differential equation (1.3.4). Inserting (1.3.15) we obtain

$$a_1[-12 + 2(x^2 + y^2)] + a_2[-20 - 36(x^2 + y^2)$$
$$+ 2(x^4 + y^4) + 24x^2y^2] = -1. \qquad (1.3.16)$$

For this expression with two unknowns we have to choose two collocation points within the square domain. We can define $x_1 = 0.5$, $y_1 = 0.5$ and $x_2 = 0.75$, $y_2 = 0.5$. Inserting this into (1.3.16) we can calculate a_1 and a_2 to obtain an approximate solution.

Since collocation will be discussed later on in detail, we postpone problems.

1.4 Differential equations as models for nature

In the last sections we have discussed boundary conditions, but where are the differential equations coming from? There are apparently two methods to derive differential equations as models for phenomena in nature and engineering: *intuition* and *derivation* from fundamental laws of nature like the energy theorem, etc. We will give two examples, the first for *intuition*: Let us assume we want to study the spread of an epidemic disease. By $S(t)$ we designate the number of healthy persons, $I(t)$ will be the number of persons contracting the disease and let $R(t)$ be the number of persons being immune against the disease. Apparently, one then has a theorem for the conservation of the number of people if nobody dies or is born during the short time period considered. This number balance reads

$$S(t) + I(t) + R(t) = const. \qquad (1.4.1)$$

Now intuition and experience come into play. Apparently the number of persons newly infected will be proportional to the number $S(t)$ of healthy persons and to the number $I(t)$ of sick persons:

$$-\frac{dS(t)}{dt} = \alpha S(t) \cdot I(t). \qquad (1.4.2)$$

The parameter α can be called *rate of infection*. On the other hand, the number of persons becoming immune after having recovered from the disease would be

$$\frac{dR(t)}{dt} = \beta I(t), \qquad (1.4.3)$$

where β can be called *rate of immunization*. We thus have three equations to determine the three unknowns S, I, R. Building dR/dt from (1.4.1) and inserting it into (1.4.3) yields

$$\frac{\mathrm{d}R}{\mathrm{d}t} \equiv -\frac{\mathrm{d}S}{\mathrm{d}t} - \frac{\mathrm{d}I}{\mathrm{d}t} = \beta I. \tag{1.4.4}$$

Calculation of I and $\mathrm{d}I/\mathrm{d}t$ from (1.4.2) and inserting into (1.4.4) gives a nonlinear differential equation of second order

$$\frac{\mathrm{d}^2 S}{\mathrm{d}t^2} - S\left(\frac{\mathrm{d}\ln S}{\mathrm{d}t}\right)^2 + \alpha \frac{\mathrm{d}S}{\mathrm{d}t}\left(\frac{\beta}{\alpha} - S\right) = 0. \tag{1.4.5}$$

We now assume the initial condition $S(t = 0) = S_0$. Thus the number of healthy persons at $t = t_0$ is given by S_0. Neglecting $\mathrm{d}\ln S/\mathrm{d}t \approx 0$, we obtain near $t = 0$

$$\frac{\mathrm{d}^2 S}{\mathrm{d}t^2} \approx \alpha \frac{\mathrm{d}S}{\mathrm{d}t}\left(S_0 - \frac{\beta}{\alpha}\right).$$

Thus the curvature $S''(t_0)$ depends on the conditions $S_0 > \beta/\alpha$ and $S_0 < \beta/\alpha$, respectively. This represents exactly the empirical basic *theorem of epidemiology*: "An epidemic starts if the number (S_0) of healthy but predisposed persons exceeds a specific threshold (β/α)".

Another method to derive differential equations is given by a *derivation* from fundamental laws. Whereas boundary conditions describe actual situations and are used to specify an actual particular solution by determining integration components, differential equations of order n deliver the general solution containing n integration constants. For example, the differential equation describing transverse vibrations of a thin uniform plate can be derived from the (empirical) law of HOOKEAN deformation plus the energy theorem. A more elegant way of deriving differential equations is *variational calculus*. Let us assume that the eigenfrequencies of transversal vibrations of plates of varying thickness are suddenly of practical interest (e.g., for the investigation of fissures in an airplane wing).

The fundamental problem of the calculus of variation is to determine the minimum of the integral

$$J(u(x,y,t)) = \iiint F(x,y,t,u(x,y,t),u_x(x,y,t),u_y(x,y,t),u_t(x,y,t),$$

$$u_{xx}(x,y,t),u_{yy}(x,y,t),u_{xy}(x,y,t)\mathrm{d}x\mathrm{d}y\mathrm{d}t \tag{1.4.6}$$

for a given functional F. The minimum of the integral J delivers this function $u(x,y,t)$, which actually makes J a minimum. This function is determined by the EULER *equations* (which will be derived later on in section 4.7):

$$F_u - \frac{\partial}{\partial x}F_{u_x} - \frac{\partial}{\partial y}F_{u_y} - \frac{\partial}{\partial t}F_{u_t}$$

$$+ \frac{\partial^2}{\partial x^2}F_{u_{xx}} + \frac{\partial^2}{\partial x \partial y}F_{u_{xy}} + \frac{\partial^2}{\partial y^2}F_{u_{yy}} \ldots = 0. \tag{1.4.7}$$

Here the indices u, u_x etc., designate differentiation of the functional $F(x, y, t, u, u_x, u_y \ldots)$ with respect to its variables u, u_x, etc.

For a physical or engineering problem the functional F is given by the LA-GRANGE functional defined by the difference kinetic energy T minus potential energy Φ. If we assume that $u(x, y, t)$ is the local transversal deflection of a plate, the kinetic energy of a plate with modestly varying thickness $h(x, y)$ is given by

$$T = \frac{1}{2} \rho_0 \iint_G h(x, y) \left(\frac{\partial u}{\partial t} \right)^2 \mathrm{d}x \mathrm{d}y. \tag{1.4.8}$$

Here ρ_0 is the (constant) surface mass density per unit of thickness, so that $\rho_0 h(x, y)$ is the local surface mass density. The surface integral T is taken over the area G of the plate.

In order to find the minimum of (1.4.6) we have to vary the integral

$$\delta \int_{t_0}^{t_1} \left(\frac{1}{2} \rho_0 \iint_G h(x, y) \left(\frac{\partial u}{\partial t} \right)^2 \mathrm{d}x \mathrm{d}y - \Phi \right) \mathrm{d}t. \tag{1.4.9}$$

δ is the variational symbol and Φ is the total elastic energy, i.e., the local elastic potential integrated over the domain G.

Using the designations E for YOUNG's modulus and μ for POISSON's ratio, the local elastic (free) energy per unit volume of the plate is given by

$$f(x, y, z, u_{xx}, u_{yy}, u_{xy})$$

$$= z^2 \frac{E}{1 + \mu} \left\{ \frac{1}{2(1 - \mu)} \left(u_{xx}^2 + u_{yy}^2 \right)^2 + \left(u_{xy}^2 - u_{xx} u_{yy} \right) \right\}. \tag{1.4.10}$$

The total elastic energy is then given by integration over the volume of the plate. Integration first over $z^2 \mathrm{d}z$ alone from $-h(x, y)/2$ to $+h(x, y)/2$ delivers $h^3(x, y)/12$. Thus the total elastic energy Φ is then given by

$$\Phi(x, y, u_{xx}, u_{yy}, u_{xy}) = \frac{E}{24(1 - \mu^2)} \iint \Big\{ h^3(x, y)(u_{xx}^2 + u_{yy}^2)^2$$

$$+ 2(1 - \mu)h^3(x, y)(u_{xy}^2 - u_{xx} u_{yy}) \Big\} \mathrm{d}x \mathrm{d}y. \tag{1.4.11}$$

If the plate has to carry a load $p(x, y)$ then we have to add the term

$$\iint_G p(x, y) \cdot u(x, y) \mathrm{d}x \mathrm{d}y. \tag{1.4.12}$$

This term describes the work done by the external forces when the points on the plate are displaced by the displacement u. Now the total functional is given by

$$F = \frac{1}{2}\rho_0 h u_t^2 + pu$$

$$-\frac{Eh^3}{24(1-\mu^2)}\left[u_{xx}^2 + u_{yy}^2 + 2u_{xx}u_{yy} + 2(1-\mu)(u_{xy}^2 - u_{xx}u_{yy})\right]. \quad (1.4.13)$$

Since then $F_{u_x} = F_{u_y} = 0$, (1.4.7) takes the form

$$F_u - \frac{\partial}{\partial t}F_{u_t} + \frac{\partial^2}{\partial x^2}F_{u_{xx}} + \frac{\partial^2}{\partial x \partial y}F_{u_{xy}} + \frac{\partial^2}{\partial y^2}F_{u_{yy}} = 0. \quad (1.4.14)$$

From the first two terms of (1.4.13) we thus obtain

$$F_u = p, \quad F_{u_t} = \rho_0 h u_t, \quad -\frac{\partial}{\partial t}F_{u_t} = -\rho_0 h u_{tt}. \quad (1.4.15)$$

Furthermore we then get the plate equation for varying thickness $h(x, y)$ in the form

$$\frac{Eh^3}{12(1-\mu^2)}(u_{xxxx} + 2u_{xxyy} + u_{yyyy}) + \rho_0 h u_{tt}$$

$$+\frac{Eh}{12(1-\mu^2)}\{6h\left[h_x(u_{xxx} + u_{xyy}) + h_y(u_{yyy} + u_{xxy})\right]$$

$$+3u_{xx}\left(2h_x^2 + hh_{xx} + 2\mu h_y^2 + \mu hh_{yy}\right)$$

$$+3u_{yy}\left(2\mu h_x^2 + \mu hh_{xx} + 2h_y^2 + hh_{yy}\right)$$

$$+6(1-\mu)u_{xy}(h_x h_y + hh_{xy})\} = p(x, y). \quad (1.4.16)$$

This is the plate equation for weakly varying thickness $h(x, y)$.

This derivation does however not answer the question where the energy theorem like (1.4.8) or (1.4.11), a consequence of HOOKE's law, comes from. There are people believing that these fundamental laws are preexistent in nature (or have been originated by a creator). Modern natural philosophy tends to another view. So mathematicians know that differential equations are invariant under special transformations of coordinates. If, for instance, the equation of motion is submitted to a simple translation along the x coordinate axis, then the momentum mv_x remains constant: it will be conserved. EMMY NOETHER has shown that the invariance of a differential equation against a transformation has the consequence of the existence of a *conservation theorem* for a related physical quantity. Thus the energy theorem is a consequence of the invariance of the equation of motion under a translation along the time axis. Human beings assume that the laws of nature are independent against a time translation. But intelligent lizards, as cold-blooded animals, would probably have a chronometry depending on the ambient temperature, so that

$t = t(T)$. Such a dependence would not allow energy conservation (EDDING-TON). But the results of the lizard physics would be the same as in human physics. Apparently human assumptions on coordinate transformations create the laws we "find" in nature. But how could we find out if these laws are correct and true? In his "Discours de la Méthode" POINCARÉ has shown that there are always several "true" models or theories describing natural phenomena. As an example, we can mention that DIVE's theory of elliptic waves [1.8] and the special relativity theory give exactly the same results up to the order $(v/c)^2$. Why have we chosen special relativity to describe nature? When we have to decide between two fully equivalent theories we should take into account:

1. Aesthetic points of view
2. MACH's *principle* of economic thinking
3. The extensibility of a theory to broader fields of applications, like the extension of special to general relativity, respectively

Problems

1. Derive the equation for transverse vibrations $u(x, y, t)$ for a plate with constant thickness $h = const$ (see section 4.5).

2. Try to solve (1.4.5) using *Mathematica* (Not possible).

3. Type the command

   DSolve[S''[t]-α*S'[t]*(S0-β/α)==0,S[t],t]

   (gives $S[t] \rightarrow \dfrac{e^{t\,(S0\,\alpha-\beta)}\,C[1]}{S0\,\alpha - \beta} + C[2]$)

   and plot the result. But there is now a problem: the solution is not given by **S[t]= .** So it is necessary to define a new function $u(t)$, which gives a value for any arbitrary t. This is done by replacing t by $t\_$. We first select the integration constants **C[2]=0, C[1]1=S0*(S0*α − β)** to obtain $S(t = 0) = S0$ and write **u[t_]=Exp[t*(S0α − β)]*S0** This $u(t)$ may be plotted for given arbitrary values of $S0, \alpha, \beta$.

4. Learn partial derivatives. Define

   u[x,y]=x^2+a*x^3*y^4+y^3

D[u[x,y],x]	$2\,x + 3\,a\,x^2\,y^4$
D[u[x,y],{x,2}]	$2 + 6\,a\,x\,y^4$
D[u[x,y],y]	$4\,a\,x^3\,y^3 + 3\,y^2$

2

Boundary problems of ordinary differential equations

2.1 Linear differential equations

Let us first consider equations of second order. According to chapter 1, equations of motion and other models combine the acceleration $\ddot{x}(t)$ of a phenomenon with some external influence like forces. The most general linear differential equation of second order apparently has the form

$$p_0(x)y'' + p_1(x)y' + p_2(x)y = f(x). \tag{2.1.1}$$

Here p_0, p_1, p_2 and f are in most cases given not-vanishing continuous functions. If $f(x)$ is zero, the equation (2.1.1) is called *homogeneous*. We now first solve the homogeneous equation

$$p_0(x)y'' + p_1(x)y' + p_2(x)y = 0. \tag{2.1.2}$$

Let

$$y(x) = C_1 y_1(x) + C_2 y_2(x) \tag{2.1.3}$$

be the *general solution*, where y_1 and y_2 are *fundamental solutions* and C_1 and C_2 are constants of integration. In the next section we will discuss methods how one may find y_1 and y_2. Let $y_0(x)$ be a *particular solution* of the inhomogeneous equation (2.1.1), then its general solution has the form

$$y(x) = C_1 y_1(x) + C_2 y_2(x) + y_0(x). \tag{2.1.4}$$

The next step is to find a particular solution $y_0(x)$ of (2.1.1). We do this by replacing the constants C_1 and C_2 by nonconstant functions $C_1(x)$ and $C_2(x)$. This method is called the method of *variation of parameters (of constants)*, because the constants in (2.1.4) are now allowed to vary. Instead of the unknown function $y(x)$, we now have two new functions additionally. The setup

$$y_0(x) = C_1(x)y_1(x) + C_2(x)y_2(x) \tag{2.1.5}$$

delivers $y_0'(x)$ and $y_0''(x)$. Since the new functions $C_1(x)$ and $C_2(x)$ are quite arbitrary, we may require two new conditions

$$C_1'(x)y_1(x) + C_2'(x)y_2(x) = 0, \tag{2.1.6}$$

$$C_1'(x)y_1'(x) + C_2'(x)y_2'(x) = f(x)/p_0(x). \tag{2.1.7}$$

From (2.1.5) and (2.1.6) we then obtain

$$y_0' = C_1 y_1' + C_2 y_2', \tag{2.1.8}$$

and from (2.1.7), (2.1.8) and (2.1.2), (2.1.1) one gets

$$y_0'' = C_1 y_1'' + C_2 y_2'' + f(x)/p_0(x). \tag{2.1.9}$$

Since (2.1.3) is assumed to be a solution of (2.1.2), insertion of (2.1.5), (2.1.8) and (2.1.9) into (2.1.1) demonstrates that (2.1.5) is actually a solution of (2.1.1). Now we determine the two still-unknown functions $C_1(x)$ and $C_2(x)$ from (2.1.6) and (2.1.7). We obtain

$$C_1'(x) = \frac{-y_2(x)f(x)}{p_0(x)\,(y_1(x)y_2'(x) - y_1'(x)y_2(x))},$$

$$C_2'(x) = \frac{y_1(x)f(x)}{p_0(x)\,(y_1(x)y_2'(x) - y_1'(x)y_2(x))}. \tag{2.1.10}$$

Integrations yield

$$C_1(x) = \int_{x_0}^{x} \frac{-y_2(\xi)f(\xi)}{p_0(\xi)W(\xi)}\,\mathrm{d}\xi, \quad C_2(x) = \int_{x_0}^{x} \frac{y_1(\xi)f(\xi)}{p_0(\xi)W(\xi)}\,\mathrm{d}\xi. \tag{2.1.11}$$

The denominator appearing in (2.1.11) is called the Wronskian *determinant* W.

$$W(x) = y_1(x)y_2'(x) - y_1'(x)y_2(x) = \begin{vmatrix} y_1(x) & y_2(x) \\ y_1'(x) & y_2'(x) \end{vmatrix}. \tag{2.1.12}$$

If this determinant vanishes, then the two solutions $y_1(x)$ and $y_2(x)$ are linearly dependent:

$$C_1(x)y_1 + C_2(x)y_2 = 0 \tag{2.1.13}$$

(for $C_1 \neq 0$, $C_2 \neq 0$). If one knows two independent solutions $y_1(x)$ and $y_2(x)$, then also the particular solution y_0 is known:

$$y_0(x) = \int_{x_0}^{x} \frac{y_1(\xi)y_2(x) - y_1(x)y_2(\xi)}{W(\xi)} \cdot \frac{f(\xi)}{p_0(\xi)}\,\mathrm{d}\xi. \tag{2.1.14}$$

If one solution $y_1(x)$ of (2.1.2) is known, then the second solution of (2.1.2) may be found:

$$y_2(x) = y_1(x) \int \frac{\exp\left[-\int (p_1(x)/p_0(x))\mathrm{d}x\right]}{y_1^2(x)}\,\mathrm{d}x. \tag{2.1.15}$$

The form of this solution initiates the idea that the setup

$$y(x) = z(x) \exp\left(-\frac{1}{2}\int \frac{p_1(x)}{p_0(x)}dx\right) \qquad (2.1.16)$$

may transform away the $y'(x)$ term in (2.1.1). Inserting (2.1.16) into (2.1.1) one obtains

$$p_0(x)z'' + \left(p_2(x) - \frac{1}{4}\frac{p_1^2(x)}{p_0(x)}\right)z = f(x)\exp\left(\frac{1}{2}\int\frac{p_1(x)}{p_0(x)}dx\right) \qquad (2.1.17)$$

and neither y' nor z' appears. The homogeneous equation (2.1.2) now takes the form

$$z''(x) + I(x)z(x) = 0. \qquad (2.1.18)$$

The invariant

$$I(x) = \frac{p_2(x)}{p_0(x)} - \frac{1}{4}\frac{p_1^2(x)}{p_0^2(x)} \qquad (2.1.19)$$

is a means to classify differential equations of second order. Thus, the general solutions of two differential equations having the same invariant differ only by a factor.

The method just described can be applied on two examples. We consider the inhomogeneous equation

$$y'' - 3y' + 2y = \exp(5x). \qquad (2.1.20)$$

Its solution is given by

$$y = C_1\exp(x) + C_2\exp(2x) + \exp(5x)/12, \qquad (2.1.21)$$

where C_1 and C_2 are constants. The same result can be obtained by the *Mathematica* command

DSolve[y″[x]-3*y′[x]+2*y[x]==Exp[5*x],y[x],x]

We now consider the *boundary problem* (two-point problem)

$$y(0) = 0, \quad y'(1) = 3 \qquad (2.1.22)$$

of the equation

$$y'' + y' - 6y = 8\exp(3x). \qquad (2.1.23)$$

The general solution of (2.1.23) is given by

$$y = C_2\exp(2x) + C_1\exp(-3x) + 4\exp(3x)/3. \qquad (2.1.24)$$

Now the integration constants C_1 and C_2 can be determined by inserting (2.1.24) into (2.1.22). Again, the solution (2.1.24) can be obtained by the *Mathematica* command

DSolve[y″[x]+y′[x]-6*y[x]==8*Exp[3*x],y[x],x]

Problems

1. For constant C_1, C_2 insert (2.1.4) into (2.1.1) and obtain the resulting differential equations for y_0, y_1, y_2. Does the solution (2.1.21) of (2.1.20) satisfy these equations derived by you? (Answer should be yes. Try to use *Mathematica* for this calculation.)

2. Calculate the invariant I (2.1.19) for the two equations (2.1.20) and (2.1.23) (Answer: $-1/4$ and $-25/4$.) Try to reproduce the solutions (2.1.21) and (2.1.24) by using (2.1.18) and (2.1.16).

3. In *Mathematica* the WRONSKIAN (2.1.12) can be defined by a determinant. Since a determinant is an operation on a matrix, we first have to define a matrix. We use the solutions $y_1(x)$ and $y_2(x)$ contained in (2.1.21):

   M={{Exp[x],Exp[2*x]},{Exp[x],2*Exp[2*x]}}

   then **Det[M]** results in e^{3x}. Calculate the WRONSKIAN for the solution (2.1.24). The answer should read $-5e^{-x}$.

4. Calculate the WRONSKIAN for

   (a) $sin(x), cos(x)$. (Answer: $(-cos^2(x) - sin^2(x))$)

   (b) $\sin x, \sin x$. (Answer: 0. Why?)

   (c) x^2, x^3. (Answer: x^4)

5. Solve $p_0 y''(x) + p_1 y'(x) + p_2 y = 0$ for constant p_0, p_1, p_2. In order to delete previous definitions for y we use

   Clear[y];DSolve[p0*y''[x]+p1*y'[x]+p2*y[x]==0,y[x],x]

   The result looks complicated, **Simplify[%]** does not help very much. But is the result correct? Can we verify the output of the calculation by inserting it into the differential equation? To do so we bring the result into the input form by using a new function $u(x)$. The new function must have x as an independent variable guaranteeing that $u(x)$ is a global function giving values for any x.

   The following example will clear the situation:

   u[x]=4*x^2 gives **u[2]=u[2]** but **v[x_]:=4*x^2** gives **v[2]=16**

   To verify the solution of the differential equation we use again

   DSolve[p0*y''[x]+p1*y'[x]+p2*y[x]==0, y[x],x]

   and we give the commands **Clear[u];u[x_]:=InputForm[%]**

   Simplify[p0*u''[x]+p1*u'[x]+p2*u[x]]

which yields $p^2\{\text{Null}\}$, zero, our differential equation is satisfied by the solution:

$$y[x] \to \exp(\frac{(-p1 - \sqrt{p1^2 - 4\ p0\ p2})\ x}{2\ p0}) \ C[1]$$

$$+ \exp(\frac{(-p1 + \sqrt{p1^2 - 4\ p0\ p2})\ x}{2\ p0}) \ C[2].$$

2.2 Solving linear differential equations

As a first example of the solution of a boundary value problem, we consider the linear differential equation

$$y'' + y = 0, \tag{2.2.1}$$

which has the general solution

$$y(x) = A \sin x + B \cos x. \tag{2.2.2}$$

$A \sin x$ or $15 \cos x$ would be particular solutions. The general solution admits both initial or boundary value problems. If we choose the *initial conditions* (*one-point conditions*)

$$y\left(\frac{\pi}{2}\right) = 10, \quad y'\left(\frac{\pi}{2}\right) = 0.5, \tag{2.2.3}$$

then the integration constants A and B can be obtained from (2.2.2)

$$10 = A \sin \frac{\pi}{2} + B \cos \frac{\pi}{2} = A, \tag{2.2.4}$$

$$0.5 = A \cos \frac{\pi}{2} - B \sin \frac{\pi}{2} = -B. \tag{2.2.5}$$

On the other hand, if we choose the *boundary conditions* (*two point conditions*)

$$y(x_0) = y_0 = 10, \tag{2.2.6}$$

$$y_1(x_1) = y_1 = 20 \tag{2.2.7}$$

we obtain from (2.2.2)

$$A = \frac{10 \cos x_1 - 20 \cos x_0}{\sin x_0 \cos x_1 - \sin x_1 \cos x_0}, \tag{2.2.8}$$

$$B = \frac{20\sin x_0 - 10\sin x_1}{\sin x_0 \cos x_1 - \sin x_1 \cos x_0}. \tag{2.2.9}$$

Thus, the general solution of the boundary value problem (2.2.1), (2.2.6) and (2.2.7) is given by the sum of two particular solutions

$$y(x) = \frac{10\cos x_1 - 20\cos x_0}{\sin x_0 \cos x_1 - \sin x_1 \cos x_0}\sin x$$

$$+ \frac{20\sin x_0 - 10\sin x_1}{\sin x_0 \cos x_1 - \sin x_1 \cos x_0}\cos x. \tag{2.2.10}$$

A warning is now necessary: not all arbitrary boundary problems can be solved. If we assume that

$$y\left(\frac{\pi}{2}\right) = 1, \quad y'(0) = 0, \tag{2.2.11}$$

then the general solution (2.2.2) is not able to satisfy these equations. From (2.2.11) one obtains the contradiction

$$A = 1 \quad \text{and} \quad A = 0. \tag{2.2.12}$$

If we replace (2.2.11) by

$$y(0) = 1, \qquad y(\pi) = -1, \tag{2.2.13}$$

we get an infinity of solutions, since $B = 1$, but A remains undetermined.

The non-vanishing boundary conditions (2.2.6), (2.2.11) and (2.2.13) are called *inhomogeneous*. Adversely, the vanishing conditions

$$y(x_0) = 0, \qquad y(x_1) = 0 \tag{2.2.14}$$

are called *homogeneous*.

We now have the same situation that we discussed in section 1.2 for partial differential equations. A *boundary problem* is called *homogeneous*, if the differential equation and the boundary conditions are both homogeneous. If the differential equation or the boundary condition or both are inhomogeneous, then the *boundary problem* is said to be *inhomogeneous*.

Having solved the homogeneous equation (2.2.1), we now consider the inhomogeneous *equation of oscillations*

$$y'' + \alpha^2 y = f(x). \tag{2.2.15}$$

A particular solution of the homogeneous equation is given by $y(x) = A\cos\alpha x$. In order to solve the inhomogeneous equation we use the method of *variation of constants*. In analogy to (2.1.5) we write

$$y = A(x)\cos\alpha x. \tag{2.2.16}$$

Inserting into the inhomogeneous equation (2.2.15) one obtains the inhomogeneous equation for $A(x)$

$$A'' \cos \alpha x - 2A' \alpha \sin \alpha x = f(x). \tag{2.2.17}$$

Application of the *Mathematica* command

**DSolve[A''[x]*Cos[α*x]-2*A'[x]* α*Sin[α*x]
==f[x],A[x],x]**

yields an expression like

$$\left\{ \left\{ A[x] \rightarrow C[2] + \int_{K\$85}^{x} \left[C[1] * \mathrm{Sec}\,[\alpha K\$84]^2 + \right. \right. \right.$$

$$\left. \left. \left. \left[\int_{K\$66}^{K\$84} \mathrm{Cos}\,[\alpha K\$65]\, f\,[K\$65]\, \mathrm{d}K\$65 \right] \mathrm{Sec}\,[\alpha K\$84]^2 \right] \mathrm{d}K\$84 \right\} \right\} \tag{2.2.18}$$

This apparently means that *Mathematica* can't solve (2.2.17). To solve the equation step by step, we consider the corresponding homogeneous equation

$$A'' - 2\alpha A' \tan \alpha x = 0. \tag{2.2.19}$$

The substitution $u(x) = A'(x), u'(x) = A''(x)$ gives the *separable equation*

$$\frac{\mathrm{d}u}{u} = 2\alpha \tan \alpha x \mathrm{d}x. \tag{2.2.20}$$

Integration yields

$$A'(x) = u(x) = C \exp\left(2\alpha \int \tan \alpha x \mathrm{d}x \right) = C/\cos^2 \alpha x. \tag{2.2.21}$$

It seems that *Mathematica* is not able to integrate equation (2.2.19). Since we do not need $A(x)$ itself, we can now solve the inhomogeneous equation (2.2.17) directly by inserting $A'(x) = C(x)/\cos^2\alpha x$ into it. The result after some short algebra is

$$\frac{\mathrm{d}C}{\mathrm{d}x} = C'(x) = f(x) \cos \alpha x \tag{2.2.22}$$

and

$$C(x) = \int f(x) \cos \alpha x \mathrm{d}x. \tag{2.2.23}$$

This result may also be derived with the help of *Mathematica*:

**A'[x]=C[x]/Cos[α*x]^2;A''[x]=D[A'[x],x];
Simplify[A''[x]*Cos[α*x]-
2*A'[x]*α*Sin[α*x]-f[x]]**

which again yields (2.2.22), but in the form

$$-f[x] + \text{Sec}[\alpha * x] * C'[x] = 0. \tag{2.2.24}$$

Finally, we obtain

$$A(x) = \int u(x)\mathrm{d}x = \int \cos^{-2}\alpha x \left[\int f(x)\cos\alpha x\mathrm{d}x \right]\mathrm{d}x \tag{2.2.25}$$

and

$$y(x) = \int \left[\int f(x)\cos\alpha x\mathrm{d}x \right] \cos^{-2}\alpha x\mathrm{d}x \cos\alpha x + A\cos\alpha x + B\sin\alpha x. \tag{2.2.26}$$

Here the first term is the particular solution of the inhomogeneous equation (2.2.15) and the other two terms represent the general solution of the homogeneous equation. Solution (2.2.26) is a consequence of the theorem that the solution of an inhomogeneous linear equation consists of the superposition of a particular solution of the inhomogeneous equation and the general solution of the homogeneous equation.

For the special function $f(x) = A\sin\beta x + D$, where A, D and β are given constants, we now solve the boundary problem

$$y(x_0) = y_0, \quad y(x_1) = y_1. \tag{2.2.27}$$

Here y_0 and y_1 are given constant values. With the function $f(x)$ given, the solution (2.2.26) takes two forms. For *resonance* between the eigenfrequency α and the exterior excitation frequency β, i.e., for $\alpha = \beta$, the solution is

$$y(x) = \frac{C}{2\alpha}x\sin\alpha x + B\sin(\alpha x - \delta), \tag{2.2.28}$$

where C, B and δ are constant. Solutions of this type are not able to satisfy (2.2.27). They are called *secular* and play a role in approximation theory. The second form of the solution is valid for $\alpha \neq \beta$ and reads

$$y(x) = \frac{A}{\alpha^2 - \beta^2}\sin\alpha x + \frac{D}{\alpha^2} + B\sin(\alpha x - \delta), \quad \alpha \neq \beta. \tag{2.2.29}$$

If one combines this solution with the boundary conditions (2.2.27), one gets

$$y_0 = \frac{A}{\alpha^2 - \beta^2}\sin\alpha x_0 + \frac{D}{\alpha^2} + B\sin(\alpha x_0 - \delta), \tag{2.2.30}$$

$$y_1 = \frac{A}{\alpha^2 - \beta^2}\sin\alpha x_1 + \frac{D}{\alpha^2} + B\sin(\alpha x_1 - \delta). \tag{2.2.31}$$

These equations determine the integration constants B and δ.

Since we now know that inhomogeneous problems of linear equations can be reduced to a homogeneous problem, we restrict ourselves to discuss methods to solve the homogeneous equation. We write equation (2.1.2) in the form

$$y''(x) + p_1(x)y'(x) + p_2(x)y(x) = 0. \tag{2.2.32}$$

Here the functions p_1 and p_2 are the functions p_1/p_0 and p_2/p_0 from (2.1.2) renamed. The differential equation (2.2.32) is called to be of the FUCHSIAN *type*, if the functions are *regular* (*rational*) with exception of *poles* (local regular singular points). To make this clear, we consider the EULER *equation*

$$a(x - x_0)^2 y''(x) + b(x - x_0)y'(x) + cy(x) = 0, \tag{2.2.33}$$

which is a special case of (2.2.32) and where a, b and c are constants. A point x_0 is called an *ordinary point*, if $a(x - x_0) \neq 0$ and a *singular point*, if $a(x - x_0) = 0$. Near an ordinary point solutions of (2.2.32) can be found using the method of *power series* $\sum_{n=0}^{\infty} a_n x^n$. Near a singular point the FROBENIUS *method* will be used.

Instead of (2.2.33) we can consider (2.2.32). Now $b(x - x_0)/a(x - x_0)^2$ will be replaced by $p_1(x)$. Thus if $a(x - x_0) = 0$, then $p_1(x) = \infty$ (singular point pole). A function regular everywhere but with one pole at x_0 can no longer be expanded into a power series, but it can be represented by a LAURENT *series* $\sum_{n=-\infty}^{\infty} a_n(x - x_0)^n$. (All these considerations could better be done in the complex plane $z = x + iy$.)

If $a_n = 0$ for $n < -m < 0$, $a_{-m} \neq 0$, one says that the point x_0 is a pole (a regular singular point) of order m. Singular points that are not poles are called *irregular singular* or *essential singular*. Equation (2.2.32) is thus called a FUCHS *equation*, if $xp_1(x)$ and $x^2 p_2(x)$ are regular for $x \to \infty$, that means that $p_1(x)$ has a pole of first, and $p_2(x)$ of second order, respectively. These regular (rational) functions can be expanded

$$p_1(x) = \sum_{l=1}^{L} \frac{A_l}{x - a_l}, \quad p_2(x) = \sum_{l=1}^{L} \left(\frac{B_l}{(x - a_l)^2} + \frac{C_l}{x - a_l} \right), \quad \sum_{l} C_l = 0. \tag{2.2.34}$$

For $L = 1$, we may write

$$p_1(x) = \frac{1}{x} \sum_{n=0}^{\infty} \alpha_n x^n, \quad p_2(x) = \frac{1}{x^2} \sum_{n=0}^{\infty} \beta_n x^n. \tag{2.2.35}$$

According to FROBENIUS, the singularity can be split off and the solution of (2.2.32) can be rewritten as a so-called FROBENIUS *series*

$$y(x) = x^\rho \sum_{n=0}^{\infty} a_n x^n, \quad a_0 \neq 0. \tag{2.2.36}$$

ρ is called the index of the series. The series is convergent. Since the method of power series is just the special case $\rho = 0$ of the FROBENIUS method, we will discuss only the latter.

We will now solve (2.2.32) using the FROBENIUS method. (2.2.35) yields

$$y'(x) = \rho x^{\rho-1} \sum_{n=0}^{\infty} a_n x^n + x^\rho \sum_{n=0}^{\infty} a_n n x^{n-1},$$

$$y''(x) = \rho(\rho-1)x^{\rho-2} \sum_{n=0}^{\infty} a_n x^n + 2\rho x^{\rho-1} \sum_{n=0}^{\infty} a_n n x^{n-1}$$

$$+ x^{\rho} \sum_{n=0}^{\infty} n(n-1)a(n-1)a_n x^{n-2}, \tag{2.2.37}$$

since a convergent power series may be differentiated.

Inserting (2.2.35) and (2.2.36) into (2.2.32) and using (2.2.37) we obtain

$$\sum_{n=0}^{\infty} x^{n+\rho-2} \left[\rho(\rho-1)a_n + 2\rho n a_n + n(n-1)a_n + \alpha_0 \rho a_n + \alpha_0 n a_n + \beta_0 a_n \right]$$

$$+ \sum_{n=1}^{\infty} x^{n+\rho-2} \left[\alpha_n a_n + \alpha_n a_n + \beta_n a_n \right] = 0. \tag{2.2.38}$$

A power series vanishes only if all coefficients vanish. For $n = 0$, (2.2.38) reads

$$\rho^2 + \rho(\alpha_0 - 1) + \beta_0 = 0, \tag{2.2.39}$$

since $a_0 \neq 0$ cancels. The case $n \neq 0$ will be treated later. Equation (2.2.39) is the so-called *indicial equation*. We now apply the method on a special form of the EULER equation (2.2.33). We use $x_0 = 0, a = 1, b = 3, c = -3$, so that $\alpha_0 = b/a = 3, \beta = c/a = -3$,

$$x^2 y''(x) + 3xy'(x) - 3y(x) = 0. \tag{2.2.40}$$

Then the indicial equation reads:

$$\rho^2 + 2\rho - 3 = 0. \tag{2.2.41}$$

Its solutions are $\rho_1 = 1, \rho_2 = -3$, so that the solution (2.2.36) of (2.2.40) is given by the superposition of two particular solutions

$$y(x) = Ax + Bx^{-3}. \tag{2.2.42}$$

The command

DSolve[x^2*y''[x]+3*x*y'[x]-3*y[x]==0, y[x],x]

delivers the same result. It would be easy to show that this solution satisfies for instance the initial conditions (2.2.3). Also the boundary conditions (2.2.6) and (2.2.7) can be satisfied by (2.2.42).

In the case that the indicial equation has real repeated roots $\rho_1 = \rho_2$ or $\rho_2 = \rho_1 - n, n = 0, 1, 2, \ldots$, the FROBENIUS method delivers only the first solution $y_1(x)$. The second solution will then contain an essential singularity like a logarithmic term. This solution may be derived from (2.1.15). For $p_0 = 1$ this equation reads

$$y_2(x) = y_1(x) \int \frac{\exp\left(-\int p_1(x)dx\right)}{y_1^2(x)} dx. \tag{2.2.43}$$

If one inserts $p_1(x)$ from (2.2.35) into (2.2.43) one obtains after integration

$$\exp\left(-\int p_1(x)\mathrm{d}x\right) = \exp\left(-\alpha_0 \ln x - \alpha_1 x - \frac{1}{2}\alpha_2 x^2 - \ldots\right)$$

$$= x^{-\alpha_0} P_2(x), \qquad (2.2.44)$$

where P_2 is a regular power series that does not vanish for $x = 0$. Assume that the first solution $y_1(x)$ has the form

$$y_1(x) = x^{\rho_1} P_1(x), \qquad (2.2.45)$$

where $P_1(x)$ is a regular power series that does not vanish for $x = 0$. Then the integrand in (2.2.43) may be written in the form

$$\frac{1}{y_1^2(x)}\exp\left(-\int p_1(x)\mathrm{d}x\right) = x^{-2\rho_1 - \alpha_0} P_1(x)^{-2} P_2(x) = x^{-n-1} P_3(x),$$
$$\qquad (2.2.46)$$

since $2\rho_1 + \alpha_0 = n+1$. Expanding the regular power series $P_3(x) = \sum_m \gamma_m x^m$, the integral becomes

$$\int x^{-n-1} P_3(x)\mathrm{d}x = \sum_{m=0}^{\infty} \frac{\gamma_m x^{m-n}}{m-n} + \gamma_n \ln x, \quad m \neq n. \qquad (2.2.47)$$

For $x_0 = 0$, $\rho^2 + (a - 1)\rho + b$, $\rho_1 = \rho_2$ the EULER equation (2.2.33) has the solution $y = C_1 x^\rho + C_2 x^\rho \ln x$. For $\rho_1 = \rho_2 = 2$ the command

DSolve[x^2*y''[x]+2*x*y'[x]+2*y[x]==0, y[x],x]

yields an expression containing power of x with complex exponents, which is equivalent to

$$y(x) = x^{-1/2}\left[C_2 \cos\left(\frac{1}{2}\sqrt{7}\ln x\right) - C_1 \sin\left(\frac{1}{2}\sqrt{7}\ln x\right)\right].$$

If the roots of the indicial equation are complex, they must be conjugate. Then the solution of (2.2.33) may be expressed in terms of trigonometric functions.

Up to now we have investigated only the case $n = 0$. We use the BESSEL equation

$$y'' + \frac{1}{x}y' + \left(1 - \frac{n^2}{x^2}\right)y = 0 \qquad (2.2.48)$$

to demonstrate the procedure for $n > 0$. Inserting (2.2.36) into (2.2.48) we receive (for n replaced by ν)

$$\sum_{\nu=0}^{\infty} x^{\nu+\rho-2} c_\nu \left[\rho(\rho - 1) + 2\rho\nu + \nu(\nu - 1) + \rho + \nu - n^2\right]$$

$$+ \sum_{\nu=0}^{\infty} x^{\nu+\rho} c_\nu = 0. \qquad (2.2.49)$$

For $\nu = 0$ one obtains the indicial equation $c_0(\rho^2 - n^2) = 0$ and $\rho_{1,2} = \pm n$. (We can expect the existence of a logarithmic solution. This solution and the determination of c_0 will be discussed later on). Making the replacement $\nu \to \nu - 2$ we can join the two sums into one to receive

$$\sum_{\nu=0} x^{\nu+\rho-2} \left[c_\nu \left(\rho^2 + 2\rho\nu + \nu^2 - n^2 \right) + c_{\nu-2} \right] = 0. \qquad (2.2.50)$$

For $[\,] = 0$, we thus obtain the two-termed *recurrence relation*

$$c_\nu = \frac{-c_{\nu-2}}{\nu(\nu + 2\rho)} = \frac{-c_{\nu-2}}{\nu(\nu + 2n)}. \qquad (2.2.51)$$

c_0 is still unknown. If we choose $\nu = 1$, $\nu + \rho - 2 = \rho - 1$, then (2.2.49) yields $c_1 = 0$. Furthermore, we find $c_1 = c_3 = c_5 = \ldots = 0$, so that only $\nu = 0, 4, 6$ appears and the series representing the BESSEL *functions* contains only the power 0, 4, 6, ...

The command

DSolve[y''[x]+y'[x]/x+y[x]*(1-n^2/x^2)==0,y[x],x]

produces the solution

BesselJ[n,x] C[1] + BesselY[n,x] C[2]

(where the space replaces the * representing multiplication). As an exercise, the reader is invited to solve the following equations using the FROBENIUS method and *Mathematica*.

Equation	Solution	Indicial Equ.
$y'' + \omega^2 y = 0$	$y = \frac{a_1}{\omega} \sin \omega x,$	$\rho(\rho - 1) = 0$
$y'' + \frac{1}{x}y' - \left(1 + \frac{n^2}{x^2}\right)y = 0$	$y = J_n(ix) = I_n(x)$	$\rho^2 = \pm n^2$
$y'' - \frac{6}{x^2}y = 0$	$y = C_1 x^3 + C_2 x^{-2},$	$\rho^2 - \rho = 6$
$y'' - 6y/x^3 = 0$	essential singularity $x = 0,$	$-6a_0 = 0$
$y'' + \frac{1}{x}y' - \frac{a^2}{x^2}y = 0$	$y = C_1 x^a + C_2 x^{-a},$	$\rho^2 = a^2$
$y'' + \frac{x+1}{2x}y' + \frac{3}{2x}y = 0$	$c_\nu = -\frac{\rho+\nu+2}{(\rho+\nu)(2\rho+2\nu-1)}c_{\nu-1}, \nu \geq 1,$	$\rho(2\rho - 1) = 0$
$y'' - \frac{1}{2x}y' + \frac{x^2+1}{2x^2}y = 0$	$c_\nu = -\frac{c_{\nu-2}}{(\rho+\nu)(2\rho+2\nu-3)+1}, \nu \geq 2, (\rho-1)(\rho-2) = 0$	
$y'' + \frac{2}{3x}y' + \frac{xy}{3} = 0$	$c_\nu = -\frac{c_{\nu-3}}{(\rho+\nu)(3\rho+3\nu-1)}, \nu \geq 3,$	$\rho(3\rho - 1) = 0$
$y'' - xy' + ny = 0$	$c_{\nu+2} = \frac{-(n-\nu)c_\nu}{(\nu+1)(\nu+2)}, \nu = 0, 1, \ldots,$	$\rho = 0$

Due to the recurrence formulae one has $c_{\nu+2}/c_\nu \to 0$ for $\nu \to \infty$ and therefore convergence.

Apparently, the singularities appearing in differential equations help to classify linear differential equations of second order. If the coefficients are given by

$$p_1(z) = \frac{1}{2}\left[\frac{m_1}{z-a_1} + \frac{m_2}{z-a_2} + \ldots + \frac{m_{n-1}}{z-a_{n-1}}\right], \qquad (2.2.52)$$

$$p_2(z) = \frac{1}{4}\left[\frac{A_0 + A_1 z + \ldots + A_l z^l}{(z-a_1)^{m_1}(z-a_2)^{m_2}\ldots(z-a_{n-1})^{m_{n-1}}}\right], (2.2.53)$$

then the equation (2.2.32) is called a BÔCHER *equation*[1.5]. Nearly all boundary value problems one may come across in physics, engineering and applied mathematics are of this type ($n \leq 4$, $l \leq 3$, $m_3 \leq 2$, $m_1 = 1$).

We now discuss some special cases. *Four* singularities are to be found in (2.2.54) to (2.2.56):

$$y''(z) + \frac{1}{2}\left[\frac{1}{z-a_1} + \frac{2}{z-a_2} + \frac{2}{z-a_3}\right]y'(z)$$
$$+ \frac{1}{4}\left[\frac{A_0 + A_1 z + A_2 z^2 + A_3 z^3}{(z-a_1)(z-a_2)^2(z-a_3)^2}\right]y(z) = 0. \qquad (2.2.54)$$

(HEINE *equation*, one pole of first order, two poles of second order and one pole at infinity),

$$y'' + \frac{1}{2}\left[\frac{1}{z} - \frac{1}{z-a_2} + \frac{1}{z-a_3}\right]y'$$
$$+ \frac{1}{4}\left[\frac{(a_2^2 + a_3^2)q - p(p+1)z + \kappa^2 z^2}{z(z-a_2)(z-a_3)}\right]y = 0. \qquad (2.2.55)$$

(LAMÉ *wave equation* or LAMÉ *equation for* $\kappa = 0$) for $\kappa = 0$, and

$$y'' + \frac{1}{2}\left[\frac{1}{z-a_1} + \frac{1}{z-a_2} + \frac{2}{z-a_3}\right]y'$$
$$+ \frac{1}{4}\left[\frac{A_0 + A_1 z + A_2 z^2}{(z-a_1)(z-a_2)(z-a_3)^2}\right]y = 0, \qquad (2.2.56)$$

WANGERIN *equation*. *Three* singularities are contained in (2.2.57) to (2.2.60)

$$y'' + \frac{1}{2}\left[\frac{1}{z-a_1} + \frac{1}{z-a_2}\right]y' + \frac{1}{4}\left[\frac{A_0 + A_1 z + A_2 z^2}{(z-a_1)(z-a_2)}\right]y = 0 \qquad (2.2.57)$$

(two poles of first order in the finite domain and one pole of fourth order in infinity), and in

$$y'' + \frac{1}{2}\left[\frac{2}{z-a_1} + \frac{2}{z-a_2}\right]y'$$
$$+ \frac{1}{4}\left[\frac{A_0 + A_1 z + A_2 z^2 + A_3 z^3 + A_4 z^4}{(z-a_1)^2(z-a_2)^2}\right]y = 0. \qquad (2.2.58)$$

(two poles of second order and one pole of fourth order). Also the MATHIEU *equation* $y'' + (\lambda - 2q \cos 2x)y = 0$ has three singularities, two are essential. The *hypergeometric equation* is the "grandmother" of many equations used in physics and engineering. It reads

$$y'' + \frac{c - (a + b + 1)z}{z(1 - z)}y' - \frac{ab}{z(1 - z)}y = 0. \tag{2.2.59}$$

This important equation has poles at 0, 1 and ∞. The values of the index ρ are $\rho_1 = 0$, $\rho_2 = 1 - c$ at the location $x = 0$, $\rho_1 = 0$, $\rho_2 = c - a - b$ at $x = 1$ and $\rho_1 = a$, $\rho_2 = b$ at infinity. The LEGENDRE *wave equation*

$$y'' + \frac{2z}{z^2 - 1}y' + \left[\frac{\kappa^2 a^2 (z^2 - 1) - p(p + 1)}{z^2 - 1} - \frac{q^2}{(z^2 - 1)^2}\right] y = 0, \tag{2.2.60}$$

(and the LEGENDRE *equation* $\kappa = 0$) are children of (2.2.59) exhibiting three singularities.

Two singularities are found in (2.2.61) - (2.2.70):

$$y'' + \frac{1}{2}\left[\frac{4}{z - a_1}\right]y' + \frac{1}{4}\left[\frac{A_2 z^2}{(z - a_1)^4}\right] y = 0 \tag{2.2.61}$$

and

$$y'' + \frac{1}{2}\left[\frac{2}{z - a_1}\right]y' + \frac{1}{4}\left[\frac{A_0 + A_2 z^2 + A_4 z^4}{(z - a_1)^2}\right] y = 0, \tag{2.2.62}$$

but even the simple equation

$$y'' + \frac{1 + a}{z}y' = 0, \quad y_1 = 1, \quad y_2 = z^{-a}, \tag{2.2.63}$$

has two poles at 0 and ∞. Furthermore, some well-known and important equations have two singularities: the *confluent hypergeometric equation* (KUMMER *equation*)

$$y'' + \frac{c - z}{z}y' - ay = 0, \tag{2.2.64}$$

which is a "daughter" of (2.2.59). It has one pole at $z = 0$ and an essential singularity at ∞; $\rho(\rho + c - 1) = 0$ is its indicial equation. The BESSEL *equation* (2.2.48) and

$$y'' + \frac{2}{z}y' - \frac{p(p + 1)}{z^2}y = 0, \tag{2.2.65}$$

the BESSEL *wave equation*

$$y'' + \frac{1}{z}y' + \left(\kappa^2 z^2 + q^2 - p^2/z^2\right) y = 0, \tag{2.2.66}$$

as well as the *generalized* BESSEL *equation*

$$y'' + \frac{1 - 2\alpha}{z}y' + \left[\left(\beta\gamma z^{\gamma - 1}\right)^2 + \frac{\alpha^2 - p^2\gamma^2}{z^2}\right] y = 0, \tag{2.2.67}$$

which has the solution

$$y = z^\alpha Z_p(\beta z^\gamma), \tag{2.2.68}$$

(Z_p is a cylinder function like J_p), have also two singularities. Other children of (2.2.59) are the WHITTAKER *equation*

$$y'' + \left(-\frac{1}{4} + \frac{\kappa}{z} + \frac{1/4 - \mu^2}{z^2}\right) y = 0, \tag{2.2.69}$$

solved by WHITTAKER *functions*, or the GEGENBAUER *equation*

$$y'' + \frac{2(\alpha + 1)}{z^2 - 1} z y' - \frac{n(n + 2\alpha + 1)}{z^2 - 1} y = 0. \tag{2.2.70}$$

One singularity will be found in

$$y'' + \frac{1}{2}\left[\frac{m_1}{z - a_1}\right] y' + \frac{1}{4}\left[\frac{A_0}{(z - a_1)^{m_1}}\right] y = 0, \tag{2.2.71}$$

which comprises the EULER equation. The most simple linear differential equation of second order is given by

$$y'' = 0. \tag{2.2.72}$$

The equation $y'' + 2y'/(z - a)$ has a pole at $z = a$ and the solutions $y_1 = 1$, $y_2 = 1/(z - a)$. Also the WEBER *equation*

$$y'' + \left[q^2\left(p + \frac{1}{2}\right) - q^4 z^2/4\right] y = 0, \tag{2.2.73}$$

which is a "grandchild" of (2.2.59) has one pole, but $y'' - ky = 0$ has an essential singularity at $z = \infty$. Also (2.2.1) and (2.2.72) have one singularity. This is a consequence of the LIOUVILLE *theorem*, which expresses the fact that all functions $y(z)$ of a complex variable $z = x + iy$ must either have one (or more) singularities or be a constant.

Problems

1. Solve (2.2.1) **y[x]→C[1] Cos[x]+C[2] Sin[x]**

2. Now solve the initial value problems (2.2.1), (2.2.3) using

   **DSolve[{y″[x]+y[x]==0,y[Pi/2]==10,y′[Pi/2]==0.5},
   y[x],x]**

   which gives **y[x]->0.5 Cos[x]...** Another possibility is numerical integration

```
NDSolve[{y''[x]+y[x]==0,y[1.5]==6.,y'[1.5]==-20.},
y,{x,0,Pi}]
```

InterpolatingFunction[{{0., 3.14159}}, <>]

In order to plot the result we use now

```
Plot[Evaluate[y[x]/.%],{x,0,Pi}]
```

In order to plot, the values of $y(x)$ must be known. **Evaluate** replaces the definition of a new function (as $v(x)$ in problem 5 of section 2.1). The phrase **y[x]/.%** has the meaning: "replace $y(x)$ by the result of the last calculation, i.e., the solution of the initial value problem."

3. Now use *Mathematica* to solve the inhomogeneous boundary value problem (2.2.6) numerically for $x_0 = 0, x_1 = 2$.
```
bsol=NDSolve[{y''[x]+y[x]==0,y[0]==1.,y[2.]==2.},
y[x],{x,0,Pi}]
```
Here we have given a name to the calculation. Plotting is now possible using **Plot[Evaluate[y[x]/.bsol],{x,0,Pi}]**

4. Now solve the homogeneous boundary problem (2.2.14)
$x_0 = 0, x_1 = 3.14159, y_0 = y(x_0) = 0, y_1 = y(x_1) = 0$.
```
Clear[y];
ts=NDSolve[{y''[x]+y[x]==0,y[0]==0,y[Pi]==0},
y[x],{x,0,Pi}]
```

and plot the result. If this does not work, look at the values $y(x)$ by **Table[ts,{x,0,Pi,0.4}]**

5. Find the indicial equation (2.2.39) or (2.2.38) for the following equations, solve them according to the FROBENIUS method and verify the result with *Mathematica*. Take some of the equations on page 38:
```
DSolve[y''[x]+m^2*y[x]==0,y[x],x]
```
$\{\{y[x] \to C[1] \operatorname{Cos}[mx] + C[2] \operatorname{Sin}[mx]\}\}$
```
DSolve[y''[x]+y'[x]/x-(1+n^2/x^2)*y[x]==0,y[x],x]
```
$\{\{y[x] \to \operatorname{BesselJ}[n, -ix] \, C[1] + \operatorname{BesselY}[n, -ix] \, C[2]\}\}$
(modified BESSEL function)
```
DSolve[y''[x]-6*y[x]/x^2==0,y[x],x]
```
$\{\{y[x] \to x^3 \, C[1] + \frac{C[2]}{x^2}\}\}$
```
DSolve[y''[x]-6*y[x]/x^3==0,y[x],x]
```

$$y[x] \to \frac{\operatorname{BesselI}\left[1, 2\sqrt{6}\sqrt{\frac{1}{x}}\right] C[1]}{\sqrt{6}\sqrt{\frac{1}{x}}} + \frac{\sqrt{\frac{2}{3}} \operatorname{BesselK}\left[1, 2\sqrt{6}\sqrt{\frac{1}{x}}\right] C[2]}{\sqrt{\frac{1}{x}}}$$

```
DSolve[y''[x]+y'[x]/x-a^2*y[x]/x^2 ==0,y[x],x]
```

$$\{\{y[x] \rightarrow C[1] \text{ Cosh}[a \text{ Log}[x]] + i \ C[2] \text{ Sinh}[a \text{ Log}[x]]\}\}$$

NDSolve[y''[x]+(x+1)*y'[x]/2*x+3*y[x]/2*x==0,y[x],x]

This does not work. Why? (For numerical solution a range must be given, see problem 3).

6. Solve the initial value problem of an equation of third order:

NDSolve[{y'''[x]+y''[x]+y[x]==0,y[0]==5,y'[0]==-10,
y''[0]==80},y,{x,0,1}]

7. Solve (2.2.59), (2.2.63), (2.2.64), (2.2.69), (2.2.72) and (2.2.73) giving

DSolve[y''[x]+(c-(a+b+1)*x)*y'[x]/(x*(1-x))-
a*b*y[x]/(x*(1-x))==0,y[x],x]

$$\{\{y[x] \rightarrow C[1] \text{ Hypergeometric2F1}[a, b, c, x] +$$
$$(-1)^{1-c} \ x^{1-c} \ C[2] \text{ Hypergeometric2F1}[1 + a - c, 1 + b - c, 2 - c, x]\}\}$$

DSolve[y''[x]+(a+1)*x*y'[x]/x==0,y[x],x]

$$\left\{\left\{y[x] \rightarrow \frac{\exp(-(1 + a) \ x) \ C[1]}{-1 - a} + C[2]\right\}\right\}$$

DSolve[y''[x]+(c-x)*y'[x]-a*y[x]==0,y[x],x]

$$\left\{\left\{y[x] \rightarrow C[1] \text{ HermiteH}\left[-a, -\frac{c}{\sqrt{2}} + \frac{x}{\sqrt{2}}\right] +\right.\right.$$
$$\left.\left. C[2] \text{ Hypergeometric1F1}\left[\frac{a}{2}, \frac{1}{2}, \left(-\frac{c}{\sqrt{2}} + \frac{x}{\sqrt{2}}\right)^2\right]\right\}\right\}$$

DSolve[y''[x]+(-0.25+κ/x+(0.25-μ^2)/x^2)+y[x]==0,
y[x],x]

$$\{\{y[x] \rightarrow C[1] \text{ Cos}[1. \ x] + C[2] \text{ Sin}[1. \ x] +$$

$$\frac{1}{x}(0.25 \ x \text{ Cos}[1. \ x]^2 + 0.25 \ x \text{ Cos}[1. \ x] \text{ CosIntegral}[1. \ x]-$$

$1. \ x \ \mu^2 \ Cos[1. \ x] \text{ CosIntegral}[1. \ x] - 1. \ x \ \kappa \text{ CosIntegral}[1. \ x] \text{ Sin}[1. \ x]+$

$0.25 \ x \text{ Sin}[1. \ x]^2 + 1. \ x \ \kappa \text{ Cos}[1. \ x] \text{ SinIntegral}[1. \ x]+$

$0.25 \ x \text{ Sin}[1. \ x] \text{ SinIntegral}[1. \ x] - 1. \ x \ \mu^2 \text{ Sin}[1. \ x] \text{ SinIntegral}[1. \ x])\}\}$

compare to (2.2.69).

DSolve[y''[x]==0,y[x],x]

$$\{\{y[x] \rightarrow C[1] + x \ C[2]\}\}$$

```
DSolve[y''[x]+(q^2*(p+0.25)-q^4*x^2/4.)*y[x]==0,
y[x],x]
```

$y[x] \to \exp(-0.25\, q^2\, x^2)\, C[1]$

$\mathrm{HermiteH}\left[\dfrac{-0.25\, q^6 + 1.\, p\, q^6}{q^6}, 0.707107\, q\, x\right] + \exp(-0.25\, q^2 x^2)\, C[2]$

$\mathrm{Hypergeometric1F1}\left[\dfrac{-0.25\, q^6 + 1.\, p\, q^6}{2\, q^6}, \dfrac{1}{2}, 0.5\, q^2 x^2\right]$

2.3 Differential equations of physics and engineering

Prior to the discussion of boundary problems, it seems to be useful to investigate some of the differential equations of physics and engineering in more detail. A large class of partial differential equations allows separation into ordinary differential equations. Many of these ordinary differential equations are children or grandchildren of the hypergeometric differential equation.

In spherical problems the separation of the pertinent partial differential equation like, e.g., HELMHOLTZ equation (1.1.35) leads to the LEGENDRE *equation*

$$\frac{d^2 y(\vartheta)}{d\vartheta^2} + \cot\vartheta\, \frac{dy(\vartheta)}{d\vartheta} + l(l+1)y(\vartheta) - \frac{m^2}{\sin^2\vartheta}y(\vartheta) = 0, \qquad (2.3.1)$$

where ϑ is the polar angle in spherical coordinates. The solutions of (2.3.1) are usually called *spherical functions*. The substitution $\cos\vartheta = x$ gives rise to the equation

$$y''(x) - \frac{2x}{1-x^2}y'(x) + \frac{l(l+1)}{1-x^2}y(x) - \frac{m^2}{(1-x^2)^2}y(x) = 0. \qquad (2.3.2)$$

This is the special case $\alpha = 0$ of the GEGENBAUER *equation* (2.2.70). It is easy to see that this equation has poles at the location $x = \pm 1$. For $m = 0$ it has the recurrence relation

$$c_{\nu+2}(\nu+2)(\nu+1) = c_\nu(\nu^2 + \nu - l^2 - l), \quad \nu = 0, 1, \ldots. \qquad (2.3.3)$$

c_0 is defined by (2.3.17). The case $m \neq 0$ will be treated later. With the substitution $x = 1 - 2\xi$, $\xi = -(x-1)/2$ one obtains from (2.3.2) the new equation

$$\xi(1-\xi)y'' + (1-2\xi)y' + l(l+1)y = 0. \qquad (2.3.4)$$

This equation has a pole at $\xi=0$ and is the special case $a = l+1, b = -l, c = 1$ of the hypergeometric equation (2.2.59). The solutions of equation (2.3.4) for arbitrary l are usually called LEGENDRE *functions of the first kind* or *transcendental spherical functions* $\mathcal{P}_n$. In order to write down the general solution of (2.3.4), we need a second solution. But due to the relations $\rho_1 = 0$, $\rho_2 = 1-c$ and $\rho_1 = 0$, $\rho_2 = c-a-b$, which are valid for the hypergeometric equation, one obtains $\rho_1 = 0, \rho_2 = 0$ for both poles. This means that the second solution is identical with $\mathcal{P}_n$. Due to the mother, the hypergoemetric differential equation, its solutions are closely related. So the function cos is the "elliptic", and cosh the "hyperbolic" child. Whereas the function of the first kind $\mathcal{P}_n$ corresponds to cos, the hyperbolic part is given by the LEGENDRE *function* $\mathcal{Q}_n$ of the second kind

$$Q_n(x) = \int\limits_0^\infty \left(x + \sqrt{x^2 - 1} \cosh t\right)^{-n-1} dt, \quad x \geq 1. \tag{2.3.5}$$

This expression is a consequence of the possibility to represent the members of the hypergeometric family by integrals, see later. The general solution of the LEGENDRE equation is now given by

$$y(x) = C_1 P_n(x) + C_2 Q_n(x). \tag{2.3.6}$$

If the parameter l is a natural number (a positive integer), then the $\mathcal{P}_n$ degenerate into polynomials and the $\mathcal{Q}_n$ go over into elementary transcendental functions. For $n = l$, the recurrence relation (2.3.3) breaks down and the solutions are given by the LEGENDRE *polynomials*

$$P_0(x) = 1, \quad P_1(x) = x = \cos\vartheta,$$
$$P_2(x) = \frac{1}{2}\left(3x^2 - 1\right) = \frac{1}{4}\left(3\cos 2\vartheta + 1\right),$$
$$P_3(x) = \frac{1}{2}\left(5x^3 - 3x\right) = \frac{1}{8}\left(5\cos 3\vartheta + 3\cos\vartheta\right). \tag{2.3.7}$$

These polynomials as well as other polynomials are important for physical and engineering problems and may be represented by

a) a RODRIGUEZ *formula*,
b) using a *generating function*,
c) or by an *integral representation*.

These possibilities offer many practical applications. The RODRIGUEZ formula for the LEGENDRE polynomial is given by

$$P_l(x) = \frac{1}{2^l l!} \frac{d^l}{dx^l} \left(x^2 - 1\right)^l \tag{2.3.8}$$

and their generating function f is

$$f(x, u) \equiv \left(1 - 2ux + u^2\right)^{-1/2} = \sum_{l=0}^\infty P_l(x)u^l. \tag{2.3.9}$$

The integral representation comes from the complex function theory, from the CAUCHY *integral*

$$g(z) = \frac{1}{2\pi i} \oint_C \frac{g(t)}{t - z} dt. \tag{2.3.10}$$

For $g(t) = (1 - t^2)^n$ one obtains

$$P_n(z) = \frac{(-1)^n}{2^n n!} \frac{d^n}{dz^n} \left[\left(1 - z^2\right)^n \right] = \frac{(-1)^n}{2^{n+1}\pi i} \oint \frac{\left(1 - t^2\right)^n}{(t - z)^{n+1}} dt. \tag{2.3.11}$$

Using $z \to x = \cos\vartheta$, $0 < \vartheta < \pi$, $t = \cos\vartheta + \sin\vartheta \exp(i\varphi)$, $-\pi \leq \varphi \leq \pi$, $t - x = i\sin\vartheta \exp(i\varphi)$, $dt = -\sin\vartheta \exp(i\varphi)$, etc., one obtains the LAPLACE *integral representation*,

$$P_n(\cos\vartheta) = \frac{1}{\pi} \int_0^\pi (\cos\vartheta + i \sin\vartheta \cos\varphi)^n d\varphi. \tag{2.3.12}$$

Using the integral representation of the solution of the hypergeometric equation (2.2.59), the transcendental functions may be represented by

$$\mathcal{P}_n(x) = \frac{1}{\pi} \int_0^\pi \left(x + \sqrt{x^2 - 1} \cos t \right)^n dt, \quad x \geq 1, \tag{2.3.13}$$

compare (2.3.5).

Up to now we have considered only the special case $m = 0$. For $m \neq 0$ we consider (2.3.2). The FROBENIUS method creates the solutions that are called *associated* LEGENDRE *polynomials* $P_l^m(x)$ and *associated* LEGENDRE *functions*, respectively. *Mathematica* defines all these various functions

$$
\begin{aligned}
P_l(x) &= \textbf{LegendreP[l,x]} \\
P_l^m(x) &= \textbf{LegendreP[l,m,x]} \\
\mathcal{P}_n(z) &= \textbf{LegendreP[n,z]} \\
\mathcal{P}_n^m(z) &= \textbf{LegendreP[n,m,z]} \\
Q_n(z) &= \textbf{LegendreQ[n,z]} \\
Q_n^m(z) &= \textbf{LegendreQ[n,m,z]}
\end{aligned}
\tag{2.3.14}
$$

but has not been able to produce these functions by solving equations (2.3.2).

We give some formulae for the associated polynomials

$$P_l^m = \frac{\left(1 - x^2\right)^{m/2}}{2^l l!} \frac{d^{l+m} \left(x^2 - 1\right)^l}{dx^{l+m}}, \tag{2.3.15}$$

$$P_1^1(x) = \left(1 - x^2\right)^{1/2} = \sin\vartheta,$$

$$P_2^1(x) = 3\left(1 - x^2\right)^{1/2} x = \frac{3}{2}\sin 2\vartheta,$$

$$P_2^2(x) = 3\left(1 - x^2\right) = \frac{3}{2}\left(1 - \cos 2\vartheta\right), \tag{2.3.16}$$

$$P_3^1(x) = \frac{3}{2}\left(1 - x^2\right)^{1/2}\left(5x^2 - 1\right) = \frac{3}{8}\left(\sin\vartheta + 5\sin 3\vartheta\right).$$

There exists an important attribute of these polynomials. This feature is very important for applications. The attribute is called *orthogonality* and is described by

$$\int_{-1}^{1} P_l(x)P_k(x)\mathrm{d}x = \frac{2}{2l+1}\delta_{lk}, \tag{2.3.17}$$

$$\int_{-1}^{1} P_l^m(x)P_k^m(x)\mathrm{d}x = 0, \quad \text{for} \quad l \neq k, \tag{2.3.18}$$

$$\int_{-1}^{1} P_l^m(x)P_l^m(x)\mathrm{d}x = \frac{2}{2l+1}\frac{(l+m)!}{(l-m)!}. \tag{2.3.19}$$

Other children of the hypergeometric equation are the CHEBYSHEV polynomials (and CHEBYSHEV functions, if x is replaced by complex z)

ChebyshevT[n,x] $= T_n(x)$
ChebyshevU[n,x] $= U_n(x)$,

which satisfy the equation

$$\left(1 - x^2\right)T_n''(x) - xT_n'(x) + n^2T_n(x) = 0, \tag{2.3.20}$$

and are given by $T_n(1) = 1$, $T_n(-1) = (-1)^n$, which fixes the c_0 in their recurrence relation, and explicitly by

$$\begin{array}{lll} T_0(x) = 1, & T_1(x) = x, & T_2(x) = 2x^2 - 1, \\ T_3(x) = 4x^3 - 3x, & T_4(x) = 8x^4 - 8x^2 + 1, & T_5(x) = 16x^5 - 20x^3 + 5x. \end{array} \tag{2.3.21}$$

CHEBYSHEV *polynomials* of first kind are often used in making numerical approximation to functions. They satisfy the orthogonality relations

$$\int_{-1}^{1} \frac{T_m(x)T_n(x)}{\sqrt{1 - x^2}}\mathrm{d}x = \begin{cases} 0 & \text{for} \quad m \neq n, \\ \pi/2 & \text{for} \quad m = n \neq 0, \\ \pi & \text{for} \quad m = n = 0. \end{cases} \tag{2.3.22}$$

The polynomials $U_n(x)$ of the second kind and the CHEBYSHEV functions are of minor importance for practical applications.

The important attribute of orthogonality belongs also to other polynomials by LAGUERRE or HERMITE. In contrast to the orthogonality relations (2.3.17) of the LEGENDRE polynomials, the orthogonality relations (2.3.22) contain a *weighting function* $(1-x^2)^{-1/2}$. The same is true for the associated LAGUERRE *polynomials*. They satisfy the LAGUERRE *differential equation*

$$xL_n^{k''}(x) + (1 + k - \alpha x)L_n^{k'}(x) + \alpha n L_n^k(x) = 0, \qquad (2.3.23)$$

and the RODRIGUEZ formula

$$L_n^k(x;\alpha) = L_n(x;\alpha,k) = \frac{e^{\alpha x} x^{-k}}{n!} \frac{d^n}{dx^n}\left(e^{-\alpha x} x^{n+k}\right). \qquad (2.3.24)$$

For $k = 0$ and $\alpha = 1$ the polynomials $L_n^k(x)$ are designated by $L_n(x)$. For $\alpha = 1$, they are given by [3.7]

$$
\begin{aligned}
&L_0(x) = 1, &&L_0^k(x) = 1,\\
&L_1(x) = -x + 1, &&L_1^k(x) = -x + k + 1,\\
&L_2(x) = \frac{x^2}{2} - 2x + 1, &&L_2^k(x) = \frac{x^2}{2} - (k+2)x + \frac{(k+2)(k+1)}{2},
\end{aligned}
$$

$$L_n^k(x) = \sum_{m=0}^{\infty} (-1)^m \frac{(n+k)!}{(n-m)!(k+m)!m!} x^m, \quad k = 0, 1, \ldots n. (2.3.25)$$

The orthogonality relations again contain a weighting function. They read

$$\int_0^{\infty} e^{-x} x^k L_n^k(x) L_m^k(x) dx = \frac{(n+k)!}{n!} \delta_{mn}. \qquad (2.3.26)$$

This has the important practical consequence that all functions $f(x)$ that are quadratic integrable can be expanded into a LAGUERRE *series*

$$f(x) = \sum_{n=0}^{\infty} c_n \exp(-x/2) x^{k/2} L_n^k(x), \qquad (2.3.27)$$

with

$$c_n = \frac{1}{n!(n+k)!} = \int_0^{\infty} f(x) \exp(-x/2) x^{k/2} L_n^k(x) dx. \qquad (2.3.28)$$

Application of **DSolve** on (2.3.23) yields the result in the form of a *confluent hypergeometric* (KUMMER) *function*. For $\alpha = 1, k = 0$ one obtains an analogous result, but for $n = 0, 1, 2, 3, \ldots$, the results (2.3.25) are retrieved. Additionally, a transcendental function appears, that seems to have no practical applications.

HERMITE *polynomials* $H_n(x)$ have similar important properties. They satisfy the HERMITE equation

$$H_n'' - 2\alpha x H_n' + 2\alpha n H_n = 0. \qquad (2.3.29)$$

If one applies **Dsolve**, one obtains again a special form of the confluent hypergeometric function. Solution by the FROBENIUS methods leads to the recurrence relation

$$c_{\nu+2}(\nu + 2)(\nu + 1) = -2c_\nu(n - \nu),\qquad(2.3.30)$$

and the RODRIGUEZ formula reads

$$H_n(x; \alpha) = (-1)^n \exp(\alpha x^2)\frac{\mathrm{d}^n}{\mathrm{d}x^n}\exp(-\alpha x^2),\quad n = 0, 1, 2\ldots.\qquad(2.3.31)$$

Application of this formula for $\alpha = 1$ yields

$$\begin{array}{lll} H_0 = 1, & H_1 = 2x, & H_2 = 4x^2 - 2, \\ H_3 = 8x^3 - 12x, & H_4 = 16x^4 - 48x^2 + 12, & H_5 = 32x^5 - 160x^3 + 120x. \end{array}\qquad(2.3.32)$$

Mathematica defines **HermiteH[n,x]** and creates the expressions given in (2.3.32). For practical purposes it is useful to define orthogonal HERMITE polynomials $h_n(\xi)$, where $\xi = \sqrt{\alpha}\, x$. The substitution of

$$H_n(x; \alpha) = \sqrt{(2\alpha)^n n! \sqrt{\pi}}\exp\left(\xi^2/2\right)h_n(\xi)\qquad(2.3.33)$$

into (2.3.29) results in

$$h_n''(\xi) + \left(1 + 2n - \xi^2\right)h_n(\xi) = 0,\qquad(2.3.34)$$

which is solved by

$$h_n(\xi) = \frac{\exp\left(-\xi^2/2\right)}{\sqrt{2^n n!\sqrt{\pi}}}H_n(\xi; 1).\qquad(2.3.35)$$

These h_n are special cases of the confluent hypergeometric function. These functions are of practical interest because quadratic integrable functions $f(x)$ may be expanded.

$$f(x) = \sum_{n=0}^{\infty} c_n h_n(x),\quad c_n = \int_{-\infty}^{+\infty} f(x)h_n(x)\mathrm{d}x,\qquad(2.3.36)$$

since

$$\int_{-\infty}^{\infty} h_n(x)h_m(x)\mathrm{d}x = 0\quad\text{for}\quad n \neq m.\qquad(2.3.37)$$

The orthogonality relations for the H_n polynomials are given by

$$\int_{-\infty}^{\infty} H_m(x)H_n(x)\exp(-x^2)\mathrm{d}x = 0\quad\text{for}\quad n \neq m.\qquad(2.3.38)$$

We would like to conclude these deliberations on polynomials by discussing the so-called *harmonic polynomials*. They are solutions of the Laplacian operator (1.2.3). They can be found by calculating the real P and the imaginary Q parts of $(x + iy)^n$ respectively:

$$\begin{array}{ll} P_1 = x, & Q_1 = y, \\ P_2 = x^2 - y^2, & Q_2 = 2xy, \\ P_3 = x^3 - 3xy^2, & Q_3 = 3x^2y - y^3, \\ P_4 = x^4 - 6x^2y^2 + y^4, & Q_4 = 4x^3y - 4xy^3. \end{array}\qquad(2.3.39)$$

The solutions (2.3.39) cannot be found using *Mathematica* since the commands **Im** and **Re** work only on numbers and not on expressions. Since the Laplacian is linear, the superposition principle holds for its solutions. Any linear combination of the polynomials (2.3.39) is also a solution. Each partial solution can be multiplied by a factor. So $ax^3 + bx^2y - 3axy^2 - by^3/3$ is also a solution.

Problems

1. Solve the equation (2.3.2):
   ```
   DSolve[y''[x]-2*x*y'[x]/(1-x^2)+1*(1+1)*y[x]/(1-x^2)
   -m^2*y[x]/(1-x^2)^2==0,y[x],x](*(2.3.2)*)
   ```
   $y[x] \to C[1]$ LegendreP$[l, m, x] + C[2]$ LegendreQ$[l, m, x]$

2. Solve the equation (2.3.4):
   ```
   DSolve[y''[x]*x*(1-x)+(1-2*x)*y'[x]+1*(1+1)*y[x]
   ==0,y[x],x](*(2.3.4)*)
   ```
   $y[x] \to C[1]$ LegendreP$[l, -1 + 2x] + C[2]$ LegendreQ$[l, -1 + 2\ x]$

3. Solve the equation (2.3.20):
   ```
   DSolve[(1-x^2)*T''[x]-x*T'[x]+n^2*T[x]==0,
   T[x],x](*(2.3.20)*)
   ```
   $T[x] \to C[1]$ Cosh$[n$ Log$[x + \sqrt{1 - x^2}]] + i\ C[2]$ Sinh$[n$ Log$[x + \sqrt{1 - x^2}]]$

4. Solve the equation (2.3.23):
   ```
   DSolve[x*L''[x]+(1+k-α*x)*L'[x]+α *n*L[x]==0,
   L[x],x](*2.3.23*)
   ```
   $L[x] \to C[1]$ HypergeometricU$[-n, 1 + k, x\ \alpha] + C[2]\ *$ LaguerreL$[n, k, x\ \alpha]$

5. Solve the equation (2.3.29):
   ```
   DSolve[H''[x]-2*α*x*H'[x]+2*α *n*H[x]==0,H[x],x]
   (*2.3.29*)
   ```
   $H[x] \to C[1]$ HermiteH$[n, x\ \sqrt{\alpha}]$
   $+ C[2]$ Hypergeometric1F1$[-\frac{n}{2}, \frac{1}{2}, x^2\ \alpha]$

6. Calculate the harmonic polynomials for $n = 5, 6$ and 7 and *Legendre polynomial*
   $P_5=$ **LegendreP[5,x]** $=\ 15x/8 - 35x^3/4 + 63x^5/8$
   $Q_5=$ **LegendreQ[5,x]** $=\ -8/15 + 49x^2/8 - 63x^4/8+$
   $15x(1 - 14x^2/3 + 21x^4/5)Log\left(\dfrac{1 + x}{1 - x}\right)/16$

7. Calculate P_0, P_1, P_2 and P_3 using (2.3.8) and compare to (2.3.14).

8. Verify (2.3.21) and (2.3.32).

2.4 Boundary value problems and eigenvalues

Prior to solving special boundary value problems and calculate eigenvalues, we make some general remarks. Let us assume we want to solve the equation

$$y''(x) + p_1(x)y'(x) + p_2(x)y(x) = f(x), \qquad (2.4.1)$$

with the very general linear boundary conditions

$$\alpha_1 y(x_0) + \alpha_2 y'(x_0) = a_1, \qquad (2.4.2)$$

$$\beta_1 y(x_1) + \beta_2 y'(x_1) = a_2, \qquad (2.4.3)$$

where α_i, β_i and a_i are given constant parameters. Since (2.4.1) is inhomogeneous, the general solution has the form (2.1.4) or

$$y(x) = C_1 y_1(x) + C_2 y_2(x) + y_0(x). \qquad (2.4.4)$$

Substitution into the boundary conditions (2.4.2) and (2.4.3) yields

$$\begin{aligned}
C_1(\alpha_1 y_1(x_0) &+ \alpha_2 y_1'(x_0)) + C_2(\alpha_1 y_2(x_0) + \alpha_2 y_2'(x_0)) \\
&= a_1 - \alpha_1 y_0(x_0) - \alpha_2 y_0'(x_0), \qquad (2.4.5) \\
C_1(\beta_1 y_1(x_1) &+ \beta_2 y_1'(x_1)) + C_2(\beta_1 y_2(x_1) + \beta_2 y_2'(x_1)) \\
&= a_2 - \beta_1 y_0(x_1) - \beta_2 y_0'(x_1). \qquad (2.4.6)
\end{aligned}$$

These two equations represent an inhomogeneous system of linear equations for the two unknown quantities C_1 and C_2. This system is only then soluble in a unique way if the determinant

$$\begin{vmatrix} \alpha_1 y_1(x_0) + \alpha_2 y_1'(x_0) & \alpha_1 y_2(x_0) + \alpha_2 y_2'(x_0) \\ \beta_1 y_1(x_1) + \beta_2 y_1'(x_1) & \beta_1 y_2(x_1) + \beta_2 y_2'(x_1) \end{vmatrix} \neq 0, \qquad (2.4.7)$$

does not vanish. If, on the other hand, the system of linear equations determining the unknown C_1, C_2 would be homogeneous, then the determinant (2.4.7) must vanish, so that a non-trivial solution exists. Then the following rules are obvious:

1. An inhomogeneous boundary problem (2.4.1) to (2.4.3) can be solved, if the accompanying homogeneous problem only has the trivial solution.

2. If an homogeneous boundary problem ($a_1 = a_2 = 0$) is solvable, then the inhomogeneous problem can be only solved for special boundary values and it will not have a unique solution.

Let us consider a simple example. The homogeneous equation

$$y''(x) + y(x) = 0 \tag{2.4.8}$$

is solved by $\sin x$ for the homogeneous conditions

$$y(0) = 0, \qquad y(2\pi) = 0, \tag{2.4.9}$$

but the inhomogeneous problem

$$y(0) = 0, \qquad y(2\pi) = 1 \tag{2.4.10}$$

can not be solved. Whereas an inhomogeneous problem consisting of an inhomogeneous equation (2.4.1) and inhomogeneous boundary conditions (2.4.2), (2.4.3) never can be transformed into an homogeneous problem, either an inhomogeneous equation or inhomogeneous conditions can be homogenized. Let $\mathcal{L}\{y\} = g$ be an inhomogeneous equation, then the setup

$$y(x) = z(x) + \eta(x), \tag{2.4.11}$$

where $\eta(x)$ is a particular solution, $\mathcal{L}\{\eta\} = g$, yields the homogeneous equation

$$\mathcal{L}\{z\} = 0 \tag{2.4.12}$$

with inhomogeneous boundary conditions

$$\mathcal{R}\{z\} \neq 0. \tag{2.4.13}$$

We give an example. The inhomogeneous equation

$$y''(x) - y(x) = 1 \tag{2.4.14}$$

can be transformed into a homogeneous equation

$$z''(x) - z(x) = 0 \tag{2.4.15}$$

by the substitution $y(x) = z(x) - 1$. If the boundary conditions belonging to (2.4.14) are inhomogeneous

$$y(0) = 1, \qquad y(1) = 2, \tag{2.4.16}$$

then the substitution gives rise to new inhomogeneous conditions

$$z(0) = 2, \quad z(1) = 3. \tag{2.4.17}$$

On the other hand, the substitution

$$y(x) = z(x) + x + 1 \tag{2.4.18}$$

transforms the inhomogeneous equation (2.4.14) into the inhomogeneous equation

$$z''(x) - z(x) = x + 2, \tag{2.4.19}$$

and the conditions (2.4.16) become homogeneous

$$z(0) = 0, \qquad z(1) = 0. \tag{2.4.20}$$

Now we return to the solution of (2.4.1) and simplify the boundary conditions (2.4.2), (2.4.3) by choosing $\alpha_1 = 1$, $\alpha_2 = 0$, $\beta_1 = 1$, $\beta_2 = 0$ so that they read

$$y(x_0) = a_1, \qquad y(x_1) = a_2. \tag{2.4.21}$$

Then (2.4.7) delivers the *solvability condition*

$$M \equiv y_1(x_0)y_2(x_1) - y_2(x_0)y_1(x_1) \neq 0. \tag{2.4.22}$$

Considering the solution (2.4.4) and the boundary conditions (2.4.5), (2.4.6), we have a system of 3 linear equations for the two unknowns C_1 and C_2. This system can only be solved, if the determinant

$$\begin{vmatrix} y_1(x) & y_2(x) & y_0(x) - y(x) \\ y_1(x_0) & y_2(x_0) & y_0(x_0) - a_1 \\ y_1(x_1) & y_2(x_1) & y_0(x_1) - a_2 \end{vmatrix} \tag{2.4.23}$$

vanishes. Using (2.4.22), this condition yields

$$y(x) = a_1 \frac{y_1(x)y_2(x_1) - y_2(x)y_1(x_1)}{M} - a_2 \frac{y_1(x)y_2(x_0) - y_2(x)y_1(x_0)}{M}$$

$$+ y_0(x) + y_0(x_1) \frac{y_1(x)y_2(x_0) - y_2(x)y_1(x_0)}{M} \tag{2.4.24}$$

$$- y_0(x_0) \frac{y_2(x)y_1(x_1) - y_1(x)y_2(x_1)}{M}.$$

This solution satisfies the boundary conditions (2.4.21), but we have not yet obtained the general solutions $y_1(x), y_2(x)$ and the particular solution $y_0(x)$. Considering now the more general form (2.1.1) instead of (2.4.1), the latter can be found by using (2.1.14). Thus, the general solution of the inhomogeneous equation (2.1.1) is given by

$$y(x) = \frac{a_1 y_2(x_1) - a_2 y_2(x_0)}{y_1(x_0)y_2(x_1) - y_2(x_0)y_1(x_1)} y_1(x)$$

$$+ \frac{a_2 y_1(x_0) - a_1 y_1(x_1)}{y_1(x_0)y_2(x_1) - y_2(x_0)y_1(x_1)} y_2(x) + \int_{x_0}^{x_1} G(x, \xi) f(\xi) \mathrm{d}\xi. \tag{2.4.25}$$

The function $G(x, \xi)$ that appears in the particular solution is called GREEN *function*. It can be found by comparing (2.4.25) with (2.4.24). We first insert the boundary locations x_0 and x_1 into (2.1.14) for x in the upper value of the integration interval. This gives

$$y_0(x_0) = \int_{x_0}^{x_0} \frac{y_1(\xi)y_2(x_0) - y_1(x_0)y_2(\xi)}{W(\xi)} \frac{f(\xi)}{p_0(\xi)} d\xi = 0,$$

$$y_0(x_1) = \int_{x_0}^{x_1} \frac{y_1(\xi)y_2(x_1) - y_1(x_1)y_2(\xi)}{W(\xi)} \frac{f(\xi)}{p_0(\xi)} d\xi. \tag{2.4.26}$$

For $W(\xi)$ see (2.1.12) We then are able to rewrite the last three terms of (2.4.24) in the form

$$\int_{x_0}^{x} \frac{y_1(\xi)y_2(x) - y_1(x)y_2(\xi)}{W(\xi)} \frac{f(\xi)}{p_0(\xi)} d\xi + \frac{y_1(x)y_2(x_0) - y_2(x)y_1(x_0)}{M}$$

$$\cdot \int_{x_0}^{x_1} \frac{y_1(\xi)y_2(x_1) - y_1(x_1)y_2(\xi)}{W(\xi)} \frac{f(\xi)}{p_0(\xi)} d\xi. \tag{2.4.27}$$

Splitting the integral $\int_{x_0}^{x_1}$ into $\int_{x_0}^{x} + \int_{x}^{x_1}$, and using the abbreviation

$$f(\xi)/(W(\xi)p_0(\xi)) = A,$$

we arrive at the expression

$$\int_{x_0}^{x} [y_1(\xi)y_2(x) - y_1(x)y_2(\xi)] A d\xi$$

$$+ \int_{x_0}^{x} [y_1(x)y_2(x_0) - y_2(x)y_1(x_0)] \cdot [y_1(\xi)y_2(x_1) - y_1(x_1)y_2(\xi)] A d\xi/M$$

$$+ \int_{x}^{x_1} [y_1(x)y_2(x_0) - y_2(x)y_1(x_0)] \cdot [y_1(\xi)y_2(x_1) - y_1(x_1)y_2(\xi)] A d\xi/M.$$

Combining the two integrals $\int_{x_0}^{x}$ into one, one obtains the same integrand for the integral $\int_{x_0}^{x}$ and the integral $\int_{x}^{x_1}$, so that an integral $\int_{x_0}^{x_1}$ can be formed. In order to achieve this, one has to use the equation

$$[\,y_1(x)y_2(x_1) - y_1(x_1)y_2(x)][y_1(\xi)y_2(x_1) - y_1(x_1)y_2(\xi)]$$

$$+ M[y_1(\xi)y_2(x) - y_1(x)y_2(\xi)] \tag{2.4.28}$$

$$= [y_1(x)y_2(x_1) - y_1(x_1)y_2(x)] \cdot [y_1(x)y_2(x_0) - y_1(x_0)y_2(\xi)].$$

We thus may read G off from (2.4.25)

$$G(x,\xi) = \begin{cases} [y_1(x)y_2(x_1) - y_1(x_1)y_2(x)][y_1(\xi)y_2(x_0) \\ -y_1(x_0)y_2(\xi)]/MW(\xi)p_0(\xi) \quad \text{for} \quad x_0 \leq \xi \leq x \leq x_1, \\ \\ [y_1(x)y_2(x_0) - y_1(x_0)y_2(x)][y_1(\xi)y_2(x_1) \\ -y_1(x_1)y_2(\xi)]/MW(\xi)p_0(\xi) \quad \text{for} \quad x_0 \leq x \leq \xi \leq x_1. \end{cases} \quad (2.4.29)$$

This GREEN function is continuous in the domains $x_0 \leq \xi$, $\xi \leq x_1$ and represents a solution of the homogeneous differential equations. At the boundaries one has $G(x_0,\xi) = G(x_1,\xi) = 0$. The derivative $\partial G/\partial x$ is, however, discontinuous at $x = \xi$ and satisfies the *discontinuity condition*

$$\left.\frac{\partial G}{\partial x}\right|_{x=\xi+0} - \left.\frac{\partial G}{\partial x}\right|_{x=\xi-0} = \frac{1}{p_0(\xi)}. \quad (2.4.30)$$

The GREEN function depends on the boundary conditions. If one replaces (2.4.21) by

$$y(x_0) = \alpha, \qquad y'(x_1) = \beta \quad (2.4.31)$$

then the solution of the differential equation reads

$$y(x) = \frac{[\alpha y_2'(x_1) - \beta y_2(x_0)]y_1(x) - [\alpha y_1'(x_1) - \beta y_1(x_0)]y_2(x)}{y_1(x_0)y_2'(x_1) - y_1'(x_1)y_2(x_0)}$$

$$+ \int_{x_0}^{x_1} G(x,\xi)\, f(\xi)\mathrm{d}\xi, \quad (2.4.32)$$

where the GREEN function is now given by

$$G(x,\xi) = \begin{cases} [y_1(x)y_2'(x_1) - y_1'(x_1)y_2(x)][y_1(\xi)y_2(x_0) \\ -y_1(x_0)y_2(\xi)]/NW(\xi)p_0(\xi) \quad \text{for} \quad x_0 \leq \xi \leq x \leq x_1 \\ \\ [y_1(x)y_2(x_0) - y_1(x_0)y_2(x)][y_1(\xi)y_2'(x_1) \\ -y_1'(x_1)y_2(\xi)]/NW(\xi)p_0(\xi) \quad \text{for} \quad x_0 \leq x \leq \xi \leq x_1. \end{cases} \quad (2.4.33)$$

Here we used N as an abbreviaition for the denominator in (2.4.32).

GREEN functions connected with boundary value problems of partial differential equations will be discussed extensively in later chapters.

Discussing equation (1.1.49), we have found that solutions of ordinary differential equations may exist that contain a special parameter called *eigenvalue*. It has been demonstrated that such an eigenvalue is determined by boundary conditions. This connection shall now be discussed more deeply. We consider the homogeneous linear ordinary differential equation of second order (2.1.2). The differential equation

$$\frac{\mathrm{d}^2}{\mathrm{d}x^2}(p_0(x)y(x)) - \frac{\mathrm{d}}{\mathrm{d}x}(p_1(x)y(x)) + p_2(x)y(x)\} = 0 \equiv$$

$$p_0 y'' - p_1 y' + p_2 y + 2p_0' y' + p_0'' y - p_1' y = 0 \qquad (2.4.34)$$

is called the differential equation adjoint to (2.1.2). If it happens that $2p_0' - p_1 = p_1$ and $p_0'' - p_1' + p_2 = p_2$ or if

$$\frac{\mathrm{d}p_0(x)}{\mathrm{d}x} = p_1(x), \qquad (2.4.35)$$

then (2.1.2) and (2.4.34) are identical and (2.1.2) is called *self-adjoint*. It is interesting that homogeneous differential equations of the form (2.1.2) can be transformed into a self-adjoint equation by the substitution

$$y(x) = p_0(x) z(x) \exp \left(- \int \frac{p_1(x)}{p_0(x)} \mathrm{d}x \right). \qquad (2.4.36)$$

Defining a self-adjoint operator $\mathcal{L}$ by

$$\mathcal{L}y(x) = \frac{\mathrm{d}}{\mathrm{d}x} \left[p(x) \frac{\mathrm{d}y}{\mathrm{d}x} \right] + q(x) y(x), \qquad (2.4.37)$$

we are now able to define a STURM-LIOUVILLE *eigenvalue problem*. These problems appear in many practical applications and are defined by

$$\mathcal{L}y(x) + \lambda w(x) y(x) = 0. \qquad (2.4.38)$$

Here the parameter λ will turn out to be the *eigenvalue* and $w(x)$ a *weighting function*. It can be shown that a self-adjoint operator leads to real eigenvalues even if all other quantities are complex and that orthogonality relations

$$\int_{x_0}^{x_1} w(x) y_i(x) y_k(x) \mathrm{d}x = 0, \qquad i \neq k \qquad (2.4.39)$$

exist for two solutions $y_i(x), y_k(x)$ of the self-adjoint differential equation (2.4.38). Replacement of the weighting function $w(x)$ by $a_i^2 w(x)$ leads to *normalization* of the solutions

$$\int_{x_0}^{x_1} a_i^2 w(x) y_i^2(x) \mathrm{d}x = 1, \qquad (2.4.40)$$

where now

$$\frac{1}{a_i} = \sqrt{ \int_{x_0}^{x_1} w(x) y_i^2(x) \mathrm{d}x }. \qquad (2.4.41)$$

The combination of (2.4.39) and (2.4.40) gives rise to *orthonormalized eigenfunctions* $y_i(x)$ defined by

$$\int_{x_0}^{x_1} w(x) y_i(x) y_j(x) \mathrm{d}x = \delta_{ij}/a_i^2. \qquad (2.4.42)$$

These orthonormalized functions are able to exactly represent continuous and differentiable functions (what we used in (2.3.27) and (2.3.37)). The GREEN function of a self-adjoint boundary problem is symmetric: $G(x, \xi) = G(\xi, x)$.

As an example for an eigenvalue problem we consider the BESSEL equation (2.2.48) in the form

$$y''(x) + \frac{1}{x} y'(x) + \left(\kappa^2 - \frac{n^2}{x^2} \right) y(x) = 0. \qquad (2.4.43)$$

Comparing (2.4.43) with (2.4.38) we find that the BESSEL equation (2.4.43) is self-adjoint and of the STURM-LIOUVILLE type ($p = x$, $w = x$, $q = -n^2(x)$). The transformation $\xi = \kappa x$ leads to the form (2.4.48) and **DSolve** yields the solution

$$\texttt{BesselJ[n,x]} = J_n(x) = \sum_{\nu=0}^{\infty} \frac{(-1)^\nu}{\nu!(n+\nu)!} \left(\frac{x}{2} \right)^{n+2\nu}, \qquad (2.4.44)$$

see also (2.2.51). c_0 is now determined by (2.4.42). A boundary condition

$$y(\xi = R) \to y(\kappa R) = 0 \to J_n(\kappa R) = 0 \qquad (2.4.45)$$

determines the eigenvalue κ through the zeros of the BESSEL function J_n. These can be found in numerical tables [2.1]

	1st zero	2nd zero
$n = 0$	$2.404\,825 = j_{01}$	$5.520\,078 = j_{02}$
$n = 1$	$3.831\,715 = j_{11}$	$7.015\,59. = j_{12}$.

Mathematica yields

FindRoot[BesselJ[0,x]==0, {x,1.}] $= 2.40483$,
FindRoot[BesselJ[0,x]==0,{x,4.5}] $= 5.52008$,
FindRoot[BesselJ[1,x]==0, {x,3.}] $= 3.83171$,
FindRoot[BesselJ[1,x]==0, {x,6.}] $= 7.01559$.

The orthogonality relations are

$$\int_0^1 x J_n\left(j_{n\nu} x \right) J_n\left(j_{n\mu} x \right) \mathrm{d}x = \delta_{\nu\mu} \frac{1}{2} \left[J_{n+1}(j_{n\nu}) \right]^2, \qquad (2.4.46)$$

where the constants $j_{n\nu}$ are given by the ν-th zero of J_n. The BESSEL *series* expansion of a function $f(x)$ is given by

$$f(x) = \sum_{\nu=1}^{\infty} c_\nu J_n(j_{n\nu} x), \qquad (2.4.47)$$

where

$$c_\nu = \frac{2}{\left[J_n'(j_{n\nu}) \right]^2} \int_0^1 x f(x) J_n(j_{n\nu} x) \mathrm{d}x. \qquad (2.4.48)$$

(2.4.47) is a uniform convergent series. It might be of interest to recall that boundary conditions may be satisfied by the general solution of an homogeneous equation simply by adapted integration constants, compare equations (2.2.1) to (2.2.10).

Problems

1. Using $y(x) = z(x) + b$ transforms the inhomogeneous boundary value problem $y''(x)+y'(x)+y(x) = a, y(0) = 0, y(\pi) = 0$ into an homogeneous equation for z with two inhomogeneous boundary conditions. (a, b are constant parameters, it will turn out $b = a$).

2. Now use inhomogeneous boundary conditions $y(0) = 2, y(\pi) = 1$ and try to homogenize both for $b \neq 0$. ($z'' + z' + z + b = a, z(0) = 0, b = 2$ but $z(\pi) = -1$.

3. `DSolve[y''[x]-y[x]-1==0,y[x],x]`

   $y[x] \rightarrow -1 + e^x \ C[1] + e^{x-1} \ C[2]$

   and satisfy (2.4.16) and then an homogeneous boundary condition.

4. `DSolve[y''-y[x]-x-2==0,y[x],x]`

   $y[x] \rightarrow -2 - x + e^x \ C[1] + e^{x-1} \ C[2]$

   Is it possible to satisfy (2.4.20)?

5. `DSolve[{y''[x]+k^2*y[x]==a*x,y[x],x]`

   $$\left\{\left\{y[x] \rightarrow \frac{a\,x}{k^2} + C[1] \ \text{Cos}[k\,x] + C[2] \ \text{Sin}[k\,x]\right\}\right\}$$

6. `DSolve[{y2''[x]+k^2*y2[x]==0,y2[0]==y0,y2'[0]==1.},`
   `y2[x],x]`

   $$\left\{\left\{y2[x] \rightarrow \frac{1.\ (k\ y0\ \text{Cos}[k\,x] + 1.\ \text{Sin}[k\,x]}{k}\right\}\right\}$$

7. `NDSolve[{y''[x]+y[x]==x,y[0]==0,y'[0]==1},`
   `y[x],{x,0,1.}]`

   $\{\{y[x] \rightarrow \text{InterpolatingFunction}[\{\{0., 1.\}\}, <>][x]\}\}$

8. `a=1.0;`

   `NDSolve[{y''[x]+(1+x^2)*y[x]==-1.,y[-1.]==0,`
   `y'[-1.]==a},y[x],{x,-1.,1.3}]`

   $\{\{y[x] \rightarrow \text{InterpolatingFunction}[\{\{-1., 1.3\}\}, <>][x]\}\}$

   `Table[Evaluate[y[x]/.%],{x,0.1,1.5,0.5}]`

   $\{\{0.303265\}, \{-0.0894297\}, \{-0.681535\}\}$

2.5 Boundary value problems as initial value problems

Due to the validity of the superposition principle for the solutions of linear differential equations, it is possible to transform boundary value problems into initial value problems. This fact is very important because many numerical methods for the solution of differential equations are based on initial value problems only.

Let us consider the inhomogeneous boundary value problem

$$y(a) = y_a, \quad y(b) = y_b \tag{2.5.1}$$

of an inhomogeneous or homogeneous linear differential equation of second (or n-th) order

$$\mathcal{L}\{y(x)\} = g(x). \tag{2.5.2}$$

This boundary value problem may be transformed into the double initial value problem

$$y_1(a) = y_a, \qquad y_1'(a) = 0, \tag{2.5.3}$$

where the solution $y_1(x)$ satisfies

$$\mathcal{L}\{y_1(x)\} = g(x) \tag{2.5.4}$$

and

$$y_2(a) = 0, \qquad y_2'(a) = 1, \tag{2.5.5}$$

where the solution $y_2(x)$ satisfies the homogeneous equation

$$\mathcal{L}\{y_2(x)\} = 0. \tag{2.5.6}$$

Then the solution $y(x)$ of (2.5.2) for the initial problem for $x = a$

$$y(a) = y_a, \qquad y'(a) = y_a' \tag{2.5.7}$$

is given by

$$y(x) = y_1(x) + y_a' y_2(x), \tag{2.5.8}$$

(since $y_2(a) = 0$, $y_1'(a) = 0$, $y_2'(a) = 1$). The parameters y_a, y_b, y_a' are constants, y_a and y_b are given. The calculation of the parameter y_a' from (2.5.8) in the form

$$y(b) = y_b = y_1(b) + y_a' y_2(b) \tag{2.5.9}$$

gives rise to the solution of the boundary problem (2.5.1), (2.5.2)

$$y(x) = y_1(x) + y_2(x)\frac{y_b - y_1(b)}{y_2(b)}. \tag{2.5.10}$$

It is easy to show that (2.5.10) satisfies the boundary problem (2.5.1), (2.5.2):

$$y(a) = y_1(a) + y_2(a)\frac{y_b - y_1(b)}{y_2(b)} = y_1(a) = y_a, \qquad (2.5.11)$$

$$y(b) = y_1(b) + y_2(b)\frac{y_b - y_1(b)}{y_2(b)} = y_b. \qquad (2.5.12)$$

We consider in detail a very simple example. We assume the equation

$$y''(x) + k^2 y(x) = ax \qquad (2.5.13)$$

and apply the boundary conditions

$$y(0) = y_0, \qquad y(1) = y_1, \qquad (2.5.14)$$

where y_0 and y_1 are known constants. The *Mathematica* command

```
DSolve[y''[x]+k^2*y[x]==a*x,y[x],x]
```
(2.5.15)

yields the solution

$$y(x) = ax/k^2 + C_2 \sin(kx) + C_1 \cos(kx). \qquad (2.5.16)$$

Inserting this solution into the two boundary conditions (2.5.14) we obtain the integration constants

$$C_2 = (y_1 - a/k^2 - y_0 \cos k) / \sin k, \qquad (2.5.17)$$

$$C_1 = y_0, \qquad (2.5.18)$$

so that the solution of the boundary problem (2.5.13), (2.5.14) assumes the form

$$y(x) = ax/k^2 + y_0 \cos kx + \sin kx \left(y_1 - a/k^2 - y_0 \cos k\right) / \sin k. \qquad (2.5.19)$$

It satisfies the two boundary conditions (2.5.14). On the other hand, the initial value problem defined by (2.5.13) and the initial conditions (2.5.7) in the form

$$y(0) = y_0, \qquad y'(0) = y'_a \qquad (2.5.20)$$

lead to equation (2.5.9). This gives

$$y'_a = (y_1 - y_1(1)) / y_2(1). \qquad (2.5.21)$$

We thus need the two solutions $y_1(x), y_2(x)$. They are determined by (2.5.13) together with the initial conditions (2.5.3)

$$y_1(0) = y_0, \qquad y'_1(0) = 0, \qquad (2.5.22)$$

and by

$$y''_2(x) + k^2 y_2(x) = 0, \qquad (2.5.23)$$

with the initial conditions in the form

$$y_2(0) = y_0, \qquad y_2'(0) = 1. \tag{2.5.24}$$

We thus obtain

$$y_2(x) = \frac{\sin kx}{k \cos k}. \tag{2.5.25}$$

For $y_1(x)$ one gets

$$y_1(x) = \frac{a}{k^2} \left(x - \frac{1}{k} \sin kx \right) + y_0 \cos kx, \tag{2.5.26}$$

which satisfies (2.5.22) and (2.5.13). From (2.5.20) we now obtain

$$y_a' = \frac{y_1 - a(1 - \sin k/k)/k^2 + y_0 \cos k}{\sin k/(k \cos k)}. \tag{2.5.27}$$

According to (2.5.10) we now have the initial value solution

$$y(x) = \frac{a}{k^2} \left(x - \frac{1}{k} \sin kx \right) + y_0 \cos kx$$
$$+ \frac{\sin kx}{k \cos kx} \cdot \frac{y_1 - \frac{a}{k^2} \left(1 - \frac{1}{k} \sin k \right) - y_0 \cos k}{\sin k/(k \cos k)}, \tag{2.5.28}$$

which satisfies the boundary condition (2.5.14).

The method that we discussed can be used only if the equations (2.5.2), (2.5.4) and (2.5.6) can be solved by closed expressions. If this is not the case the equation (2.5.2) has to be solved by numerical methods. So it is possible to solve numerically (2.5.13) for given values of k, a for an initial value problem. For $k = a = 1$ and the initial conditions

$$y(0) = 0, \qquad y'(0) = 1, \tag{2.5.29}$$

this can be done using the *Mathematica* command

```
NDSolve[{y''[x]+y[x]==x,y[0]==0,y'[0]==1},
y[x],{x,0,1.}]
```
(2.5.30)

The result $y(x)$ may be plotted by the command

```
Plot[Evaluate[y[x]/.%],{x,0,1.}]
```

But what can we do if instead of the initial conditions the boundary conditions (2.5.1) are given? We could vary the slope $y'(0)$ to match $y_1(1) = 1$. This method is called the *shooting method* because it resembles an artillery problem. The elevation of the gun, the slope $y'(0)$, is set and a first fire does not hit the target. One then corrects the slope for several successive shots until the target $y(1) = 1$ is hit. We give an example. COLLATZ [1.7] describes the bending of a strut and solves the equation by a finite difference method. We will use the shooting method with *Mathematica*. The bending of a strut with flexural rigidity $EJ(\xi)$ and axial compressive load P by a distributed transverse load $p(\xi)$ is described by

$$\frac{d^2 M(\xi)}{d\xi^2} + \frac{P}{EJ(\xi)} M(\xi) = -p(\xi). \tag{2.5.31}$$

Here ξ is the coordinate along the axis of the strut and $M(\xi)$ is the local bending moment. Assuming that the transverse load is a constant p and that the flexural rigidity is given by

$$EJ(\xi) = \frac{EJ_0}{1 + (\xi/l)^2}, \tag{2.5.32}$$

where $2l$ is the length of the strut, one may use

$$P = EJ_0/l^2, \quad x = \xi/l, \quad y = M/l^2 p \tag{2.5.33}$$

and obtains

$$y''(x) + (1 + x^2) y(x) = -1. \tag{2.5.34}$$

Due to smoothly hinged end supports, we assume $M = 0$ at each end. Then the boundary conditions read

$$y(-1) = 0, \qquad y(+1) = 0. \tag{2.5.35}$$

We now write a *Mathematica* program for the shooting method

```
a=1.0;
NDSolve[{y''[x]+(1+x^2)*y[x]==-1.,y[-1.]==0,
y'[-1.]==a},y[x],{x,-1.,1.3}];
```
(2.5.36)

and pick values of $y(x)$ in the interval $0 < x < 1.5$ with a step 0.5

```
Table[Evaluate[y[x]/.%],{x,0.,1.5,0.5}]
```

yielding 0.345, 0.011, -0.559, 313.35.

The first line in (2.5.36) defines the (variable) value a of the initial condition $y'(-1) = a$. The semicolon informs *Mathematica* not to print the value a on the screen. Notice that in (2.5.36) the numbers 1. etc., are written as floating point (decimal) numbers. The second line in (2.5.36) executes the numerical integration of the initial value problem (2.5.34), (2.5.35) and the next command picks up the values $u(1)$, based (/.) on the result of the preceding calculation (%). We now recalculated (2.5.36) many times with different values to obtain a good approximation to the exact value **a** which satisfies (2.5.35) using first the commands

```
Clear[a,T,y,u];a=1.0;
T=NDSolve[{y''[x]+(1+x^2)*y[x]==-1.,y[-1.]==0,y'[-1.]==a},
y[x],{x,-1.,1.3}];
Table[Evaluate[y[x]/.%],{x,0,1.5,0.5}]
u[x_]=y[x]/.First[T];
Plot[u[x],{x,-1.,1.3}];   u[1.]
```
(2.5.37)

we obtain -0.559557 for $u(1.)$ and Figure 2.1 $u(x)$ for wrong a.

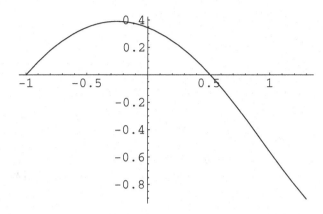

Figure 2.1
$u(x)$ **for wrong** a

Repeating the commands for $a = 1.0, 1.2\ldots$, we obtain Table 2.1.

Table 2.1. The dependence of $u(1)$ on a

a	1.0	1.2	1.5	1.6	1.7	1.8
$u(1)$	−0.5596	−0.4076	−0.1797	−0.1037	−0.0277	+0.0483

From Table 2.1 we conclude $1.7 < a < 1.8$. The command **First** picks out **u[1.]** of **T**.

In some simple cases *Mathematica* is able to solve a numerical boundary value problem directly, that means without use of the shooting method.

The *Mathematica* command

```
Sol=NDSolve[{y''[x]+(1+x^2)*y[x]==-1.,y[-1.]==0,
   y[1.]==0},y,{x,-1.,1.}]
```
$\qquad(2.5.38)$

solves the boundary value problem (2.5.34), (2.5.35) directly and delivers

$$y- > \text{InterpolatingFunction}[\{\{-1.,\}\}, <>] \qquad (2.5.39)$$

which may be plotted using the command

```
Plot[Evaluate[y[x]/.Sol],{x,-1.,1}]
```
$\qquad(2.5.40)$

This plot is shown in Figure 2.2. Evaluate causes *Mathematica* to calculate values of the interpolation function.

Since we have given the name **Sol** to the result of (2.5.38) we now write **Sol** instead of **%** as in (2.5.36). In equation (2.5.40) one finds the replacement operator. It is able to replace x by value in the expression $expr : expr$ **\.** x / value.

In many cases one has, however, an eigenvalue problem and the parameter k is not known; it depends on the boundary conditions. Let us consider the eigenvalue problem

$$y''(x) + k^2 y(x) = 0, \quad y(0) = 0, \quad y(1) = 0. \tag{2.5.41}$$

Its solution is given by

$$y = A \sin kx, \ k = n\pi, \ n = 0, 1, 2 \ldots, \ k = 3.14159, 6.28318 \ldots \tag{2.5.42}$$

The value $y'(0) = a$ necessary to reproduce (2.5.41) by the shooting method is easily calculated

$$y'(x) = Ak \cos kx, \quad y'(0) = Ak = a, \tag{2.5.43}$$

so that for $A = 1/n$, $a = \pi$. It is clear that for $a = \pi$, $k = \pi$ the shooting method will deliver a numerical result that is identical with (2.5.42).

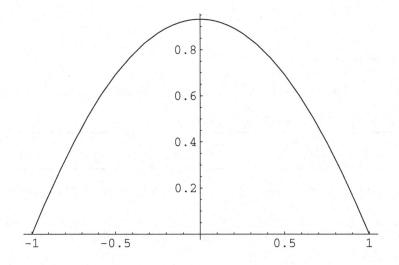

Figure 2.2
Plot of $y(x)$ according to (2.5.40)

A double loop *Mathematica* program could be created $(0 \le k \le \ldots, 0 \le a \le \ldots)$ but would probably be of no great use.

Another possibility of solving initial (and boundary) value problems of ordinary (and partial) linear and nonlinear differential equations is offered by GROEBNER'S LIE *series method* [2.2], [2.3]. In various papers GROEBNER has shown that an autonomous system of ordinary differential equations

$$\frac{dy_m(p)}{dp} = \vartheta_m(y_1, y_2, \ldots y_n, x), \quad \frac{dx}{dp} = 1, \ m = 1, 2, \ldots, n \tag{2.5.44}$$

with the initial conditions

$$y_m(p_0) = \bar{y}_m, \quad x(p_0) = x_0 \qquad (2.5.45)$$

has an absolutely convergent LIE series solution. Defining the LIE *operator*

$$D = \sum_{m=1}^{m} \vartheta\,(y_1, y_2, \ldots y_n, x)\,\frac{\partial}{\partial y_m} + \frac{\partial}{\partial x}, \qquad (2.5.46)$$

the solution of (2.5.44) reads

$$y_m(p) = \sum_{\nu=1}^{\infty} \frac{(p - p_0)^\nu}{\nu!} D^\nu \bar{y}_m,$$

$$x = \sum_{\nu=1}^{\infty} \frac{(p - p_0)^\nu}{\nu!} D^\nu x_0 = x_0 + p - p_0. \qquad (2.5.47)$$

Substitution for p yields

$$y_m(x) = \sum_{\nu=1}^{\infty} \frac{(x - x_0)^\nu}{\nu!} D^\nu \bar{y}_m, \ \nu = 1, 2, \ldots, n, \ y_0 = x. \qquad (2.5.48)$$

The prescription (2.5.48) says: assume that the parameters $\bar{y}_m$ are the same variables as the y_m. Therefore the LIE operator works on these variables, but only after formation of the LIE series (2.5.48), i.e., after application of D, the y_m may be considered to be the fixed and therefore const initial values $\bar{y}_m$.

The method seems to be quite complicated, especially when applied on simple examples, but its advantages become clear when difficult initial or boundary value problems have to be solved.
We consider

$$y''(x) + y(x) = 0 \quad \text{or} \quad y_2 = y_1', \quad y_2' = -y_1,$$
$$\vartheta_1 = y_2, \quad \vartheta_2 = -y_1. \qquad (2.5.49)$$

This equation has the solution

$$y = y_1 + y_2 = \sin x + \cos x, \qquad (2.5.50)$$

where

$$\sin x = x - \frac{x^3}{3!} + \frac{x^5}{5!} - \frac{x^7}{7!} + \ldots \qquad (2.5.51)$$

and

$$\cos x = 1 - \frac{x^2}{2!} + \frac{x^4}{4!} - \frac{x^6}{6!} + \ldots. \qquad (2.5.52)$$

Since a function has only one series representation, it is to be expected that the LIE series solution of (2.5.49) gives rise to the series

$$\sin x + \cos x = 1 + x - \frac{x^2}{2!} - \frac{x^3}{3!} + \frac{x^4}{4!} + \frac{x^5}{5!} - \frac{x^6}{6!} - \frac{x^7}{7!} \ldots. \qquad (2.5.53)$$

This result can be obtained by the *Mathematica* command

`Series[Sin[x]+Cos[x],{x,0,7}]` $\qquad$ (2.5.54)

This command expands a given function into a power series up to terms of degree 7. Thus the initial conditions realized by the solution (2.5.50) are

$$y_1(0) = 0, \quad y_2(0) = 1. \qquad (2.5.55)$$

According to (2.5.46) the LIE operator associated with (2.5.49) is given by

$$D = y_2 \frac{\partial}{\partial y_1} - y_1 \frac{\partial}{\partial y_2} + \frac{\partial}{\partial x},$$

$$D^2 = y_2^2 \frac{\partial^2}{\partial y_1^2} - y_1 \frac{\partial}{\partial y_1} - y_1 y_2 \frac{\partial^2}{\partial y_2 \partial y_1} + \frac{\partial y_2}{\partial x}\frac{\partial}{\partial y_1} + y_2 \frac{\partial^2}{\partial x \partial y_1}$$

$$-y_2 \frac{\partial}{\partial y_2} - y_2 y_1 \frac{\partial^2}{\partial y_1 \partial y_2} + y_1^2 \frac{\partial^2}{\partial y_2^2} - \frac{\partial y_1}{\partial x}\frac{\partial}{\partial y_2} - y_1 \frac{\partial^2}{\partial x \partial y_2}$$

$$+y_2 \frac{\partial^2}{\partial y_1 \partial x} - y_1 \frac{\partial^2}{\partial y_2 \partial x} + \frac{\partial^2}{\partial x^2},$$

$$Dy_1 = y_2 + y_1',$$

$$Dy_2 = -y_1 + y_2',$$

$$D^2y_1 = -y_1 + y_2' + y_1''$$

$$D^2y_2 = -y_2 - y_1' + y_2''. \qquad (2.5.56)$$

Since after application of D, the variables y_i become the constant initial values (2.5.55), the derivatives vanish and we obtain from (2.5.48) for $x_0 = 0$ the first three terms of (2.5.53).

Applications of the LIE series method for boundary problems of linear and nonlinear ordinary differential equations in perturbation theory and for partial differential equations will be discussed later.

Boundary problems solved by the LIE series method can either be reformulated as initial value problems or can be solved as boundary problems using numerical methods, LIE series perturbation theory and methods to accelerate the rate of convergence of the series [2.3]. A simple example is the eigenvalue problem [2.4] of equation

$$y''(x) + \lambda(1 + \sin x)y(x) = 0, \; y(0) = 0, \; y(\pi) = 0. \qquad (2.5.57)$$

This eigenvalue problem can be written in the form of a boundary value problem

$$\begin{aligned}
\dot{x}_1(x) &= x_2(x) & x_1(0) &= 0 \\
\dot{x}_2(x) &= -x_3(1 + \sin x)x_1 & x_2(\pi) &= 0 \\
\dot{x}_3 &= 0 & x_2(0) &= 1.
\end{aligned} \qquad (2.5.58)$$

The initial value of $x_3 = \lambda = const$ is to be calculated. This has been done with the shooting method (arbitrary initial conditions) and by splitting the operator $D = D_1 + D_2$ (perturbation theory). The result was $\lambda = 0.54031886$, [2.2], [2.3].

Problems

1. The tedious method to find the correct a by the shooting method can be improved by a loop. Learn to write a program for a loop:

```
Clear[a,l]
a[1]=1.6;
For[l=1,l<4,
Print[{"iteration=",l,"value of a=",a[l]}];
a[l]=a[l]+0.01*l
l++];
```
(2.5.59)

resulting for $l = 1, 2, 3$ in
{iteration =,   1, value of $a =$, 1.6}
{iteration =,   2, value of $a =$, 1.601}
{iteration =,   3, value of $a =$, 1.604}

2. Consider a LIE series solution of
$\dot{Z}(t) = \dfrac{1}{2(Z(t) + a)}$ with the initial condition $Z(t = 0) = z$ so that

$D = \dfrac{1}{2(z + a)} \dfrac{d}{dz}.$

Now

$D^\nu z = \binom{1/2}{\nu} \dfrac{\nu!}{(z + a)^{2\nu - 1}}, \quad \nu = 1, 2, \ldots$

and the solution reads

$Z = z + \displaystyle\sum_{\nu}^{\infty} = 1 \binom{1/2}{\nu} \dfrac{t^\nu}{(z + a)^{2\nu - 1}}$

For $t = 0$ one obtains $Z(0) = z$.

3. Show that the invariant (2.1.19) of BESSEL equations given by $1 - (4n^2 - 1)/4x^2$ and that for $n = \pm 1/2$ the BESSEL function is given by

$J_{\pm 1/2}(x) = \dfrac{1}{\sqrt{x}}(C_1 \cos x + C_2 \sin x).$

Using *Mathematica* show that
$J_{1/2}(x) = \sqrt{(2/\pi x)} \sin x, J_{-1/2}(x) = \sqrt{(2/\pi x)} \cos x.$

4. Consider

$\dot{Z}_1 = a_{11} Z_1 + a_{12} Z_2, \qquad \dot{Z}_2 = a_{21} Z_1 + a_{22} Z_2,$

where a_{ik} are constants. Show that these equations may be solved by a setup $Z_1 = (Az_1 + Bz_2) \exp(\lambda_1 t) + (Cz_1 + Dz_2) \exp(\lambda_2 t)$ and similar for Z_2, A, B, C, D etc., are constants and the eigenvalue λ is determined by

$$Det \begin{pmatrix} \lambda - a_{11} & -a_{12} \\ -a_{21} & \lambda - a_{22} \end{pmatrix} = 0.$$

What initial conditions are satisfied?

2.6   Nonlinear ordinary differential equations

Even the first differential equation in this book, namely (1.1.1), and several others have been nonlinear. The problem with nonlinear equations is that there is no general theory for the integration. There are only four possible cases of simple integration of differential equations of the type

$$F(y''(x), y'(x), y(x), x) = f(x). \tag{2.6.1}$$

Case 1: The variable x is missing explicitly and the equation is homogeneous. Such equations are called *autonomous*.

$$F(y''(x), y'(x), y(x)) = 0. \tag{2.6.2}$$

In this case it is possible to reduce the order of the equation. The ansatz $p(y) = y'(x(y))$ creates

$$y''(x) = \frac{dp(y(x))}{dx} = \frac{dp}{dy}\frac{dy}{dx} = \frac{dp}{dy}p(y) \tag{2.6.3}$$

and produces

$$F\left(p(y), \frac{dp}{dy}, y, p\right) = 0, \tag{2.6.4}$$

which is a differential equation of first order for the function $p(y)$. If a solution $p(y) = f(y, C) \neq 0$ can be found then separation of variables and integration are possible:

$$dx = \frac{dy}{f(y, C)}, \qquad x = \int \frac{dy}{f(y, C)} + C_1. \tag{2.6.5}$$

Here C and C_1 are integration constants. If a solution $y(p) = g(p, C)$ has been found, then

$$x = \int \frac{dy}{p(y)} = \int \frac{dg}{dp} \cdot \frac{dp}{p(y)}, \qquad x = \frac{y}{p} + \int g(p, C)\frac{dp}{p^2} \tag{2.6.6}$$

are solutions. As an example for (2.6.2) we consider the equation of oscillation with *quadratic damping*

$$y''(x) + ay'^2(x) + by(x) = 0. \tag{2.6.7}$$

Using (2.6.3) we obtain (2.6.4) in the form

$$p'(y) + ap(y) + by/p(y) = 0. \tag{2.6.8}$$

The *Mathematica* command $\tag{2.6.9}$

`StandardForm[DSolve[p'[y]+a*p[y]+b*y/p[y]==0,p[y],y]]`
gives

$$p(y) = \pm\sqrt{-2b\left(-\frac{1}{4a^2} + \frac{y}{2a}\right) + \exp(-2ay)C}, \tag{2.6.10}$$

what we define as function

`p[y_]=Sqrt[-2*b*(-1/(4*a^2)+y/(2*a))+C[1]Exp[-2*a*y];`

The command **StandardForm** produces a result written in usual mathematical notation. The command **InputForm** forces *Mathematica* to show the result in the input form, that means that the resulting expression will be in such a form that it can be used as the input for the next *Mathematica* command. C is the integration const. To check on the correctness of the solution (2.6.10) of the equation (2.6.8) we use the command

`Simplify[p'[y]+a*p[y]+b*y/p[y]]` $\tag{2.6.11}$

This gives zero what proves the correctness of (2.6.10).

If y' does not appear in (2.6.2) the integration is simpler. We multiply an *equation of motion*

$$\ddot{x}(t) = f(x) \tag{2.6.12}$$

by $\dot{x}(t)$ and receive after integration

$$\dot{x}^2(t) = \int f(x)\mathrm{d}x. \tag{2.6.13}$$

Separation of variables yields $x(t)$.

Case 2: The variable y is missing in (2.6.1). Using $y'(x) = p(x)$ one gets

$$F(p'(x), p(x), x) = 0. \tag{2.6.14}$$

Let us consider the example

$$x^2y''(x) - y'^2(x) + 2xy'(x) - 2x^2 = 0. \tag{2.6.15}$$

Using $y'(x) = p(x)$, $y''(x) = p'(x)$ one obtains

$$x^2p'(x) - p^2(x) + 2xp - 2x^2 = 0. \tag{2.6.16}$$

Mathematica gives

$$p(x) = \frac{\mathrm{d}y}{\mathrm{d}x} = \frac{-2x + \exp(C)x^2}{-1 + \exp(C)x}. \tag{2.6.17}$$

This allows integration. We thus have to apply

```
Integrate[(x*(C*x-2)/(C*x-1)),x]
```
(2.6.18)

on (2.6.17) giving the solution of (2.6.15)

$$y(x) = -\frac{x}{C} + \frac{x^2}{2} - \frac{\ln(Cx - 1)}{C^2} + const.$$
(2.6.19)

In some cases a solution of (2.6.14) is given in the form $x(p) = g(p, C)$. Then

$$y = \int p \mathrm{d}x = \int pg'(p, C)\mathrm{d}p + const$$
(2.6.20)

or

$$y = \int p \mathrm{d}x = px - \int x \mathrm{d}p = pg(p, C) - \int g(p, C)\mathrm{d}p + const$$
(2.6.21)

give x or y as function of p.

Case 3: Both variables x and y are missing. Then the equation is again homogeneous. If one has

$$y'' = f(y') = f(p)$$
(2.6.22)

then the solution reads

$$x = \int \frac{\mathrm{d}p}{f(p)} + const, \quad y = \int \frac{p}{f(p)} \mathrm{d}p + const.$$
(2.6.23)

Case 4: If all terms of an equation have the same power of $y, y'y''$, the equation is called uniform (e.g., of the second degree)

$$xyy'' - xy'^2 + yy' = 0.$$
(2.6.24)

Then the introduction of a new variable $u = y'/y$ reduces the order of the equation.

A great class of nonlinear ordinary differential equations of second order can be subsumed under the very general form

$$y''(x) + f(y(x), y'(x))h(x)y'(x) + g(y(x)) \cdot p(x) = r(x).$$
(2.6.25)

The physical meaning of the terms in (2.6.25) can be classified as follows:
1) exterior excitation $r(x)$, compare (2.2.15),
2) damping $f(y(x), y'(x))$, compare (2.6.7),
3) nonlinearity of the oscillator $g(y)$,
4) selfexcitation by *negative damping*, compare the VAN DER POL *equation*

$$y'' - \mu(1 - y^2)y' + y = 0,$$
(2.6.26)

5) *parametric excitation* $f(x)$ in a non-autonomous equation like

$$y'' + f(x)y = 0,$$
(2.6.27)

compare the MATHIEU equation or the similar equation (2.5.57). If $f(x)$ is a periodic function, the equation (2.6.27) is called HILL equation. For $f(x) \sim \sin x$ (2.6.27) becomes a MATHIEU equation (2.7.53).

We give some examples.

$$y''(t) + \rho y'(t) + \left(\frac{g}{l} + \frac{a}{l}\omega^2 \cos \omega t\right) \sin y(t) = 0 \qquad (2.6.28)$$

is the oscillation equation of a damped parametrically excited mathematical pendulum.

More general are the LIÉNARD *equation*

$$y'' + h(y)y' + f(y) = 0, \text{ where } y = y(x), \qquad (2.6.29)$$

and the LEVINSON-SMITH *equation*

$$y'' + f(y, y')y' + g(y) = 0. \qquad (2.6.30)$$

It helps to understand how to solve equations of these types if the behavior of the solutions is known qualitatively. In the case of ordinary differential equations of second order $y''(x)$, the splitting into two equations of first order for $x_1(t) = y(t), x_2(t) = y'(t)$, where t is a parameter, allows the study of the *trajectories* $x_1(t), x_2(t)$ or $y'(y)$ in the phase plane x_1, x_2 or y, y'. These trajectories are also called a *phase portrait* of the differential equation of second order. In order to investigate the phase portrait $y'(y)$ of (2.6.29), we replace the independent variable x by a new independent variable t. We then have two dependent variables $x_1(t)$ and $x_2(t)$. Using new designations for x and y by $x_1 \to y$, $x_2 \to y''$ two new quantities, $P(x, y)$ and $Q(x, y)$ will then be defined by

$$y'(x) \to \dot{x}(t) = y(t) = Q(x, y) = y \qquad (2.6.31)$$

and

$$y''(x) \to \ddot{x}(t) = \dot{y}(t) = -h(x(t))\dot{x}(t) - f(x(t))$$
$$= -h(x(t))y(t) - f(x(t)) = P(x, y). \qquad (2.6.32)$$

The starting point for the portrait analysis will then be

$$\frac{dx}{dt} = \dot{x} = Q(x, y), \qquad (y) \qquad (2.6.33)$$

$$\frac{dy}{dt} = \dot{y} = P(x, y) \qquad (y'') \qquad (2.6.34)$$

and the differential equation for the phase portrait trajectories reads

$$\frac{dy}{dx} = \frac{P(x, y)}{Q(x, y)}. \qquad (2.6.35)$$

Special points x_0, y_0, which are always at rest, satisfy $\dot{x} = 0, \dot{y} = 0$. They are called *equilibrium points* or *singular points* and satisfy

$$P(x_0, y_0) = 0, \quad Q(x_0, y_0) = 0. \qquad (2.6.36)$$

Since the investigation of phase portraits is mainly interested in the surroundings of equilibrium points, we can expand $P(x,y)$ and $Q(x,y)$ into TAYLOR *series*

$$P(x,y) = \frac{\partial P}{\partial x}(x - x_0) + \frac{\partial P}{\partial y}(y - y_0),$$

$$Q(x,y) = \frac{\partial Q}{\partial x}(x - x_0) + \frac{\partial Q}{\partial y}(y - y_0). \tag{2.6.37}$$

Since it is possible to dislocate equilibrium points, we chose $x_0 = 0$, $y_0 = 0$ and insert the ansatz

$$x(t) = A\exp(\lambda t), \quad y(t) = B\exp(\lambda t) \tag{2.6.38}$$

into (2.6.33), (2.6.34), (2.6.37). The result is

$$A\lambda = \frac{\partial Q}{\partial x}A + \frac{\partial Q}{\partial y}B, \quad B\lambda = \frac{\partial P}{\partial x}A + \frac{\partial P}{\partial y}B. \tag{2.6.39}$$

So that these two linear homogeneous equations possess nontrivial solutions for the unknown constant amplitudes A and B, the determinant of the system has to vanish. This allows to calculate the value of the parameter λ

$$\lambda_{1,2} = \frac{Q_x + P_y}{2} \pm \sqrt{\frac{(Q_x + P_y)^2}{4} - (Q_x P_y - P_x Q_y)}. \tag{2.6.40}$$

Six cases appear:
   a) both solutions λ_1, $\lambda_2 \neq \lambda_1$ are real: no oscillations
1) $\lambda_1 < 0$, $\lambda_2 < 0$: attracting (stable) node,
2) $\lambda_1 > 0$, $\lambda_2 > 0$: repelling (unstable) node,
3) $\operatorname{sign}\lambda_1 \neq \operatorname{sign}\lambda_2$: saddle (unstable) point,
   b) the solutions λ_1, λ_2 are complex: oscillations possible
4) $\operatorname{Re}\lambda_1 > 0, \operatorname{Re}\lambda_2 > 0$: unstable focus (repelling spiral),
5) $\operatorname{Re}\lambda_1 < 0, \operatorname{Re}\lambda_2 < 0$: stable focus (attracting spiral),
6) λ_1, λ_2: imaginary: stable center.

As an example we consider (2.6.25) with the specifications $p = 1$, $h = 1$, $f = -c_1 - c_2 y^2$, $g = -c_3 y - c_4 y^3$. This gives

$$y''(x) + \left(-c_1 - c_2 y^2(x)\right) y'(x) - c_3 y(x) - c_4 y^3(x) = 0. \tag{2.6.41}$$

This equation comprises not only the VAN DER POL equation (2.6.26) ($c_1 = \mu$, $c_2 = 1$, $c_3 = -1$, $c_4 = 0$) but also the DUFFING *equation* for a nonlinear oscillator

$$y''(x) + ay(x) + by^3(x) = 0 \tag{2.6.42}$$

and the LASHINSKY *equation*

$$y''(x) + ay'(x) - ay(x) + by^3(x) = 0 \tag{2.6.43}$$

of plasma physics [2.5]. From (2.6.41) we can read after change of designation

$$P(x, y) = c_1 y + c_2 y x^2 + c_3 x + c_4 x^3,$$

$$Q(x, y) = y. \tag{2.6.44}$$

According to (2.6.36) the equilibrium points are then found from $Q = 0$ and $P = 0$. There are two of them:

$$x_0 = 0, \quad y_0 = 0 \tag{2.6.45}$$

and

$$x_0 = \pm\sqrt{-c_3/c_4}, \quad y_0 = 0. \tag{2.6.46}$$

According to the linearized stability analysis (2.6.37), we now have to calculate

$$P_x = \frac{\partial P}{\partial x} = 2c_2 yx + c_3 + 3c_4 y^2, \qquad P_y = \frac{\partial P}{\partial y} = c_1 + c_2 x^2,$$

$$Q_x = \frac{\partial Q}{\partial x} = 0, \qquad\qquad Q_y = \frac{\partial Q}{\partial y} = 1. \tag{2.6.47}$$

For the equilibrium point (2.6.45) we then find from (2.6.40)

$$\lambda_{1,2} = \frac{c_1}{2} \pm \sqrt{\frac{c_1^2}{4} + c_4}. \tag{2.6.48}$$

We can now conclude:

	$c_3 > 0$	$c_3 < 0$	$c_3 = 0$
$c_1 < 0$	unstable	stable	stable
$c_1 > 0$	unstable	unstable	unstable
$c_1 = 0$	unstable	stable	indifferent.

There are more unstable than stable solutions. But what happens with the other equilibrium point (2.6.46)? Since our linearized anaysis is valid only very near zero, we have to shift this singular point into the origin. This is done by the transformation

$$v = x \pm \sqrt{-c_3/c_4}, \ y = y_0 = 0, \tag{2.6.49}$$

compare (2.6.46). We then obtain another differential equation for v and

$$\lambda_{1,2} = \frac{(c_1 - c_2 c_3/c_4)}{2} \pm \sqrt{\frac{(c_1 - c_2 c_3/c_4)^2}{4} - 2c_3}. \tag{2.6.50}$$

This indicates stability for

$$c_1 - c_2 c_3/c_4 \leq 0, \quad c_3 \geq 0. \tag{2.6.51}$$

For

$$(c_1 - c_2 c_3/c_4) - 8c_3 < 0 \tag{2.6.52}$$

one has an oscillatory solution. These calculations demonstrate *stabilization due to nonlinear terms*. $c_1 = 1$, $c_2 = -1$, $c_3 = c_4 = 0$ produces $y'' - y' + y = 0$, which has an unstable oscillatory solution. Now if we add the nonlinear terms we have

$$y'' - y' + y - c_2 y^2 y' - c_4 y^3 = 0, \qquad (2.6.53)$$

and $c_2 = +1$, $c_4 = -0.5$ give a stable nonoscillatory solution. For the LASHINKY equation (2.6.43) one obtains a damped oscillatory solution for $\alpha^2 < 8a$ and a stable nonoscillatory solution for $\alpha^2 > 8a$. For $\alpha = 0$ JACOBI elliptic functions satisfy the equation. Negative restoring forces like in $y'' + y' - y = 0$ - similar to the BÉNARD problem [2.5] - can also be stabilized by parametric effects, see section 2.7, equation (2.7.61).

We now treat some examples.

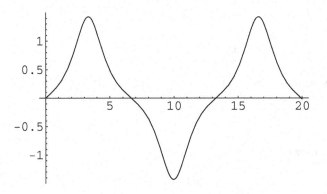

Figure 2.3
Stable nonlinear oscillation $y'' - y + y^3 = 0$ for $y(0) = 0, y'(0) = 0.2$

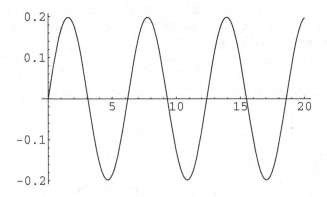

Figure 2.4
Stable nonlinear oscillation$y'' + y + y^3 = 0$ for $y(0) = 0, y'(0) = 0.2$

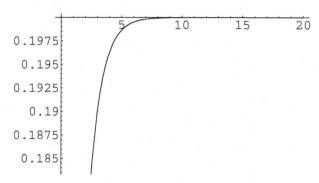

Figure 2.5
Creeping solution$y'' + y' = 0$ for $y(0) = 0, y'(0) = 0.2$

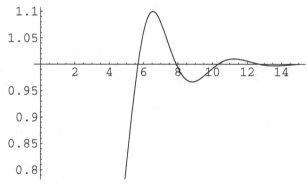

Figure 2.6
Damped nonlinear oscillation $y'' + y' - y + y^3 = 0$ for $y(0) = 0, y'(0) = 0.1$

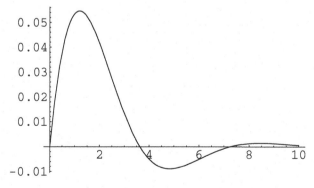

Figure 2.7
Damped nonlinear oscillation $y'' + y' + y + y^3 = 0$ for $y(0) = 0, y'(0) = 0.1$

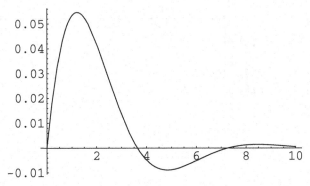

Figure 2.8
Damped nonlinear oscillation $y'' + y' + y - y^3 = 0$ for $y(0) = 0, y'(0) = 0.1$

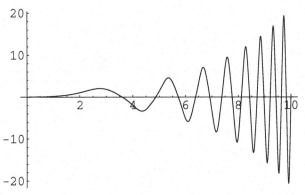

Figure 2.9
Unstable nonlinear oscillation $y'' - y' - y + y^3 = 0$ for $y(0) = 0, y'(0) = 0.1$

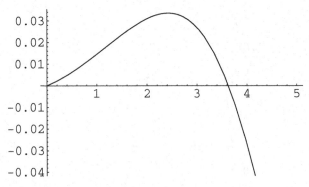

Figure 2.10
Strongly unstable nonlinear oscillation $y'' - y' + y - y^3 = 0$ for $y(0) = 0, y'(0) = 0.01$

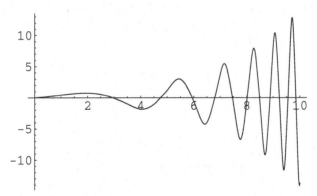

Figure 2.11
Increasing nonlinear oscillation $y'' - y' + y + y^3, y(0) = 0, y'(0) = 0.3$

Similar results can be obtained from a LYAPUNOV *analysis*. We now may consider the LEVINSON-SMITH equation (2.6.30). First we have to construct a LYAPUNOV *function*

$$V(y, y') = \frac{1}{2}y'^2 + \int_0^y g(y)\mathrm{d}y. \tag{2.6.54}$$

As one can see, this function describes energy. Apparently $\mathrm{d}V/\mathrm{d}x = 0$ or $\mathrm{d}V/\mathrm{d}t$ for $y(t)$ conserves energy (stability). We build

$$\frac{\mathrm{d}V}{\mathrm{d}x} = y'y'' + \frac{\mathrm{d}y}{\mathrm{d}x}\frac{\mathrm{d}}{\mathrm{d}y}\int_0^y g(y)\mathrm{d}y = y'y'' + y'g(y). \tag{2.6.55}$$

Inserting for y'' one obtains

$$\frac{\mathrm{d}V}{\mathrm{d}x} = y'\left[g(y) - f(y, y')y' - g(y)\right] = -y'^2 f(y, y'). \tag{2.6.56}$$

An equilibrium point is given by $y = 0, y' = 0$. In its environment one has $V = const$. For $f > 0$ one has $\mathrm{d}V/\mathrm{d}x < 0$ (asymptotically stable) and for $f < 0$ one has $\mathrm{d}V/\mathrm{d}x > 0$, unstable (near the origin).

Solutions of some equations have the property that their portrait curves never leave a closed domain in the phase space y, y'. These domains are called *limit cycle*. The VAN DER POL equation has such a limit cycle. Solutions inside such a cycle are periodic. BENDIXON has shown that such a limit cycle may exist, if

$$G = P_y + Q_x = 0. \tag{2.6.57}$$

If G never changes its sign and does not vanish, then no limit cycle can exist. The periodic solutions belonging to such a limit cycle may be stable or unstable.

We now investigate the VAN DER POL equation in detail. Splitting (2.6.26) into two equations of first order by using $y = x_1(t)$, $y' = x_2(t)$ one gets the system (2.6.33), (2.6.34)

$$y' \to \frac{\mathrm{d}x_1}{\mathrm{d}t} = x_2, \tag{2.6.58}$$

$$y'' \to \frac{\mathrm{d}x_2}{\mathrm{d}t} = -x_1 + \mu(1 - x_1^2)x_2. \tag{2.6.59}$$

Integration of this system gives an expression for the phase portrait trajectories $x_2(x_1)$ or $y'(y)$. If we rewrite the system in the form

$$P(x, y) = -y + \mu(1 - y^2)y', \quad Q(x, y) = y, \tag{2.6.60}$$

then we see that the BENDIXON criterion (2.6.57) is satisfied by the VAN DER POL equation. The phase portrait will be a limit cycle. In order to find it, we may integrate (2.6.58), (2.6.59) using the *Mathematica* commands

```
Clear[mu,n];mu=1.;n=2.; Sol=NDSolve[{x1'[t]==x2[t],x2'[t]
==-x1[t]+mu*(1-(x1[t])^2)*x2[t],x1[0]==n,x2[0]==0},
{x1,x2},{t,0,8.*Pi}];                                    (2.6.61)
```

```
ParametricPlot[Evaluate[{x1[t],x2[t]}/.Sol],{t,0,4.*Pi}]
                                                         (2.6.62)
```

By this way we produced Figure 2.12. All phase portraits of the six cases mentioned after (2.6.40) could be created by the same method.

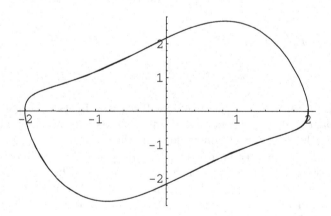

Figure 2.12
Limit cycle of the Van der Pol equation

We see that the phase portrait is a good way to understand the behavior of a second-order ordinary differential equation.

Problems

1. We now consider the parametric resonance (2.6.27), which is of industrial importance [2.12]. The charge $Q(t)$ of a variable capacitor $C(t)$, which is connected in a series resonance circuit with a constant inductance L_0 and an ohmic resistance R_0, is described by

$$C(t)L_0\frac{d^2Q(t)}{dt^2} + R_0\frac{dQ(t)}{dt}C(t) + Q(t) = 0. \qquad (2.6.63)$$

Let us assume $C(t) = C_0\left(1 + \frac{\Delta C}{C_0}\cos\frac{2\pi pt}{\tau}\right),$ $\qquad (2.6.64)$

where $\Delta C = (C_{\max} - C_{\min})/2$. $C_0 = (C_{\max} - C_{\min})/2$ and $p = \omega\tau/\pi$. (The pumping angular frequency is designated by ω). Using the *Mathematica* command

`G[t]=1/C[t]=Series[1/(1+C0*Cos[2*ω*t]),{t,0,5}`

one obtains a symbolic series representation of $G(t)$ up to the order 5.

For `Clear[G,F];C0=1.;` and $2\omega t = u$ one has
`G[u]=Series[1/(1+C0*Cos[u]),{u,0,5}]` yielding
$0.5 + 0.125\ u^2 + 0.0208333\ u^4 + O[u]^6$.

However this expression cannot be used to calculate or plot the function $G(t)$. It is necessary to use the *Mathematica* function **Normal** to generate a polynomial without the reminder $O[u]^6$. Now one defines

`F[u_]=Normal[G[u]];` and one may pick out special values by
`Table[F[u],{u,0,Pi,Pi/8}]`
This allows to `Plot[F[u],{u,-Pi,Pi}]`
the function. A polynomial representation of the pump function is however not suitable. One can obtain a FOURIER series representation by loading the package and

`<<Calculus`FourierTransform`` and
`Clear[F];`
`F[u_]=FourierTrigSeries[1/(1+Cos[4*u]),u,3];`
`Plot[F[u],{u,-Pi/2,Pi/2}]`

yields an approximative picture of the pumping function. For a better plot one can replace $3 \to 100$. One now can integrate (2.6.63) numerically.

2. Plot the phase portrait of the DUFFING *equation* (2.6.42). You should give these commands:

`Clear[a,b,n,Duff,x1,x2];a=1.;b=0.1;n=2.;`
`Duff=NDSolve[{x1'[t]==x2[t],x2'[t]==-a*x1[t]-`
`b*x1[t]^3,x1[0]==n,x2[0]==0},{x1,x2},{t,0,8*Pi}];`

```
ParametricPlot[Evaluate[{x1[t],x2[t]}
/.Duff],{t,0,4.*Pi}]
```

The figure should be an oval. Play around with varying values of a, b, n. Is the BENDIXON criterion satisfied?

3. Plot the phase portrait of the MATHIEU *equation*

```
Clear[λ,n,Ma,x1,x2];λ=0.5;n=2.;
Ma=NDSolve[{x1'[t]==x2[t],x2'[t]==-λ*(1+Sin[t])*
x1[t],
x1[0]==n,x2[0]==0},{x1,x2},{t,0,8*Pi}];
ParametricPlot[Evaluate[{x1[t],x2[t]}/.Ma],
{t,0,4.*Pi}]
```
Play with varying values of λ and n. Is the BENDIXON criterion satisfied? Is the solution of $y'' + (\lambda - 2q\cos 2x)y = 0$ stable for $0 < \lambda < 10$ and $q = 1$? Compare [1.4].

4. Solve the equation and plot the Figures 2.3 to 2.11.

2.7   Solutions of nonlinear differential equations

There are very few nonlinear equations that have a closed solution. One of these equations is the *pendulum equation*

$$\varphi''(t) + \frac{g}{l}\sin\varphi(t) = 0; \tag{2.7.1}$$

here φ is the angle of deflection, l the length of the pendulum and g is the acceleration of gravity. The pendulum mass has been assumed $m = 1$. Multiplying (2.7.1) by $\dot\varphi(t)$ we obtain

$$\dot\varphi\ddot\varphi = \frac{d}{dt}\frac{\dot\varphi^2}{2} = -\frac{d\varphi}{dt}\frac{g}{l}\sin\varphi, \tag{2.7.2}$$

or after integration one has

$$\dot\varphi = \frac{d\varphi}{dt} = \sqrt{\frac{2g}{l}}\sqrt{\cos\varphi - \cos\alpha}, \tag{2.7.3}$$

where $\cos\alpha$ is the integration constant. Using the identity

$$\cos\varphi - \cos\alpha = 2\left(\sin^2\frac{\alpha}{2} - \sin^2\frac{\varphi}{2}\right) \tag{2.7.4}$$

we rewrite (2.7.3) in the form

$$\int_0^\varphi \frac{d(\varphi/2)}{\sqrt{\sin^2(\alpha/2) - \sin^2(\varphi/2)}} = \sqrt{\frac{g}{l}}t. \tag{2.7.5}$$

The expression $\sqrt{g/l}$ is usually called eigenfrequency ω of the pendulum. The integral in (2.7.5) is usually called an elliptic integral because similar integrals appear in the calculation of the curve length of an ellipse. In order to transform (2.7.5) into the form of a LEGENDRE *elliptic integral*, we use the following abbreviations

$$\sin(\varphi/2) = \sin(\alpha/2)\sin u = k\sin u, \tag{2.7.6}$$

$$\cos(\varphi/2) = \sqrt{1 - \sin^2(\varphi/2)} = \sqrt{1 - k^2 \sin^2 u}, \tag{2.7.7}$$

$$d\sin(\varphi/2) = \cos(\varphi/2)d(\varphi/2) = kd\sin u = k\cos u du, \tag{2.7.8}$$

$$d(\varphi/2) = \frac{k\cos u du}{\cos(\varphi/2)} = \frac{k\cos u du}{\sqrt{1 - k^2 \sin^2 u}}. \tag{2.7.9}$$

The new parameter k is called *elliptic modulus* and u is the *amplitude*. Now we may write (2.7.5) in the form

$$\sqrt{\frac{g}{l}}t = \int\limits_0^u \frac{du}{\sqrt{1 - k^2 \sin^2 u}} = F(k; u). \tag{2.7.10}$$

The new function $F(k; u)$ is called the *incomplete elliptic integral of the first kind*. The adjective "incomplete" is often omitted: it refers to the fact that the upper limit in the integral is variable. If the upper limit is given by $\pi/2$, then the integral is called *complete elliptic integral* (of the first kind).

$$K(k) = \int\limits_0^{\pi/2} \frac{du}{\sqrt{1 - k^2 \sin^2 u}} \approx \frac{\pi}{2}\left(1 + \frac{k^2}{4} + \frac{9k^4}{64} + \ldots\right) = \sqrt{\frac{g}{l}}\frac{\tau}{4}. \tag{2.7.11}$$

Here $\tau = 2\pi/\omega$ is the *period of oscillation* of the pendulum ($0 \le \varphi \le 2\pi$). It depends on the initial maximum deflection. $k(\alpha)$ represents the energy that has been put into the pendulum at $t = 0$. We are now interested in the inverse function $u(t)$ of (2.7.10). It is called the *elliptic amplitude*

$$u(t) = \text{am}\left(\sqrt{\frac{g}{l}}t, k\right) \tag{2.7.12}$$

and

$$\sin u = \sin\text{am}\left(\sqrt{\frac{g}{l}}t, k\right) = \text{sn}\left(\sqrt{\frac{g}{l}}t, k\right) \tag{2.7.13}$$

is called JACOBI (or *elliptic*) *sine function*. Now we are in the position to write down the solution $\varphi(t)$ to (2.7.1)

$$\varphi(t) = 2\arcsin\left[k\,\text{sn}\left(\sqrt{\frac{g}{l}}t, k\right)\right]. \tag{2.7.14}$$

Mathematica can solve (2.7.1); the solution reads

```
EllipticF[u,k];  EllipticK[k];
```
$\mathrm{am}(x,k)$=`JacobiAmplitude[x,k]`; $\mathrm{sn}[x,k]$
```
=JacobiSN[x,k]
```
                                                                    (2.7.15)

In *Mathematica* the arguments of elliptic functions may be given in the opposite order from what is used normally. The commands
```
a=3.;
pen=NDSolve[{y''[t]+a*Sin[y[t]]==0,y[0]==Pi/8.,
y'[0]==0.1},y,{t,0,Pi}]
```
                                                                    (2.7.16)
```
Plot[Evaluate[y[t]/.pen],{t,0,Pi}]
```
gives a nice plot $y(t)$. Of more interest is however the phase portrait. We use the commands
```
InputForm[Integrate[x''[t]*x'[t]+Sin[x[t]]
*x'[t],t]/.{x[t]->x,x'[t]->y}]
```
                                                                    (2.7.17)
on (2.7.2) for $g/l = a = 1$. This gives $y^2 - 2\cos(x)$. Now we define the energy function F which we obtained in **InputForm**
```
F[x_,y_]=(y^2-2*Cos[x])/2
```
                                                                    (2.7.18)
This energy function produces a portrait by
```
ContourPlot[F[x,y],{x,-5.,5.},{y,-4.,4.},
CountourShading->False,Contours->Range[-5.,5.,0.45],
PlotPoints->100]
```
                                                                    (2.7.19)
This plot is shown in Figure 2.13.

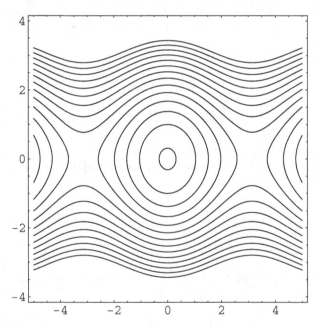

Figure 2.13
Phase portrait of the pendulum

Now we learn more *Mathematica* details.

ContourPlot

plots curves $F(x, y) = const$ in the domain given by $-5. \leq x \leq 5., -4. \leq y \leq 4$.

ContourShading->False avoids shading of the plot, so that the curves may be seen more clearly.

The command **Contours->Range[-5.,5.,0.45]**

orders the plotting of contour curves in the interval from -5 to $+5$ with a step 0.45 and

PlotPoints->100

fixes the number of evaluation points in each direction.

There exist few nonlinear equations that can be solved by transformations. So the equation

$$yy'' + ay'^2 + f(x)yy' + g(x)y^2 = 0 \qquad (2.7.20)$$

may be transformed into a linear equation

$$u'' + f(x)u' + (a+1)g(x)u = 0 \qquad (2.7.21)$$

with the use of the transformation

$$y = u^{1/(1+a)}. \qquad (2.7.22)$$

Another equation is

$$y'' + f(x)y' + g(y)y'^2 = 0 \qquad (2.7.23)$$

which may be transformed by

$$y' = u(x)v(y). \qquad (2.7.24)$$

Looking for a closed solution to a nonlinear equation is eased by a search in KAMKE [2.6]. But very often it is necessary to resort to approximation methods. According to POINCARÉ, a nonlinear differential equation of the type

$$\ddot{x} + x = f(x, \dot{x}, t) \qquad (2.7.25)$$

can be solved by a power series

$$x(t) = x_0(t) + x_1(t) + x_2(t), \qquad (2.7.26)$$

where the $x_n(t)$ satisfy linear equations. As an example we consider

$$m\ddot{x} + kx - \varepsilon x^2 = K_0 \cos \omega t. \qquad (2.7.27)$$

ε is a small parameter < 1. We set up the *method of successive approximation*

$$x(t) = x_0(t) + \varepsilon x_1(t) + \varepsilon^2 x_2(t). \qquad (2.7.28)$$

Insertion into (2.7.27) and sorting according to powers of ε results in the following system

$$\varepsilon^0 : \quad m\ddot{x}_0 + kx_0 = K_0 \cos \omega t, \quad x_0 = \frac{K_0}{k - m\omega^2} \cos \omega t = A \cos \omega t \qquad (2.7.29)$$

$$\varepsilon^1 : \quad m\ddot{x}_1 + kx_1 = x_0^2 = A^2 \cos 2\omega t = A^2 \left(\frac{1}{2} + \frac{1}{2} \cos 2\omega t \right). \qquad (2.7.30)$$

We just see that the first harmonic 2ω is $\sim \cos 2\omega t$.

Another method has been proposed by KRYLOV and BOGOLYUBOV. They consider

$$\ddot{x} + \omega^2 x = -\mu f(x, \dot{x}), \tag{2.7.31}$$

where μ is a small parameter.

The so-called *averaging method* starts from the setup *generating solution*

$$x(t) = A(t) \sin(\omega t + \varphi(t)), \tag{2.7.32}$$

$$\dot{x}(t) = \omega A(t) \cos(\omega t + \varphi(t)), \tag{2.7.33}$$

which contains two unknown functions $A(t), \varphi(t)$. Differentiation of (2.7.32) and insertion into (2.7.33) results in

$$\dot{A}(t) = -A(t)\dot{\varphi}(t) \cos(\omega t + \varphi(t)) / \sin(\omega t + \varphi(t)). \tag{2.7.34}$$

Differentiation of (2.7.33) and insertion into (2.7.31) yield

$$\dot{A}(t) \cos(\omega t + \varphi(t)) - A(t)\dot{\varphi}(t) \sin(\omega t + \varphi(t)) = -\frac{\mu}{\omega} f(x, \dot{x}). \tag{2.7.35}$$

If one inserts $\dot{A}$ and takes $\sin^2 + \cos^2 = 1$ into account one obtains the two equations determining $A(t)$ and $\varphi(t)$. They read

$$\frac{d\varphi}{dt} = \frac{\mu}{A\omega} f(x, \dot{x}) \sin(\omega t + \varphi), \tag{2.7.36}$$

$$\frac{dA}{dt} = -\frac{\mu}{\omega} f(x, \dot{x}) \cos(\omega t + \varphi). \tag{2.7.37}$$

Apparently, the KRYLOV-BOGOLYUBOV *method* considers two different time scales: a slow change of $A(t)$ and $\varphi(t)$, which is proportional to μ and a fast time scale proportional to ω. Averaging over the fast time scale results in

$$<\dot{\varphi}> = \frac{\mu}{A\omega} \cdot \frac{1}{2\pi} \int_0^{2\pi} f(x, \dot{x}) \sin(\omega t + \varphi) d\varphi. \tag{2.7.38}$$

Here one has to substitute x and $\dot{x}$ from (2.7.32) and (2.7.33). In (2.7.38) one assumes that A and φ are constant during the short time period, so that $d\varphi = \omega dt$ during the integration in (2.7.38). For $A(t)$ the averaging process gives rise to

$$<\dot{A}> = -\frac{\mu}{\omega} \cdot \frac{1}{2\pi} \int_0^{2\pi} f(x, \dot{x}) \cos(\omega t + \varphi) d\varphi. \tag{2.7.39}$$

Two interesting results are apparent:

1. If $f(x, \dot{x}) = f(\dot{x})$, one obtains $<\varphi> = const$, that means that a small nonlinear damping term does not modify the phase φ.

2. If $f(x, \dot{x}) = f(x)$, no modification of the amplitude A appears, but the phase φ will be modified.

We now apply the averaging method on the VAN DER POL *equation* (2.6.26). In this case one has $f(x, \dot{x}) = \mu(1 - x^2)\dot{x}$, so that (2.7.39) gives

$$< \dot{A} >= \frac{\mu A}{2} \left(1 - \frac{A^2}{4}\right). \qquad (2.7.40)$$

The substitution $u = A^2$ gives a differential equation of first order that is separable. Integration yields

$$A(t) = \frac{A_0 \exp(\mu t/2)}{\sqrt{1 + (A_0^2/4)[\exp(\mu t) - 1]}}. \qquad (2.7.41)$$

Here the integration constant has to be determined by the condition $A(t = 0) = A_0$. In the limit $t \to \infty$ one obtains the solution $x(t) = 2\sin(t + \varphi)$ that is independent of any initial condition.

The averaging method can only be used if an harmonic generating solution (2.7.32) exists. This is not the case for the LASHINSKY *equation* (2.6.43), which contains a large damping term. So we can try the POINCARÉ *method of successive approximations* (2.7.28). To be able to use this method, we have to make some substitutions in the LASHINSKY equation [2.7]. We replace $x \to t$, $y = a^{1/2}b^{-1/2}u(t)$, which eliminates b; furthermore, we choose $\alpha t = \xi$, $\varepsilon = a/\alpha^2$, $du/d\xi = \dot{u}(\xi)$. Then we have $\ddot{u} + \dot{u} - \varepsilon u + \varepsilon u^3 = 0$ and (2.7.28) gives rise to the following equations

$$\ddot{u}_0 + \dot{u}_0 = 0, \quad \ddot{u}_1 + \dot{u}_1 = u_0 - u_0^3, \quad \ddot{u}_2 + \dot{u}_2 = u_1 - 3u_0^2 u_1. \qquad (2.7.42)$$

The solutions of the equations (2.7.42) are given by

$$u_0(\xi) = A\exp(-\xi) + B, \qquad (2.7.43)$$

$$u_1(\xi) = -A\xi\exp(-\xi) - \frac{A^3}{6}\exp(-3\xi) - \frac{3A^2 B}{2}\exp(-2\xi)$$

$$+ 3AB^2\xi\exp(-\xi) - (-B - B^3)\xi + C\exp(-\xi) + D. \qquad (2.7.44)$$

Here A, B, C, D are integration constants. These solutions contain *secular terms* proportional to ξ, which become infinite for $\xi \to \infty$, so that the series (2.7.28) does not converge. Since we are free to determine the integration constants, we follow the LINDSTEDT *method* for the elimination of the secular terms by choosing $B = 0$ and $D = 0$ (for higher orders). We remark that terms $\xi\exp(-\xi)$ are not secular, since they tend to zero for $\xi \to \infty$.

If the damping term is small, but the nonlinear term is large, as is the case in the LANGEVIN *equation* (FROUDE equation for rolling ships)

$$\ddot{x}(t) + \omega^2 \sin lx = \varepsilon F(x, \dot{x}) \qquad (2.7.45)$$

averaging must be done starting at the exact solution of the undamped equation ($\varepsilon = 0$) that has the solution

$$x(t) = \frac{2}{l} \arcsin \left[k \operatorname{sn}(\sqrt{l}\varphi, k) \right], \tag{2.7.46}$$

where $\varphi = \omega t + \psi$, where $\psi = const$. Switching on the damping $\varepsilon \neq 0$, energy will be dissipated and the modulus k and the phase ψ change [2.8] according to

$$\langle \frac{dk}{dt} \rangle = \frac{\varepsilon \sqrt{l}}{8\omega K} \int_0^{4K} F(\operatorname{sn}(u), \operatorname{cn}(u)) du, \tag{2.7.47}$$

$$\langle \frac{d\psi}{dt} \rangle = -\frac{\varepsilon}{8\omega K} \int_0^{4K} F(\operatorname{sn}(u), \operatorname{cn}(u)) \frac{\operatorname{sn}(u) + k \dfrac{d\operatorname{sn}(u)}{dk}}{\operatorname{dn}(u)} du. \tag{2.7.48}$$

Here $u = \sqrt{l}\varphi$, K is given by (2.7.11), dn is the elliptic function defined by [2.9]

$$\operatorname{dn}(k, u) = \sqrt{1 - k^2 \operatorname{sn}^2(k, u)} \tag{2.7.49}$$

and the elliptic cosinus is defined by

$$\operatorname{cn}(k, u) = \sqrt{1 - \operatorname{sn}^2(k, u)} = \cos(\operatorname{am}(k, u)). \tag{2.7.50}$$

During the derivation of (2.7.47), (2.7.48) use has been made of the identities [2.9, 2.8]

$$\operatorname{sn}^2 + \operatorname{cn}^2 = 1, \quad \operatorname{sn}\frac{d}{dk}\operatorname{sn} = \frac{d}{dk}\frac{\operatorname{sn}^2}{2}, \quad \operatorname{sn}\frac{d\operatorname{sn}}{dk} + \operatorname{cn}\frac{d\operatorname{cn}}{dk} = 0. \tag{2.7.51}$$

In (2.7.1) the pendulum is not excited, it gains its energy from the deflection $\varphi(0) \neq 0$ at the time $t = 0$. Any energy dissipation proportional to a damping term $\dot{\varphi}$ is neglected. We now consider *parametric excitation* of the pendulum. Let be l the length of a stiff but elastic rod of mass $m = 1$, ω_0 the parametric *pump frequency* with which the location ξ of the support of the pendulum oscillates $\xi = a \cos \omega_0 t$. Assume that the pendulum rod at $t = 0$ has the initial position $\varphi = \pi$; that means that the center of gravity of the vertical rod is higher than the support (*"inverted pendulum"*). This is surely an unstable situation. Then for a small deviation $\sin \psi \approx \psi$, $\varphi = \pi - \psi$ from the vertical line $\varphi = \pi$, $\psi = 0$, the equation of an excited undamped motion reads

$$\ddot{\psi}(t) - \frac{g}{l}\psi(t) + \frac{a}{l}\omega_0^2 \psi(t) \cos \omega_0 t = 0. \tag{2.7.52}$$

The last lhs term stems from the parametric pump term $\ddot{\xi}(t) = -a\omega_0^2 \cos \omega_0 t$. With the definitions $\varepsilon = a/l$, $g/l = \omega_0^2 \varepsilon^2 k^2$, $\tau = \omega_0 t$, $\psi(t) = x(t)$ the equation

(2.7.52) will be transformed into a nonautonomous parametrically excited equation

$$\ddot{x} - \varepsilon^2 k^2 x + \varepsilon x \cos \tau = 0. \tag{2.7.53}$$

This is a MATHIEU *equation* that is solved by MATHIEU functions se, ce, which are stable only in certain domains (*stability regions*) of the parametric space ε, k [2.13]. The MATHIEU equation contains two frequencies: there is the high pump frequency ω_0 and the low eigenfrequency characterized by $\sqrt{g/l}$. We thus use the *multiple time scale method* to solve (2.7.53). Defining $\tau_0, \tau_1 = \varepsilon\tau_0$, $\tau_2 = \varepsilon^2\tau_0$, $d\tau_0/dt = 1$, $d\tau_1/dt = \varepsilon$, $d\tau_2/dt = \epsilon^2$ and

$$x(\tau) = x_0(\tau_0, \tau_1, \tau_2 \ldots) + \varepsilon x_1(\tau_0, \tau_1, \tau_2 \ldots) + \varepsilon^2 x_2(\tau_0, \tau_1, \tau_2 \ldots) + \ldots \tag{2.7.54}$$

as well as

$$\frac{\partial}{\partial \tau} = \frac{\partial}{\partial \tau_0} + \varepsilon \frac{\partial}{\partial \tau_1} + \varepsilon^2 \frac{\partial}{\partial \tau_2} + \ldots \tag{2.7.55}$$

we obtain

$$\frac{d^2}{dt^2} = \frac{\partial^2}{\partial \tau_0^2} + 2\varepsilon \frac{\partial^2}{\partial \tau_1 \partial \tau_0} + \varepsilon^2 \frac{\partial^2}{\partial \tau_1^2} + 2\varepsilon^2 \frac{\partial^2}{\partial \tau_2 \partial \tau_0} + \ldots . \tag{2.7.56}$$

Thus insertion into (2.7.53) gives rise to the following equations and solutions

$$\frac{\partial^2 x_0}{\partial \tau_0^2} = 0, \quad x_0(\tau_0, \tau_1, \tau_2 \ldots) = a(\tau_1, \tau_2, \ldots)\tau_0 + b(\tau_1, \tau_2 \ldots); \ a \to 0, \tag{2.7.57}$$

$$2\frac{\partial^2 x_0}{\partial \tau_1 \partial \tau_0} + \frac{\partial^2 x_1}{\partial \tau_0^2} + x_0 \cos \tau_0 = 0, \tag{2.7.58}$$

$$\frac{\partial^2 x_0}{\partial \tau_1^2} + 2\frac{\partial^2 x_1}{\partial \tau_2 \partial \tau_0} + 2\frac{\partial^2 x_1}{\partial \tau_1 \partial \tau_0} + \frac{\partial^2 x_2}{\partial \tau_0^2} - k^2 x_0 + x_1 \cos \tau_0 = 0. \tag{2.7.59}$$

Using $\partial b/\partial \tau_0 = 0$, $\cos \tau_0 \approx \cos \tau$ and eliminating secular terms one arrives at

$$x_0 = b = c(\tau_2) \exp \left(\sqrt{k^2 - \frac{1}{2}} \tau_1 \right),$$

$$x_1 = c(\tau_2) \exp \left(\sqrt{k^2 - \frac{1}{2}} \tau_1 \right) \sin \tau_0. \tag{2.7.60}$$

These two solutions oscillate and are bounded if

$$k^2 = \frac{gl}{\omega^2 a^2} < \frac{1}{2}, \quad \omega^2 > \frac{2gl}{a^2}. \tag{2.7.61}$$

If the pump frequency of the support is large enough the inverted pendulum will stay in the "unstable" position $\varphi \approx \pi$ (*"dynamic stabilization"* [2.5]). For a normal pendulum $\varphi \approx 0$, pump frequencies $\omega = \sqrt{g/l} \cdot 2/n$, $n = 1, 2, \ldots$ excite *parametric resonance*. Other similar methods are the *iterative* FOURIER *setup* and the *equivalent integro-differential equation* method by G. SCHMIDT

[2.10]. These methods enable one to find resonance curves (amplitude $A(\omega)$) and their backbone curves for periodic solutions for equations of the type

$$y'' + \lambda y - \gamma y \cos nx + \beta y' + by^2 y' - ey^3 = f(\omega, x). \qquad (2.7.62)$$

Many engineering problems can be described by this equation: torsional vibrations of beams and shafts, vibration of many layered sandwich plates and construction shells, sawtooth oscillations of struts etc. For equations of the type

$$y'' + f(x)y = 0 \qquad (2.7.63)$$

the WKB *method* (named after WENTZEL, KRAMERS, BRILLOUIN) is sometimes used. The setup $y = \exp(i\phi(x))$ creates

$$-(\phi')^2 + i\phi'' + f = 0. \qquad (2.7.64)$$

For $|\phi''| \ll f$, which corresponds to a geometric-optic approximation of wave phenomena, namely that inhomogeneities are small with respect to the wave length, one obtains

$$\phi'(x) = \pm\sqrt{f(x)}, \quad \phi(x) = \pm\int\sqrt{f(x)}\mathrm{d}x \qquad (2.7.65)$$

and in the second iteration

$$\phi'' \approx \frac{1}{2}f^{-1/2}f', \quad \phi'^2 \approx f \pm \frac{i}{2}\frac{f'}{\sqrt{f}}, \quad \phi \approx \pm\int\sqrt{f(x)}\mathrm{d}x + \frac{i}{4}\ln f \qquad (2.7.66)$$

the solution

$$y(x) \approx f^{-1/4}\left[C_1 \exp\left(i\int\sqrt{f(x)}\mathrm{d}x\right) + C_2 \exp\left(-i\int\sqrt{f(x)}\mathrm{d}x\right)\right]. \qquad (2.7.67)$$

More complicated problems can be handled by the LIE *series method*. If one considers a problem of *celestial mechanics* like the relative motion of Sun, Jupiter and its eighth Moon one has a system of 18 equations of first order [2.3, 2.4, 2.11]. Due to the very slow convergence of the LIE series, a transformation is necessary: first one determines approximate orbits, which are then corrected by a perturbation method. This consists in splitting of the LIE *operator*

$$D = D_1 + D_2. \qquad (2.7.68)$$

The FORTRAN *program LIESE* handles such operators and quite general ordinary differential equations of first order very satisfactorily. In celestial mechanics one step of 10 digit computation took 2 sec, whereas a 12-digit step using the COWELL method needed 10 sec on nearly equivalent mainframes.

Problems

1. Solve (2.7.1) with the help of *Mathematica*

   `DSolve[φ''[t]+g*Sin[φ[t]]/l==0,φ[t],t]`

   and compare with the solution given earlier.

2. Integrate (2.7.11) using

   `Integrate[1/(Sqrt[1-k^2*Sin[u]^2]),u]`
   yielding EllipticF$[u, k^2]$

3. Insert (2.7.28) into (2.7.27) and derive the differential equations for $x_0(t), x_1(t)$ and $x_2(t)$.

4. Does `DSolve[y''[x]+a*y'[x]-a*y[x]+b*y[x]^3==0,y[x],x]`

   solve (2.6.43)?

5. What is the *Mathematica* solution of (2.7.53)? Depending on your computer and your version of *Mathematica* something like

   $$x(\tau) = C_1 \text{ MathieuC}(-4 \ k^2 \ \epsilon^2, -2 \ \epsilon, \tau/2)$$
   $$+ C_2 \text{ MathieuS}(-4 \ k^2 \ \epsilon^2, -2 \ \epsilon, \tau/2)$$

   should appear on your screen. The traditional notation would be MathieuC = ce, MathieuS = se [1.4], [2.1].

6. Solve equation (2.7.23) using $f(x) = 1/x$, $g(y) = y$.

   `DSolve[y''[x]+y'[x]/x+y[x]*y'[x]^2==0, y[x],x]`

   InverseFunction::ifun : Inverse functions are being used. Values may be lost for multivalued inverses.

   $$y[x] \rightarrow \sqrt{2} \text{ Erfi}^{(-1)} \left[\frac{e^{-2 \ c[2]}(2 \ e^{2 \ C[2]} \ C[1] + \text{Log}[x])}{\sqrt{2} \ \pi} \right]$$

7. Integrate the following function $f(x) = ax^2/(bx + c)$. Use

   `G[x]=Integrate[a*x^2/(b*x+c),x]`

   $$a \left(-\frac{c \ x}{b^2} + \frac{x^2}{2 \ b} + \frac{c^2 \ \text{Log}[c + b \ x]}{b^3} \right).$$

8. Verify the result of the last problem by using `D[G[x],x]`

   $$a \left(-\frac{c}{b^2} + \frac{x}{b} + \frac{c^2}{b^3 \ (c + b \ x)} \right).$$

This is true but not exactly what you expected. Give the command

Simplify[%]

$$\frac{a\,x^2}{c+b\,x}$$

9. Plot the portrait of $x_1(t) = A\sin(\omega t)$, $X_2(t) = B\cos(\omega t)$.

10. The pendulum equation (2.7.1) does not contain any damping. Add a damping term $ay'(t)$ and use

DSolve[y''[t]+a*y'[t]+b*Sin[y[t]]==0,y[t],y]

Solve::ifun : Inverse functions are being used by Solve, so some solutions may not be found.

Apparently *Mathematica* cannot solve this equation since it gives this message and simply repeats the original command.

3

Partial differential equations

3.1   Coordinate systems and separability

In section 1.1 we have seen that it is possible to separate (linear) differential equations into ordinary differential equations. The setup (1.1.37) made this possible for cartesian coordinates. Since the HELMHOLTZ equation is separable in 11 coordinate systems we now investigate these systems more closely.

We consider the *line element* in cartesian coordinates

$$\mathrm{d}s^2 = \mathrm{d}x^2 + \mathrm{d}y^2 + \mathrm{d}z^2 = \sum_{i=1}^{3} \mathrm{d}x_i^2. \qquad (3.1.1)$$

Making a transformation $x_i = x_i(q_k)$ to new coordinates q_k, then

$$\mathrm{d}x_i = \sum_{k=1}^{3} \frac{\partial x_i}{\partial q_k}\mathrm{d}q_k, \quad i = 1, 2, 3. \qquad (3.1.2)$$

This expression depends only on the new coordinates q_k, since the $x_i = x_i(q_k)$ are functions of q_k. If the new coordinates q_k are such that the expression

$$\sum_{i=1}^{3} \frac{\partial x_i}{\partial q_l}\frac{\partial x_i}{\partial q_m} = 0, \quad \text{or} \quad \vec{e}_l\vec{e}_m = 0, \quad l \neq m \qquad (3.1.3)$$

vanishes, then the new coordinate system is called *orthogonal*. In this case the unit vectors $\vec{e}_l$ and $\vec{e}_m$ are orthogonal. Furthermore, if the JACOBIAN *functional determinant*

$$\det |\partial q_i(x_k)/\partial q_k| \qquad (3.1.4)$$

does not vanish, an inverse transformation exists, and

$$\mathrm{d}q_l = \sum_{i=1}^{3} \frac{\partial q_l}{\partial x_i}\mathrm{d}x_i. \qquad (3.1.5)$$

Since the cartesian coordinates are independent from each other one has $\partial x_i/\partial x_k = \delta_{ik}$ (KRONECKER *symbol*) and one can write

$$\frac{\partial x_i}{\partial q_l}\frac{\partial q_l}{\partial x_k} = \delta_{ik} = \frac{\partial q_i}{\partial q_k}. \qquad (3.1.6)$$

One sees that the value of δ_{ik} is invariant under coordinate transformations. Inserting (3.1.2) into (3.1.5) one obtains

$$\mathrm{d}q_l = \sum_{i=1}^{3} \frac{\partial q_l}{\partial x_i} \sum_{k=1}^{3} \frac{\partial x_i}{\partial q_k} \mathrm{d}q_k = \sum_{i,k=1}^{3} \frac{\partial q_l}{\partial x_i} \frac{\partial x_i}{\partial q_k} \mathrm{d}q_k = \delta_{lk} \mathrm{d}q_k = \mathrm{d}q_l. \qquad (3.1.7)$$

Raising (3.1.2) to the second power results in

$$\mathrm{d}x_i^2 = \sum_{k,l=1}^{3} \frac{\partial x_i}{\partial q_k} \frac{\partial x_i}{\partial q_l} \mathrm{d}q_k \mathrm{d}q_l. \qquad (3.1.8)$$

Defining now the *metric tensor*

$$g_{kl} = \sum_{i=1}^{3} \frac{\partial x_i}{\partial q_k} \frac{\partial x_i}{\partial q_l} \qquad (3.1.9)$$

one may write the line element in the form

$$\mathrm{d}s^2 = \sum_{k,l=1}^{3} g_{kl} \mathrm{d}q_k \mathrm{d}q_l = \sum_{i=1}^{3} \mathrm{d}x_i^2. \qquad (3.1.10)$$

Mathematica is of great help. In the standard Add On Packages (located at /usr/local/mathematica/AddOns/StandardPackages) one finds the package VectorAnalysis. The command

<<Calculus`VectorAnalysis` $\qquad\qquad\qquad\qquad\qquad\qquad\qquad$ (3.1.11)

loads the package. In this package the default coordinate system is **Cartesian** with coordinate variables Xx, Yy, Zz. The command

Coordinates[Cartesian] $\qquad\qquad\qquad\qquad\qquad\qquad\qquad\qquad$ (3.1.12)

yields $\{Xx, Yy, Zz\}$ and

Coordinates[Spherical] $\qquad\qquad\qquad\qquad\qquad\qquad\qquad\qquad$ (3.1.13)

gives $\{Rr, Ttheta, Pphi\}$.

To set a special coordinate system we type

SetCoordinates[Spherical] $\qquad\qquad\qquad\qquad\qquad\qquad\qquad$ (3.1.14)

so that now spherical coordinates represent the default system. Some more commands are useful:

CoordinateRanges[ ] $\qquad\qquad\qquad\qquad\qquad\qquad\qquad\qquad\qquad$ (3.1.15)

gives the intervals over where each of the coordinate variables of the last defined default system may range, $(0 \le Rr, 0 \le Ttheta, -\pi < Pphi \le \pi)$ and

CoordinateRanges[Cylindrical] $\qquad\qquad\qquad\qquad\qquad\qquad$ (3.1.16)

gives the result for the Cylindrical system etc. *Mathematica* offers transformations too. Thus

`CoordinatesToCartesian[{1,Pi/2,Pi/4},Spherical]` (3.1.17)

results in $(1/\sqrt{2},\ 1/\sqrt{2}, 0)$, (3.1.18)

which are the cartesian coordinates of the point whose spherical coordinates are $1, \pi/2, \pi/4$. On the other hand,

`CoordinatesFromCartesian[{x,y,z},Cylindrical]` (3.1.19)

yields

$$\left(r = \sqrt{x^2 + y^2},\ \varphi = \text{ArcTan}(y/x) \text{ or } \text{ArcTan}[x, y], z\right).$$ (3.1.20)

To obtain the formulae for the transformation between cartesian and spherical coordinates, we give the commands

`CoordinatesFromCartesian[{x,y,z},Spherical]` (3.1.21)

yielding

$$\left(\sqrt{x^2 + y^2 + z^2},\ \text{ArcCos}\left[\frac{z}{\sqrt{x^2 + y^2 + z^2}}\right],\ \text{ArcTan}[x,\ y]\right)$$ (3.1.22)

and

`CoordinatesToCartesian[{Rr,Ttheta,Pphi},Spherical]` (3.1.23)

is yielding

$$(Rr\ \text{Cos}[Pphi]\ \text{Sin}[Ttheta],\ Rr\ \text{Sin}[Pphi]\ \text{Sin}[Ttheta],\ Rr\ \text{Cos}[Ttheta]).$$ (3.1.24)

Mathematica also understands how to calculate the functional determinant. The command

`MatrixForm[JacobianMatrix[Spherical[r,theta,phi]]]` (3.1.25)

forces *Mathematica* to give the result in the form of a matrix

$$\begin{pmatrix} \text{Cos}[phi]\ \text{Sin}[theta] & r\ \text{Cos}[phi]\ \text{Cos}[theta] & -r\ \text{Sin}[phi]\ \text{Sin}[theta] \\ \text{Sin}[phi]\ \text{Sin}[theta] & r\ \text{Cos}[theta]\ \text{Sin}[phi] & r\ \text{Cos}[phi]\ \text{Sin}[theta] \\ \text{Cos}[theta] & -r\ \text{Sin}[theta] & 0 \end{pmatrix}$$ (3.1.26)

and calculate the determinant of (3.1.26)

`Simplify[Det[%]]` (3.1.27)

results in $g = r^2 \sin(\vartheta)$, (3.1.28)

where g is the determinant of the metric tensor (3.1.9). Using (3.1.9) the tensor itself is then represented by the matrix

$$g_{kl} = \begin{pmatrix} 1 & 0 & 0 \\ 0 & r^2 & 0 \\ 0 & 0 & r^2 \sin^2 \vartheta \end{pmatrix}.$$ (3.1.29)

Now the differentials of the coordinates may also be calculated using (3.1.2) and (3.1.5)

$$dx = \sin\vartheta\cos\varphi\,dr + r\cos\vartheta\cos\varphi\,d\vartheta - r\sin\vartheta\sin\varphi\,d\varphi$$

$$dy = \sin\vartheta\sin\varphi\,dr + r\cos\vartheta\sin\varphi\,d\vartheta + r\sin\vartheta\cos\varphi\,d\varphi$$

$$dz = \cos\vartheta\,dr - r\sin\vartheta\,d\vartheta$$

$$dr = \sin\vartheta\cos\varphi\,dx + \sin\vartheta\sin\varphi\,dy + \cos\vartheta\,dz$$

$$d\vartheta = \cos\vartheta\cos\varphi\,dx + \cos\vartheta\sin\varphi\,dy - \sin\vartheta\,dz$$

$$d\varphi = -\sin\varphi\,dx + \cos\varphi\,dy. \tag{3.1.30}$$

The line element now reads

$$ds^2 = dx^2 + dy^2 + dz^2 = \sum_{k,l} g_{kl} dq_l dq_k$$

$$= dr^2 + r^2 d\vartheta^2 + r^2 \sin^2\vartheta\,d\varphi^2. \tag{3.1.31}$$

The HELMHOLTZ equation

$$\Delta u(x,y,z) + k^2 u(x,y,z) = 0 \tag{3.1.32}$$

and the three-dimensional LAPLACE equation

$$\Delta u(x,y,z) \equiv u_{xx} + u_{yy} + u_{zz} = 0 \tag{3.1.33}$$

are separable in 11 coordinate systems. As long as a boundary coincides with the coordinate lines of these 11 coordinate systems, it is easy to solve the boundary value problem. For other partial differential equations and other coordinate systems the question of separability is answered by the STAECKEL *determinant* [1.2] and [1.4].

The 11 coordinate systems are in *Mathematica* notation:

Cartesian (x,y,z)	Cylindrical (r,ϑ,z)
Spherical (r,ϑ,φ)	ParabolicCylindrical (u,v,z)
Paraboloid (u,v,φ)	EllipticCylindrical (u,v,z,a)
ProlateSpheroidal (ξ,η,φ,a)	OblateSpheroidal (ξ,η,φ,a)
Conical $(\lambda,\mu,\nu;a,b)$	ConfocalEllipsoidal $(\lambda,\mu,\nu;a,b,c)$
ParabolicCylindrical (u,v,z).	

The parameters a, b, c refer on the centre of the system and other geometric properties (axes, eccentricity). Other useful coordinate systems are Bipolar $(u,v,z;a)$, Bispherical (u,v,φ,a) and various toroidal systems.

Vector analysis is also supported by *Mathematica* . In cartesian coordinates a vector $\vec{K}$ will be defined by

$$\vec{K} = K_x \vec{e}_x + K_y \vec{e}_y + K_z \vec{e}_z \tag{3.1.34}$$

and in spherical coordinates one has

$$\vec{K} = K_r \vec{e}_r + K_\vartheta \vec{e}_\vartheta + K_\varphi \vec{e}_\varphi, \tag{3.1.35}$$

where K_x, K_y, etc., and K_r are the vector components in the actual coordinate system.

A *vector* is defined as an invariant quantity. Thus the transformation from cartesian into other coordinates necessitates a transformation not only of the coordinates and the components, but also of the unit vectors $\vec{e}_i \to \vec{\tilde{e}}_l$ according to the scheme:

cartesian → new system		new system → cartesian	
$dx_i = \sum_{k=1}^{3} \dfrac{\partial x_i}{\partial q_k} dq_k,$	(3.1.2)	$dq_l = \sum_{i=1}^{3} \dfrac{\partial q_l}{\partial x_i} dx_i,$	(3.1.5)
$\vec{e}_i = \sum_{k=1}^{3} \dfrac{\partial x_i}{\partial q_k} \dfrac{1}{\sqrt{g_{kk}}} \vec{\tilde{e}}_k,$	(3.1.36)	$\vec{\tilde{e}}_l = \sum_{i=1}^{3} \dfrac{\partial q_l}{\partial x_i} \vec{e}_i,$	(3.1.37)
$K_i = \sum_{l=1}^{3} \dfrac{\partial q_l}{\partial x_i} \tilde{K}_l,$	(3.1.38)	$\tilde{K}_l = \sum_{i=1}^{3} \dfrac{\partial x_i}{\partial q_l} \dfrac{1}{\sqrt{g_{ll}}} K_i.$	(3.1.39)

Due to (3.1.10) the differentials will also be transformed. We now give an example for the transformation of a vector between cartesian and spherical coordinates.

$$\vec{e}_x = \sin\vartheta \cos\varphi \, \vec{\tilde{e}}_r + \cos\vartheta \cos\varphi \, \vec{\tilde{e}}_\vartheta - \sin\varphi \, \vec{\tilde{e}}_\varphi,$$
$$\vec{e}_y = \sin\vartheta \sin\varphi \, \vec{\tilde{e}}_r + \cos\vartheta \sin\varphi \, \vec{\tilde{e}}_\vartheta + \cos\varphi \, \vec{\tilde{e}}_\varphi,$$
$$\vec{e}_z = \cos\vartheta \, \vec{\tilde{e}}_r - \sin\vartheta \, \vec{\tilde{e}}_\vartheta. \tag{3.1.40}$$

$$\vec{\tilde{e}}_r = \sin\vartheta \cos\varphi \, \vec{e}_x + \sin\vartheta \sin\varphi \vec{e}_y + \cos\vartheta \, \vec{e}_z,$$
$$\vec{\tilde{e}}_\vartheta = \cos\vartheta \cos\varphi \, \vec{e}_x + \cos\vartheta \sin\varphi \, \vec{e}_y - \sin\vartheta \vec{e}_z,$$
$$\vec{\tilde{e}}_\varphi = -\sin\varphi \, \vec{e}_x + \cos\varphi \, \vec{e}_y. \tag{3.1.41}$$

$$K_x = K_r \sin\vartheta \cos\varphi + K_\vartheta \cos\vartheta \cos\varphi - K_\varphi \sin\varphi,$$
$$K_y = K_r \sin\vartheta \sin\varphi + K_\vartheta \cos\vartheta \sin\varphi + K_\varphi \cos\varphi,$$
$$K_z = K_r \cos\vartheta - K_\vartheta \sin\vartheta, \tag{3.1.42}$$

$$K_r = K_x \sin\vartheta \cos\varphi + K_y \sin\vartheta \sin\varphi + K_z \cos\vartheta,$$
$$K_\vartheta = K_x \cos\vartheta \cos\varphi + K_y \cos\vartheta \sin\varphi - K_z \sin\vartheta,$$
$$K_\varphi = -K_x \sin\varphi + K_y \cos\varphi. \tag{3.1.43}$$

These calculations have been made using the two matrices

$$M_{li} = \frac{\partial x_i}{\partial q_l} = \begin{pmatrix} \sin\vartheta \cos\varphi & \sin\vartheta \sin\varphi & \cos\vartheta \\ r \cos\vartheta \cos\varphi & r \cos\vartheta \sin\varphi & -r \sin\vartheta \\ -r \sin\vartheta \sin\varphi & r \sin\vartheta \cos\varphi & 0 \end{pmatrix} \tag{3.1.44}$$

$$N_{il} = \frac{\partial q_l}{\partial x_i} = \begin{pmatrix} \sin\vartheta \cos\varphi & \cos\vartheta \cos\varphi & -\sin\varphi \\ \sin\vartheta \sin\varphi & \cos\vartheta \sin\varphi & \cos\varphi \\ \cos\vartheta & -\sin\vartheta & 0 \end{pmatrix} \tag{3.1.45}$$

according to (3.1.38) in the form

$$K_i = \sum_l N_{il} K_l \tag{3.1.46}$$

and according to (3.1.39)

$$\tilde{K}_l = \sum_i M_{li}(g_{ll})^{-1/2} K_i. \tag{3.1.47}$$

Now we can elaborate vector algebra and vector analysis. The *scalar (dot) product* of two vectors v_1 and v_2 is computed in the default coordinates using <<Calculus`VectorAnalysis`

$$\vec{v}_1 \cdot \vec{v}_2 = \textbf{DotProduct[v1,v2]} \tag{3.1.48}$$

where $v_1 = \{v_{1x}, v_{1y}, v_{1z}\}$ defines a vector. If the scalar product is to be calculated in another coordinate system, the command **DotProduct[v1,v2]** is useful.

Let $\vec{v}_1 = a\vec{e}_x + 5\vec{e}_y + \vec{e}_z, \vec{v}_2 = \vec{e}_x + 10\vec{e}_y$, or

v1={a*x,5,1};   v2={1,10,0}; $\tag{3.1.49}$

then $\vec{v}_1 \cdot \vec{v}_2 = 50 + a\,x$ is obtained by **DotProduct[v1,v2]**

The *vector (cross) product* is defined by

$$\vec{v}_1 \times \vec{v}_2 = \textbf{CrossProduct[v1,v2]} \tag{3.1.50}$$

so that one obtains the new vector $\{-10,\ 1,\ -5 + 10ax\}$.

Mathematica calculates the scalar triple product using

ScalarTripleProduct[v1,v2,v3]

Next we define the del (*nabla operator*) by

$$\nabla = \vec{e}_x \frac{\partial}{\partial x} + \vec{e}_y \frac{\partial}{\partial y} + \vec{e}_z \frac{\partial}{\partial z}; \quad \nabla U = \operatorname{grad} U \tag{3.1.51}$$

or more generally

$$\nabla = \sum_{l=1}^{3} \frac{\vec{e}_l}{\sqrt{g_{ll}}} \frac{\partial}{\partial q_l}; \quad \nabla U = \operatorname{grad} U = \sum_{l=1}^{3} \frac{\vec{e}_l}{\sqrt{g_{ll}}} \frac{\partial U}{\partial q_l}. \tag{3.1.52}$$

Mathematica uses **Grad[f,coordsys]** $\tag{3.1.53}$

Thus **Grad[7*x^3+y^2-z^4,Cartesian[x,y,z]]**

delivers the vector $\{21x^2, 2y, -4z^3\}$.

The curl$\vec{v}$ is elaborated by

Curl[v,Spherical[r,theta,phi]] $\tag{3.1.54}$

Let **v={r^2*Sin[theta],r*Cos[phi]*Sin[theta],r*Sin[phi]};**

then (3.1.54) yields the vector

$\{(1 + \operatorname{Cot}[theta])\operatorname{Sin}[phi], -2\operatorname{Sin}[phi], -r\operatorname{Cos}[theta] + 2\operatorname{Cos}[phi]\operatorname{Sin}[theta]\}$

Using the palette with Greek and other special symbols we may rewrite

v={r^2*Sin[ϑ],r*Cos[φ]*Sin[ϑ],r*Sin[φ]}; $\tag{3.1.55}$

then we obtain for the curl

$$\left\{ \frac{\text{Csc}[\vartheta] \ (r^2 \ \text{Cos}[\vartheta] \ \text{Sin}[\varphi] + r^2 \ \text{Sin}[\vartheta] \ \text{Sin}[\varphi])}{r^2}, \right.$$

$$\left. -2 \ \text{Sin}[\varphi], \quad \frac{-r^2 \ \text{Cos}[\vartheta] + 2 \ r \ \text{Cos}[\varphi] \ \text{Sin}[\vartheta]}{r} \right\}. \tag{3.1.56}$$

Finally the *Laplacian operator* ($\triangle$) can be realized by the command

Laplacian[f]

and the divergence of a vector v is given by **Div[v]**.

Behind these commands there are some complicated calculations. In general coordinates q_l one has

$$\nabla = \sum_{l=1}^{3} \frac{\vec{e}_l}{\sqrt{g_{ll}}} \frac{\partial}{\partial q_l}; \quad \nabla U = \text{grad} \, U = \sum_{l=1}^{3} \frac{\vec{e}_l}{\sqrt{g_{ll}}} \frac{\partial U}{\partial q_l}. \tag{3.1.57}$$

In spherical coordinates this reads

$$\nabla = \vec{e}_r \frac{\partial}{\partial r} + \vec{e}_\vartheta \frac{1}{r} \frac{\partial}{\partial \vartheta} + \vec{e}_\varphi \frac{1}{r \sin \vartheta} \frac{\partial}{\partial \varphi}. \tag{3.1.58}$$

Applying the operator div on a vector yields

$$\text{div} \, (\vec{e} U) = U \, \text{div} \, \vec{e} + \vec{e} \, \text{grad} \, U,$$

$$\text{div} \, \vec{K} = \text{div} \sum_l (\vec{e}_l K_l) = \sum_l (K_l \text{div} \, \vec{e}_l + \vec{e}_l \, \text{grad} \, K_l)$$

or in spherical coordinates

$$\text{div} \, \vec{K} = \text{div} \, (K_r \vec{e}_r + K_\vartheta \vec{e}_\vartheta + K_\varphi \vec{e}_\varphi)$$

$$= K_r \text{div} \, \vec{e}_r + K_\vartheta \text{div} \, \vec{e}_\vartheta + K_\varphi \text{div} \, \vec{e}_\varphi$$

$$+ \vec{e}_r \, \text{grad} \, K_r + \vec{e}_\vartheta \, \text{grad} \, K_\vartheta + \vec{e}_\varphi \, \text{grad} \, K_\varphi. \tag{3.1.59}$$

It is thus clear that the basic vectors $\vec{e}$ have also to be transformed. But expressions like

$$\left(\vec{e}_r \frac{\partial}{\partial r} + \vec{e}_\vartheta \frac{1}{r} \frac{\partial}{\partial \vartheta} + \vec{e}_\varphi \frac{\partial}{\partial \varphi} \right) (\sin \vartheta \cos \varphi \vec{e}_x + \sin \vartheta \sin \varphi \vec{e}_y + \cos \vartheta \vec{e}_z) \tag{3.1.60}$$

are quite cumbersome. It is hence of advantage to use cartesian coordinates for such calculations. We can write

$$\text{div} \, \vec{e}_r = \left(\vec{e}_x \frac{\partial}{\partial x} + \vec{e}_y \frac{\partial}{\partial y} + \vec{e}_z \frac{\partial}{\partial z} \right) \left(\frac{x}{r} \vec{e}_x + \frac{y}{r} \vec{e}_\vartheta + \frac{z}{r} \vec{e}_\varphi \right). \tag{3.1.61}$$

Using $(\vec{e}_l \vec{e}_k = \delta_{lk})$ for orthogonal coordinates, one obtains

$$\operatorname{div}\vec{e}_r = \frac{\partial}{\partial x}\frac{x}{r} + \frac{\partial}{\partial y}\frac{y}{r} + \frac{\partial}{\partial z}\frac{z}{r} = \frac{2}{r} \qquad (3.1.62)$$

and

$$\operatorname{div}\vec{e}_\vartheta = \frac{\cot\vartheta}{r}, \quad \operatorname{div}\vec{e}_\varphi = 0. \qquad (3.1.63)$$

Then

$$\operatorname{div}\vec{K} = \frac{\partial K_r}{\partial r} + \frac{1}{r}\frac{\partial K_\vartheta}{\partial \vartheta} + \frac{1}{r\sin\vartheta}\frac{\partial K_\varphi}{\partial\varphi} + \frac{2}{r}K_r + \frac{1}{r}K_\vartheta\cot\vartheta \qquad (3.1.64)$$

or more generally

$$\operatorname{div}\vec{K} = \nabla\cdot\vec{K} = \sum_{l=1}^{3}\frac{1}{\sqrt{g_{ll}}}\frac{\partial K_l}{\partial q_l} + \frac{1}{\sqrt{g}}\sum_l K_l\frac{\partial}{\partial q_l}\sqrt{\frac{g}{g_{ll}}}. \qquad (3.1.65)$$

Similar situations appear for curl and $\triangle$. One has

$$\operatorname{curl}\vec{K} = \nabla\times\vec{K} = K_1\operatorname{curl}\vec{e}_1 + K_2\operatorname{curl}\vec{e}_2 + K_3\operatorname{curl}\vec{e}_3 + [\operatorname{grad}K_1\times\vec{e}_1]$$
$$+ [\operatorname{grad}K_2\times\vec{e}_2] + [\operatorname{grad}K_3\times\vec{e}_3] \qquad (3.1.66)$$

and in spherical coordinates this reads

$$(\operatorname{curl}\vec{K})_r = \frac{1}{r}\frac{\partial K_\varphi}{\partial\vartheta} - \frac{1}{r\sin\vartheta}\frac{\partial K_\vartheta}{\partial\varphi} + \frac{1}{r}K_\varphi\cot\vartheta, \qquad (3.1.67)$$

$$(\operatorname{curl}\vec{K})_\vartheta = \frac{1}{r\sin\vartheta}\frac{\partial K_r}{\partial\varphi} - \frac{\partial K_\varphi}{\partial r} - \frac{1}{r}K_\varphi, \qquad (3.1.68)$$

$$(\operatorname{curl}\vec{K})_\varphi = \frac{\partial K_\vartheta}{\partial r} - \frac{1}{r}\frac{\partial K_r}{\partial\vartheta} + \frac{1}{r}K_\vartheta \qquad (3.1.69)$$

and more generally

$$(\nabla\times\vec{K})_1 = \frac{1}{\sqrt{g_{22}}}\frac{\partial K_3}{\partial q_2} - \frac{1}{\sqrt{g_{33}}}\frac{\partial K_2}{\partial q_3}$$
$$+ \frac{K_3}{\sqrt{g_{22}g_{33}}}\frac{\partial\sqrt{g_{33}}}{\partial q_2} - \frac{K_2}{\sqrt{g_{22}g_{33}}}\frac{\partial\sqrt{g_{22}}}{\partial q_3}, \qquad (3.1.70)$$

$$(\nabla\times\vec{K})_2 = \frac{1}{\sqrt{g_{33}}}\frac{\partial K_1}{\partial q_3} - \frac{1}{\sqrt{g_{11}}}\frac{\partial K_3}{\partial q_1}$$
$$+ \frac{K_1}{\sqrt{g_{11}g_{33}}}\frac{\partial\sqrt{g_{11}}}{\partial q_3} - \frac{K_3}{\sqrt{g_{11}g_{33}}}\frac{\partial\sqrt{g_{33}}}{\partial q_1}, \qquad (3.1.71)$$

$$(\nabla \times \vec{K})_3 = \frac{1}{\sqrt{g_{11}}} \frac{\partial K_2}{\partial q_1} - \frac{1}{\sqrt{g_{22}}} \frac{\partial K_1}{\partial q_2}$$

$$+ \frac{K_2}{\sqrt{g_{11} g_{22}}} \frac{\partial \sqrt{g_{22}}}{\partial q_1} - \frac{K_1}{\sqrt{g_{11} g_{22}}} \frac{\partial \sqrt{g_{11}}}{\partial q_2}. \qquad (3.1.72)$$

For the Laplacian one finds

$$\Delta U = \operatorname{div} \operatorname{grad} U = \frac{1}{\sqrt{g}} \sum_{l=1} \frac{\partial}{\partial q_l} \frac{\sqrt{g}}{\sqrt{g_{ll}}} \frac{\partial U}{\partial q_l} \qquad (3.1.73)$$

or in spherical coordinates

$$\Delta U = \frac{\partial^2 U}{\partial r^2} + \frac{2}{r} \frac{\partial U}{\partial r} + \frac{1}{r^2 \sin^2 \vartheta} \frac{\partial U}{\partial \varphi^2} + \frac{1}{r^2} \frac{\partial^2 U}{\partial \vartheta^2} + \frac{\cos \vartheta}{r^2 \sin \vartheta} \frac{\partial U}{\partial \vartheta}. \qquad (3.1.74)$$

Now it is quite simple to separate the HELMHOLTZ equation $\Delta U + k^2 U = 0$. Using the separation setup

$$U(r, \vartheta, \varphi) = R(r)\Theta(\vartheta)\Phi(\varphi) \qquad (3.1.75)$$

one gets after division by U

$$R'' + \frac{2}{r} R' + (k^2 - a/r^2)R = 0, \qquad (3.1.76)$$

$$\Theta'' + \cot \vartheta \, \Theta' + (a - b/sin^2\vartheta)\Theta = 0, \qquad (3.1.77)$$

$$\Phi'' + b\Phi = 0, \qquad (3.1.78)$$

where a and b are separation constants. These ordinary differential equations may be solved using the methods discussed in chapter 2. Again *Mathematica* commands are of great help. Using (3.1.75) and applying

Expand[(Laplacian[U[r,ϑ,φ],Spherical[r, ϑ,φ]])
*r^2/U[r,ϑ,φ]]; (3.1.79)

results in the yet unseparated equations (3.1.76) - (3.1.78) in the form

$$\frac{2 \, r \, R'[r]}{R[r]} + \frac{\mathrm{Cot}[\vartheta] \, \Theta'[\vartheta]}{\Theta[\vartheta]} + \frac{r^2 \, R''[r]}{R[r]} + \frac{\Theta''[\vartheta]}{\Theta[\vartheta]} + \frac{\mathrm{Csc}[\vartheta]^2 \, \Phi''[\varphi]}{\Phi[\varphi]}.$$

The command **Expand** is used to expand out products (or positive integer powers) in an expression. (Please remember to load the package VectorAnalysis.)

The *vector* HELMHOLTZ *equation*, which can be found in electromagnetism presents difficulties. In general coordinate systems it is defined by

$$\Delta \vec{K} + k^2 \vec{K} = \operatorname{grad} \operatorname{div} \vec{K} - \operatorname{curl} \operatorname{curl} \vec{K} + k^2 \vec{K} = 0. \qquad (3.1.80)$$

If one uses cartesian coordinates, this vector equation can be separated into three equations,

$$(\Delta \vec{K})_x + k^2 K_x = \frac{\partial^2 K_x}{\partial x^2} + \frac{\partial^2 K_x}{\partial y^2} + \frac{\partial^2 K_x}{\partial z^2} + k^2 K_x = 0, \qquad (3.1.81)$$

$$(\Delta \vec{K})_y + k^2 K_y = \frac{\partial^2 K_y}{\partial x^2} + \frac{\partial^2 K_y}{\partial y^2} + \frac{\partial^2 K_y}{\partial z^2} + k^2 K_y = 0, \qquad (3.1.82)$$

$$(\Delta \vec{K})_z + k^2 K_z = \frac{\partial^2 K_z}{\partial x^2} + \frac{\partial^2 K_z}{\partial y^2} + \frac{\partial^2 K_z}{\partial z^2} + k^2 K_x = 0, \qquad (3.1.83)$$

but in other coordinate systems problems appear - the equivalent equations (3.1.81) - (3.1.83) are coupled! In spherical coordinates one finds

$$(\Delta \vec{K})_r + k^2 K_r = \frac{\partial^2 K_r}{\partial r^2} + \frac{1}{r^2} \frac{\partial^2 K_r}{\partial \vartheta^2} + \frac{1}{r^2 \sin^2 \vartheta} \frac{\partial^2 K_r}{\partial \varphi^2} + \frac{2}{r} \frac{\partial K_r}{\partial r}$$

$$+ \frac{\cot \vartheta}{r^2} \frac{\partial K_r}{\partial \vartheta} - \frac{2 K_r}{r^2} \qquad (3.1.84)$$

$$- \frac{2}{r^2} \frac{\partial K_\vartheta}{\partial \vartheta} - \frac{2 K_\vartheta \cot \vartheta}{r^2} - \frac{2}{r^2 \sin \vartheta} \frac{\partial K_\varphi}{\partial \varphi} + k^2 K_r = 0.$$

But a trick may help: use the cartesian vector components as functions of the general (spherical) coordinates, so that $(\Delta \vec{K})_l = \Delta \vec{K}_l$. Solve the three equations and then transform the cartesian coordinates into the general (spherical etc.) coordinates, which had been chosen originally to fit a boundary surface.

In two dimensions, the problem of separability of partial differential equations looks nicer. Due to *conformal mapping* the LAPLACE equation becomes separable in an infinite number of two-dimensional systems and for axially symmetric problems (*three-dimensional conformal mapping*). There are also some separable two-dimensional toroidal systems:

$$\begin{aligned}
x &= a \sinh \eta \cos \psi / T, \quad y = a \sinh \eta \sin \psi / T, \\
z &= a \sin \vartheta / T, \qquad\qquad T = \cosh \eta - \cos \vartheta, \qquad (3.1.85) \\
0 &\le \eta \le \infty, \quad -\pi < \vartheta \le +\pi, \quad 0 \le \psi < 2\pi,
\end{aligned}$$

or quasitoroidal systems like the helical coordinate system [2.5], [3.1] or magnetic field lines coordinate according to BOOZER [3.2] or HAMADA [3.3].

It might be of interest to state that there are also nonlinear partial differential equations that are separable. So the equation

$$u_{xx} u_{tt} - u_{xt}^2 = k u_x^n u_{xx}^{1+\gamma} \qquad (3.1.86)$$

is separable by the setup

$$u(x,t) = x^m G(t) = F(x)G(t). \qquad (3.1.87)$$

Using

$$n(m-1) = 2\gamma - (\gamma - 1)m \tag{3.1.88}$$

one obtains

$$m(m-1)GG'' - m^2 G'^2 = km^{n+\gamma+1}(m-1)^{1+\gamma}G^{n+\gamma+1}. \tag{3.1.89}$$

On the other hand

$$u_t = (u^n u_x)_x = nu^{n-2}u_x^2 + u^n u_{xx} \tag{3.1.90}$$

can be separated by $u(x,t) = T(t)X(x)$ into

$$\begin{aligned} T' - \lambda T^{n+1} &= 0 \\ X''X' + XX'' - \lambda X &= 0, \end{aligned} \tag{3.1.91}$$

where λ is the separation const. Another possibility to separate (3.1.90) is given by $u = F(t) + x^p G(t)$. Another separable nonlinear partial differential equation (of first order) is given by the JACOBI *equation* of mechanics

$$\left(\frac{\partial S}{\partial x}\right)^2 + \left(\frac{\partial S}{\partial y}\right)^2 + \left(\frac{\partial S}{\partial z}\right)^2 = f(x) + g(y) + h(z) \tag{3.1.92}$$

which can be separated by $S_1(x) + S_2(y) + S_3(z)$.

Problems

1. Find the transformation formulae between cartesian and cylindrical coordinates using (3.1.21) and (3.1.23)

   CoordinatesFromCartesian[{x,y,z},Cylindrical] (3.1.93)

   which should give

   $\sqrt{x^2 + y^2}$, ArcTan$[x, y]$, z

   and

   CoordinatesToCartesian[{Rr,Ttheta,Pphi},Cylindrical]
   (3.1.94)

   should yield

   Rr Cos[Ttheta], Rr Sin[Ttheta], Pphi.

   What happens if you replace $Rr \to r$, Ttheta $\to \vartheta$, Pphi $\to \varphi$?

   r Cos$[\vartheta]$, r Sin$[\vartheta]$, φ.

2. Derive the formulas for the transformation between cartesian and elliptic cylinder coordinates. First we look up the elliptic cylindrical system by

   Coordinates[EllipticCylindrical]

resulting in Uu, Vv, Zz, and then we type

CoordinatesFromCartesian[{x,y,z},
EllipticCylindrical]

The result is astonishing: $\mathrm{Re}[\mathrm{ArcCosh}[x + iy]], \mathrm{Im}[\mathrm{ArcCosh}[x + iy]], z$
but becomes clear by the command

CoordinatesToCartesian[{u,v,z},EllipticCylindrical]

giving $\mathrm{Cos}[v]\,\mathrm{Cosh}[u], \mathrm{Sin}[v]\,\mathrm{Sinh}[u], z$.

Thus the coordinate surfaces are given by elliptic cylinders $(x/a\cosh\eta)^2 + (y/(a\sinh\eta)^2 = 1$, $\eta = const$, by hyperbolic cylinders $(x/a\cos\psi)^2 - (y/(a\sin\psi)^2 = 1$, $\psi = const$, and planes $z = const$. The default value of $a = 1$.

Now we would like to know the interval over which the coordinates u, v, z may vary. The command

CoordinateRanges[EllipticCylindrical]   (3.1.95)

gives the answer. $0 \le u \le \infty, -\pi < v \le \pi, -\infty < z, \infty$.

3. Let us now consider conical coordinates. Typing

CoordinateRanges[Conical]

$\{-\infty < Llambda < \infty, 1 < Mmu^2 < 4, Nnu^2 < 1\}$

CoordinateRanges[Toroidal]

which gives $0 \le Vv < \infty, -\pi < Uu \le \pi, -\pi < Pphi \le \pi$ but

CoordinatesFromCartesian[{x,y,z},Toroidal]

$\{2\,\mathrm{Re}[\mathrm{ArcCoth}[\sqrt{x^2 + y^2} + i\,z]], -2\,\mathrm{Im}[\mathrm{ArcCoth}[\sqrt{x^2 + y^2} + i\,z]],$
$\mathrm{ArcTan}[x, y]\}$

4. Investigate cylindrical coordinates (r, ϑ, z). We first try

CoordinateRanges[Cylindrical]

which yields $0 \le Rr < \infty, -\pi < Ttheta \ge \pi, -\infty < Zz < \infty$.

Now let us solve the LAPLACE equation in these coordinates. We define a separation setup

V[r_,ϑ_,z_]=R[r]*Θ[ϑ]*Z[z]

and give the command

Expand[(Laplacian[V[r,ϑ,z],
Cylindrical[r,ϑ,z]])/V[r,ϑ,z]]

This yields

$$\frac{R'r}{r\,R[r]} + \frac{R''[r]}{R[r]} + \frac{Z''[z]}{Z[z]} + \frac{\Theta''[\vartheta]}{r^2\,\Theta[\vartheta]} = 0.$$

We now read off the three ordinary equations

$$\frac{R''}{R} + \frac{R'}{rR} = b + \frac{a}{r^2} \quad \text{or} \quad R'' + \frac{R'}{r} - \left(\frac{a}{r^2} + b\right)R = 0,$$

$$\frac{Z''}{Z} = -b \quad \text{or} \quad Z'' + bZ = 0,$$

$$\frac{\Theta''}{r^2\Theta} = -\frac{a}{r^2} \quad \text{or} \quad \Theta'' + a\Theta = 0,$$

which can be solved according to the methods given earlier.

5. Investigate now the two-dimensional LAPLACE equation. If one inserts the ansatz

   `W[x_,y_]=X[x]*Y[y];` into `Expand[(Laplacian[W[x,y],`
   `Cylindrical[x,y,z]])*x^2/W[x,y]]`

   one gets the result

   $$\frac{x\,X'[x]}{X[x]} + \frac{x^2\,X''[x]}{X[x]} + \frac{Y''[y]}{Y[y]} = 0$$

   or `Laplacian[W[x,y]]` $= 0$

6. Solve the two ordinary equations just obtained for $X(x)$ and $Y(y)$

   `DSolve[X''[x]+X'[x]/x-a^2*X[x]/x^2==0,X[x],x]`

   The result should be

   $$\{\{X[x] \to C[1]\ \text{Cosh}[a\ \text{Log}[x]] + i\ C[2]\ \text{Sinh}[a\ \text{Log}[x]]\}\}$$

   and the solution of the two-dimensional Laplacian reads

   `M[x_,y_]=X[x]*Y[x]`

   Proof: `Laplacian[M[x,y]]` $= 0$

7. Demonstrate that the *harmonic polynomial*

   `x^4-6*x^2*y^2+y^4` is a solution of the two-dimensional LAPLACE equation.

8. The operation `Div` on a vector may formally be regarded as the `DotProduct` of the two vectors to calculate this divergence.

   Try: `K={5*x^2*y, y^2*x*z^2, 5*x*y};Div[K]` (result?)
   Try: `f[x_,y_,z_]=x^2+y^2+2*z^2; Div[f*K]`

3.2 Methods to reduce partial to ordinary differential equations

In the last section we have seen that many, but not all partial differential equations can be reduced to ordinary differential equations by the separation method. There exist, however, other methods to reach this goal. Some of them are very useful for engineering problems. The LAPLACE transform theory has become a basic part of electronic engineering study [3.4]. Like in all other integral transformations [3.5, 3.6, 3.7] of ordinary differential equations

$$p_0(z)w''(z) + p_1(z)w'(z) + p_2(z)w(z) = 0 \tag{3.2.1}$$

a setup

$$w(z) = \int_{t_0}^{t_1} K(z,t)y(t)\mathrm{d}t \tag{3.2.2}$$

is made. t is a new complex (or real) variable and $K(z,t)$ is the *kernel* of the transformation. Various such kernels have been used:

LAPLACE-Transformation	$\int_0^\infty \exp(-zt),$	$\dfrac{1}{2\pi i}\int_{\gamma-i\infty}^{\gamma+i\infty}\exp(zt),$
FOURIER transformation	$\int_{-\infty}^{+\infty}\exp(-izt),$	$\dfrac{1}{2\pi}\int_{-\infty}^{+\infty}\exp(izt),$
HANKEL transformation	$\int_0^\infty t\mathrm{J}_n(zt),$	$\int_0^{+\infty} z\mathrm{J}_n(zt),$
MELLIN transformation	$\int_0^\infty t^{z-1},$	$\dfrac{1}{2\pi i}\int_{\gamma-i\infty}^{\gamma+i\infty} z^{-t},$
EULER transformation	$\int_{-\infty}^{+\infty}(z-t)^{\mu}.$	

The right column presents the inverse transformation. Now we will demonstrate the application of the LAPLACE *transformation* on partial differential equations. As an example, we choose the *heat conduction equation* for heat conductivity $\lambda = 1$

$$\frac{\partial^2 T}{\partial x^2} = \frac{\partial T}{\partial t}, \tag{3.2.3}$$

and the initial condition

$$T(x,0) = \psi(x). \tag{3.2.4}$$

Here $T(x,t)$ represents the temperature at point x at time t. Multiplication of (3.2.4) by the LAPLACE kernel $\exp(-zt)$ and integration results in

$$\int\limits_0^\infty \exp(-zt)\frac{\partial^2 T}{\partial x^2}\mathrm{d}t = \frac{\partial^2}{\partial x^2}\int\limits_0^\infty \exp(-zt)T(x,t)\mathrm{d}t = \int\limits_0^\infty \exp(-zt)\frac{\partial T}{\partial t}\mathrm{d}t. \tag{3.2.5}$$

Due to (3.2.2) one has the LAPLACE transform $w(z,x)$ of $T(x,t)$

$$w(z,x) = \int\limits_0^\infty \exp(-zt)T(x,t)\mathrm{d}t. \tag{3.2.6}$$

Partial integration of the rhs term of (3.2.5) $\int u\mathrm{d}v = uv - \int v\mathrm{d}u$ yields

$$\int\limits_0^\infty \exp(-zt)\frac{\partial T}{\partial t}\mathrm{d}t = \left| \exp(-zt)T(x,t)\right|_0^\infty + \int\limits_0^\infty zT(x,t)\exp(-zt)\mathrm{d}t \tag{3.2.7}$$

the LAPLACE transform of $\partial T/\partial t$ due to $\exp(0) = 1$ and $\exp(-\infty) = 0$. Since z and t are independent variables, we may rewrite (3.2.7) as

$$\int\limits_0^\infty \exp(-zt)\frac{\partial T}{\partial t}\mathrm{d}t = -T(x,0) + z\int\limits_0^\infty T(x,t)\exp(-zt)\mathrm{d}t.$$

Using the definitions (3.2.6), (3.2.4) we thus have

$$\int\limits_0^\infty \exp(-zt)\frac{\partial T}{\partial t}\mathrm{d}t = zw(x;z) - \psi(x). \tag{3.2.8}$$

Now we consider $\partial^2 T/\partial x^2$. The next step is

$$\int\limits_0^\infty \exp(-zt)\frac{\partial T}{\partial x}\mathrm{d}t = \frac{\mathrm{d}}{\mathrm{d}x}\int\limits_0^\infty \exp(-zt)T(x,t)\mathrm{d}t = \frac{\mathrm{d}}{\mathrm{d}x}w(x,z) \tag{3.2.9}$$

and thus

$$\int\limits_0^\infty \exp(-zt)\frac{\partial^2 T}{\partial x^2}\mathrm{d}t = \frac{\mathrm{d}^2 w(x;z)}{\mathrm{d}x^2}. \tag{3.2.10}$$

Collecting the terms and inserting we obtain from (3.2.3)

$$\frac{\mathrm{d}^2 w(x;z)}{\mathrm{d}x^2} = zw(z,x) - \psi(x), \tag{3.2.11}$$

which is an ordinary differential equation for $w(x; z)$. z is a parameter. After solving (3.2.11) one needs to apply the inverse transformation $w(z, x)$ to obtain the solution $T(x, t)$ of (3.2.3).

It is remarkable that the initial condition $\psi(x)$ became the inhomogeneous term of the ordinary differential equation. For $\psi = 0$, the solution of (3.2.11) reads

$$w(x; z) = C_1(z) \exp(x\sqrt{z}) + C_2(z) \exp(-x\sqrt{z}). \tag{3.2.12}$$

C_1, C_2 are the integration constants depending on z. They allow to satisfy the boundary conditions of the partial differential equation (3.2.3). They might be

$$T(0, t) = \varphi(t), \, T(\infty, t) = 0 \tag{3.2.13}$$

or

$$w(0; z) = \int_0^\infty \exp(-zt)T(0, t)\mathrm{d}t = \int_0^\infty \exp(-zt)\varphi(t)\mathrm{d}t = \bar{w}(z) \tag{3.2.14}$$

so that $C_1(z) = 0$, $C_2(z) = \bar{w}(z)$. To obtain the correct full solution of the partial differential equation one has to apply the inverse transformation on $w(x, z)$ with respect to z.

Mathematica may help with these calculations [2]. It has a special package that is to be loaded into the memory of your computer by the command

`<<Calculus`LaplaceTransform``

(In some newer versions of *Mathematica* the package is autoloading and this command no longer needed.)

To understand how *Mathematica* handles the problem, we start step by step with (3.2.6), which now reads

`w[z,x]=LaplaceTransform[D[T[x,t],t],t,z]` (3.2.15)

yielding

z LaplaceTransform$[T[x, t], t, z] - T[x, 0]$ (3.2.16)

what is equivalent with (3.2.8). The next step is

`LaplaceTransform[D[T[x,t],{x,2}],t,z]` (3.2.17)

results in

LaplaceTransform$[T^{(2,0)}[x, t], t, z]$, (3.2.18)

what is equivalent with (3.2.10). *Mathematica* sometimes uses the notation `w[z][x]` for $w(z, x)$ to point to the difference between the variable and the parameter z.

The integral transformations work only for linear ordinary or partial differential equations. Nonlinear equations may be handled by *similarity transformations*. It seems that BOLTZMANN was first to suggest this method. Taking into account the heat conductivity λ of various materials, which may depend itself on temperature, BOLTZMANN suggested the *pseudo Laplacian operator*

$$\frac{\partial u}{\partial t} = \frac{\partial}{\partial x}\left[f(u(x,t))\frac{\partial u}{\partial x}\right] = \frac{\partial u}{\partial t} \qquad (3.2.19)$$

to replace (3.2.3). Now a similarity variable

$$\eta(x,t) = x^\alpha t^\beta \qquad (3.2.20)$$

has been defined. Insertion of (3.2.20) into (3.2.19) results in

$$u_x = \frac{du}{d\eta}\frac{\partial \eta}{\partial x} = u'\alpha x^{\alpha-1}t^\beta, \quad u_t = u'\beta x^\alpha t^{\beta-1}, \qquad (3.2.21)$$

$$u_{xx} = \alpha(\alpha-1)x^{\alpha-2}t^\beta u' + \alpha^2 x^{2(\alpha-1)}t^{2\beta}u'', \qquad (3.2.22)$$

and assuming the transition $u(x,t) \to u(\eta)$

$$\beta\frac{x^2}{t}\eta u' = \alpha(\alpha-1)\eta f(u)u' + \alpha^2\eta^2(f(u)u')'. \qquad (3.2.23)$$

In this expression all terms depend on η with the exception of x^2/t. To continue, one assumes

$$x^2/t = \eta \quad \text{or} \quad x^2/t = \eta^2 \qquad (3.2.24)$$

which yields α and β. The physical interpretation and the usefulness of (3.2.22) even for the linear equation ($\lambda = const$) will be discussed later on (problems 10 and 11 in section 3.3).

The disadvantage of similarity transformations is hidden in the fact that similarity solutions may only satisfy very restricted boundary conditions. BIRKHOFF [3.8] has developed several methods to find similarity transformations [3.9], [3.10], [3.11], [3.12]. He uses transformations of two independent variables occurring in linear or nonlinear partial differential equations

$$\begin{aligned}\bar{x}_i &= a^{\alpha_i}x_i, \quad i = 1\ldots m,\\ \bar{y}_j &= a^{\gamma_j}y_j, \quad j = 1\ldots n.\end{aligned} \qquad (3.2.25)$$

Let us look at an example. We consider the nonlinear partial differential equation of second order for $u(x,y,t)$.

$$u^p(u_{xx} + u_{yy}) + pu^{p-1}(u_x^2 + u_y^2) - u_t = 0. \qquad (3.2.26)$$

Using $\bar{x} = a^m x$, $\bar{y} = a^n y$, $\bar{t} = a^r t$, $\bar{u} = a^s u$, one obtains the condition $2m - (p+1)s = 2n - (p+1)s = r - s$, so that all powers of a are the same. This yields

$$\xi = xt^{-\alpha}, \quad \eta = yt^{-\alpha}, \quad f(\xi,\eta) = \frac{u(x,y,t)}{t^A} \qquad (3.2.27)$$

and the new partial differential equation for $f(\xi, \eta)$

$$f^p \left[f_{\xi\xi} + f_{\eta\eta}\right] + pf^{p-1} \left[f_\xi^2 + f_\eta^2\right] - Af + \alpha\left[\xi f_\xi + \eta f_\eta\right] = 0. \qquad (3.2.28)$$

The next step $\vartheta = \eta/\xi$, $\psi(\vartheta) = f(\xi, \eta)/\xi^{2/p}$ finally results in the ordinary differential equation for $\psi(\vartheta)$

$$\psi^{p-1} \left[(1 + \vartheta^2)\, \psi'' + 2\vartheta\left(1 - \frac{2}{p}\right)\psi' + \frac{2}{p}\left(\frac{2}{p} - 1\right)\psi\right]$$

$$+ p\psi^{p-2} \left[\psi'^2 + \left(\frac{2}{p}\psi - \vartheta\psi'\right)^2\right] + \frac{1}{p} = 0. \qquad (3.2.29)$$

Another method to solve disagreable partial differential equations consists of multiple integrations. Electromagnetic waves progressing along wires are described by the two partial differential equations

$$\frac{\partial I(x,t)}{\partial x} = -C\frac{\partial U(x,t)}{\partial t}, \quad E = -\frac{\partial U}{\partial x}, \quad E = R(E)I, \qquad (3.2.30)$$

where I is the current intensity, C the (constant) capacity for unit length of the wire and U is the voltage. The (nonlinear) resistance depends on the electric field $E(x,t)$. Eliminating U yields $E = RI$, $I_{xx} = CE_t$, $E(x,t) = F(I(x,t))$. Formal double integration of I_{xx} results in

$$I(x,t) = C\int_0^x\!\!\int_0^x E_t(x',t)\mathrm{d}x'\mathrm{d}x' + A(t)x + B(t). \qquad (3.2.31)$$

Using the HILDEBRAND *formula*

$$\underbrace{\int_a^x \ldots \int_a^x}_{n\text{-times}} f(x)\,\underbrace{\mathrm{d}x \ldots \mathrm{d}x}_{n\text{-times}} = \frac{1}{(n-1)!}\int_a^x (x - x')^{n-1} f(x')\mathrm{d}x', \qquad (3.2.32)$$

one gets

$$I(x,t) = C\int_0^x (x - x')E_t(x',t)\mathrm{d}x' + A(t)x + B(t). \qquad (3.2.33)$$

Now one has to solve the *integral equation* (3.2.33) under the constraint $E(x,t) = F(I)$ using an iterative procedure

$$E_n = F(I_n), \quad I_{n+1} = C\int_0^x (x - x')E_{tn}\mathrm{d}x', \quad n = 1, 2, \ldots. \qquad (3.2.34)$$

The first approximation for this procedure may be

$$\int_0^\infty E(x,t)\mathrm{d}x = const, R(E) = R_0(E_0/E)^m, \, A(t) = B(t) = 0,$$

so that $E_n = [R_0 E_0^m I_n]^{1/(m+1)}$ results.

Even a combination of the integral equation method together with a similarity transformation is very useful. Let us consider an example. The BLASIUS *equations* for the two-dimensional *boundary layer* problem of a liquid read

$$uu_x + vu_y + \nu u_{yy} = 0, \quad u_x + v_x = 0. \tag{3.2.35}$$

Here $u(x,y)$ and $v(x,y)$ are the components of a steady two-dimensional viscous flow satisfying the boundary conditions on the wall

$$u(x,0) = 0, \quad v(x,0) = 0, \quad u(x,\infty) = u_0. \tag{3.2.36}$$

Assume a similarity transformation

$$\eta = y\sqrt{u_0/\nu x}, \quad \psi = 2\sqrt{\nu x u_0}\, f(\eta), \tag{3.2.37}$$

where $\psi(x,y)$ is the *stream function* defined by $\psi_y = u(x,y), \psi_x = -v(x,y)$. From (3.2.37) one gets

$$f''' + ff'' = 0 \tag{3.2.38}$$

subject to

$$f(0) = 0, \quad f'(0) = a, \quad a \geq 0, \quad f''(0) = 1. \tag{3.2.39}$$

Now a double integration yields

$$f(\eta) = \int_0^\eta \int_0^\eta f''(\eta)\mathrm{d}\eta\mathrm{d}\eta + b\eta + c. \tag{3.2.40}$$

By the way, there exists an extensive literature on integral equations [1.2], [1.7], [3.7], [3.13], [3.14], [3.15].

There are many fields in physics and engineering where *integro-differential equations* appear. In statistical thermodynamics [3.16], [3.17] each individual particle (molecule, electron, neutron, atom) is described by its space coordinates x, y, z and its momentum components p_x, p_y, p_z. The six-dimensional space x, y, z, p_x, p_y, p_z related to a single particle is called μ-phase space and for N particles comprising the $6N$ *phase space* is called Γ-space. The LIOUVILLE *distribution function* (also called N-particle distribution function)

$$F(x_1, y_1, z_1, x_2, y_2, z_2, \ldots$$

$$x_n, y_n, z_n, p_{x1}, p_{y1}, p_{z1}, p_{x2}, p_{y2}, p_{z2}, \ldots p_{xN}, p_{yN}, p_{zN}, t)$$

describes the probability of finding at time t particle number l at x_l, y_l, z_l carrying there a momentum p_{xl}, p_{yl}, p_{zl}. According to statistical thermodynamics this function in Γ-space satisfies the LIOUVILLE *equation*

$$\frac{\partial F}{\partial t} + \sum_{i=1}^{N} \dot{q}_i \frac{\partial F}{\partial q_i} + \sum_{i=1}^{N} \dot{p}_i \frac{\partial F}{\partial p_i} = 0, \tag{3.2.41}$$

where $q_i(t), p_i(t)$ are a new denotation for x, y, z and p_x, p_y, p_z. If interactions between particles may be neglected, μ-space distribution function is sufficient to describe the particle system. If there exists only an interaction between two particles (for instance if one assumes short range forces) (like VAN DER WAALS forces [3.18], [3.19]) then the BOLTZMANN *integro-differential equation* may be derived [3.20] from (3.2.41). It describes a one-particle distribution function $f(x, y, z, u_x, u_y, u_z, t)$ of an individual particle number 1, which is in short range interaction with a family of particles number 2

$$\frac{\partial f}{\partial t} + u_x \frac{\partial f}{\partial x} + u_y \frac{\partial f}{\partial y} + u_z \frac{\partial f}{\partial z} + K_x \frac{\partial f}{\partial u_x} + K_y \frac{\partial f}{\partial u_y} + K_z \frac{\partial f}{\partial u_z}$$

$$= \int |\vec{u}_2 - \vec{u}_1| \Big\{ f\left(\vec{x}_1, \vec{u}_1^*, t\right) f\left(\vec{x}_2, \vec{u}_2^*, t\right)$$

$$- f\left(\vec{x}_1, \vec{u}_1, t\right) f\left(\vec{x}_2, \vec{u}_2, t\right) \Big\} \sigma \mathrm{d}\vec{u}_2. \tag{3.2.42}$$

Here the subscripts 1 and 2 designate interacting particles, the star * is reserved for particles prior to the interaction and σ is the *scattering cross section* [2.5], [3.21]. $\vec{K}$ describes the interaction forces in the single-particle equation of motion $m\vec{\dot{u}} = \vec{K}$. The equation (3.2.42) has been solved by HILBERT using the *moment method* [3.22], [3.23]. This method is based on the setup

$$f = f_0 + \frac{\lambda}{l} f_1 + \left(\frac{\lambda}{l}\right)^2 f_2 + \dots. \tag{3.2.43}$$

Here λ is the *mean free path* between two particle collisions [2.5] and l designates the length over which the distribution function changes considerably. This might be the size of an apparatus, of a thermodynamic system, etc., HILBERT obtained a set of integral equations for f_0, f_1, f_2, etc., which can only be solved if five *integrability conditions* are satisfied. These conditions turn out to be exactly the basic hydrodynamic equations, vector *equation of motion*, continuity equation (conservation of mass) and the *energy theorem*. Thus fluid and gas dynamics as well as thermodynamics are subdomains of the LIOUVILLE equation and the BOLTZMANN equation (3.2.39), respectively. In order to derive the basic hydrodynamic equations from (3.2.42), we first abbreviate the rhs term by $(\delta f/\delta t)_{coll}$, then we multiply (3.2.42) in vector denotation by Θ and integrate over particle velocities:

$$\frac{\partial}{\partial t}\int \Theta f \mathrm{d}\vec{u} + \nabla \int \Theta \vec{u} f \mathrm{d}u$$

$$+ \frac{1}{m}\int \Theta \vec{K}\cdot\nabla_u f \mathrm{d}\vec{u} = \int \Theta \left(\frac{\delta f}{\delta t}\right)_{coll} \mathrm{d}\vec{u}. \qquad (3.2.44)$$

Now we define the quantity Θ: $\qquad\qquad\qquad\qquad\qquad\qquad\qquad$ (3.2.45)

1. $\Theta = m$	moment of zero order	mass
2. $\Theta m\vec{u}$	moment of first order	momentum
3. $\Theta = m\vec{u}\ \ \vec{u}$	momentum of second order	energy

Defining the normalized average over the distribution function

$$<\Theta> = \frac{\int \Theta f \mathrm{d}\vec{u}}{\int f \mathrm{d}\vec{u}} \qquad (3.2.46)$$

we immediately obtain from (3.2.44) for $\Theta = m$ the *continuity equation*

$$\frac{\partial \rho}{\partial t} + \nabla(\rho\vec{v}) = m\int \left(\frac{\delta f}{\delta t}\right)_{coll}\mathrm{d}\vec{u} = \left(\frac{\delta\rho}{\delta t}\right)_{coll}, \qquad (3.2.47)$$

where

$$\rho = m\int f(\vec{x},\vec{u},t)\mathrm{d}\vec{u} \qquad (3.2.48)$$

and

$$\vec{v} = \int \vec{u} f(\vec{x},\vec{u})\mathrm{d}\vec{u} \qquad (3.2.49)$$

is the average particle velocity, which becomes the fluid velocity. Since mass, momentum and energy will be conserved during an elastic particle-particle collision, the terms $\int \Theta(\delta f/\delta t)_{coll}\mathrm{d}\vec{u}$ vanish for the zeroth, first and second moment. The first moment gives

$$\frac{\partial}{\partial t}(\rho\vec{v}) + \nabla\cdot(\rho\vec{v}\cdot\vec{v}) + \nabla p = 0, \qquad (3.2.50)$$

where

$$p = \frac{1}{3}\rho\sum_{k=1}^{3}<\vec{u}_k\cdot\vec{u}_k> = \frac{\rho\bar{u}^2}{3} + \rho\frac{kT}{m} \qquad (3.2.51)$$

is the fluid pressure. k is the BOLTZMANN const ($1.3805\cdot 10^{-23}\,Joule/^\circ K$). Using $\mathrm{d}/\mathrm{d}t = (\partial/\partial t) + \vec{v}\nabla$ one obtains the EULER *equation of motion* of a fluid

$$\rho\left(\frac{\partial\vec{v}}{\partial t} + \vec{v}\cdot\nabla\vec{v}\right) = \rho\frac{\mathrm{d}\vec{v}}{\mathrm{d}t} = -\nabla p. \qquad (3.2.52)$$

Finally the second moment produces the energy theorem in various forms depending on the assumptions made on the fluid or gas properties (viscosity, heat conductivity, electric conductivity, anisotropy etc.) [2.5], [3.1], [3.16], [3.22].

The integro-differential equation describing the critical mass of a *nuclear reactor* or an *atomic bomb* is far more complicated [3.21], [3.24] - [3.27] than (3.2.42).

Actually, there are several such equations: for the fast neutron distribution $n(x, y, z, t, E, \vartheta, \varphi)$ and for the groups of delayed neutrons $n_i(x, y, z, t)$. E is the neutron energy and the spherical coordinates ϑ, φ describe the direction of the neutron velocity vector. Using the scattering σ_s, absorption σ_A and fission σ_f cross sections of the neutrons, it is possible to solve the set of integro-differential equations containing various integrals over space and energy by a spherical functions series with unknown coefficients [3.27]. It is then possible to calculate approximatively the critical masses of U235 and Pu239 atomic bombs for a naked sphere with $R = 8\,\mathrm{cm}$ 40 kg and 10 kg, respectively.

It should be remarked that another method to reduce partial differential equations to ordinary equations consists in special transformations that are not integral transformations, compare section 3.4.

Problems

1. Determine the LAPLACE transform of $\cos(t)$. The command to load the package ($<<$Calculus`LaplaceTransform`) may not be necessary depending on your computer and the version of *Mathematica* you are using.

   LaplaceTransform[Cos[t],t,z]

   should result in $z/(1 + z^2)$.

2. Find the LAPLACE transforms [3.28] of $\cos\sqrt{t}/\sqrt{t}$,
   $J_n(t)$,
   $\delta(t) = \mathrm{DiracDelta}[t]$,
   $(\exp(-at) - \exp(-bt))t$
   giving $\sqrt{\pi/z}\exp(-1/4z), 1$, and $1/(a + z)^2 - 1/(b + z)^2$ respectively. This will be done by

   LaplaceTransform[Cos[Sqrt[t]]/Sqrt[t],t,z]

   yields $\dfrac{e^{-\frac{1}{4z}}\sqrt{\pi}}{\sqrt{z}}$.

   InputForm[LaplaceTransform[BesselJ[n,t],t,z]]

   Here we used the command InputForm to obtain a result which could be copied and directly pasted into the inverse transformation. One obtains

   $$\mathrm{InputForm} = 2^{-n}z^{-1-n}\mathrm{Hypergeometric2F1}\left[\frac{1+n}{2}, \frac{2+n}{2}, 1+n, -\frac{1}{z^2}\right]$$

```
LaplaceTransform[DiracDelta[t],t,z]
```

```
LaplaceTransform[(Exp[-a*t]-Exp[-b*t])*t,t,z]
```
yields $\dfrac{1}{(a+z)^2} - \dfrac{1}{(b+z)^2}$.

3. Calculate some inverse transformations.

   If the LAPLACE transform of a function $f(t)$ is designated by $F(z)$, then the inverse transformation operator applied on $F(z)$ should restore $f(t)$. Calculate the inverse transform of $z/(z^2+1)$ giving $\cos(t)$.

   We first define

```
f[t]:=Cos[t];F[z]=LaplaceTransform[f[t],t,z]
```

   giving $z/(1+z^2)$. Then we type

```
InverseLaplaceTransform[F[z],z,t]
```
to obtain $\mathrm{Cos}(t)$.

```
InverseLaplaceTransform[(z^(-1-n)*
Hypergeometric2F1[(1+n)/2,(2+n)/2,1+n,
-z^(-2)])/2^n,z,t]
```

   yields

   $$2^{-n}\mathrm{InverseLaplaceTransform}\left[z^{-1-n}\ \mathrm{Hypergeometric2F1}\right.$$
   $$\left[\frac{1+n}{2},\frac{2+n}{2},1+n,-\frac{1}{z^2}\right],z,t\right].$$

```
LaplaceTransform[BesselJ[0,a*t],t,z]
```

   giving $1/\sqrt{z^2+a^2}$.

4. Here some exercises for a deeper understanding how LAPLACE-transformation transforms an ordinary differential equation into an algebraic equation.

```
LaplaceTransform[x'[t],t,z]
```
should produce

   $z\ \mathrm{LaplaceTransform}[x[t],t,z] - x[0]$, then

```
LaplaceTransform[y''[x]-y'[x]+y[x],x,z]
```
yields

   $\mathrm{LaplaceTransform}[y[x],x,z] - z\ \mathrm{LaplaceTransform}[y[x],x,z] +$
   $z^2\ \mathrm{LaplaceTransform}[y[x],x,z] + y[0] - z\ y[0] - y'[0]$

```
LaplaceTransform[D[u[x,t],{x,2}]-a*D[u[x,t],{t,2}]-
b*D[u[x,t],t]-c*u[x,t]==0,t,z]
```

   produces

   $-\mathrm{LaplaceTransform}[c*u[x,t],t,z] + \mathrm{LaplaceTransform}[u^{(2,0)}[x,t],t,z] -$
   $b\left(z\ \mathrm{LaplaceTransform}[u[x,t],t,z] - u[x,0]\right) -$
   $a\left(z^2\ \mathrm{LaplaceTransform}[u[x,t],t,z] - z\ u[x,0] - u^{(0,1)}[x,0]\right) == 0.$

5. Apply the LAPLACE operator on an ordinary differential equation of second order $y''(t) + c_1 y'(t) + c_0 y(t) = g(t)$. The result should be

$$w(z) = \tilde{g}(z)\frac{1}{z^2 + c_1 z + c_0} + y(0) \cdot \frac{z + c_1}{z^2 + c_1 z + c_0} + y'(0) \cdot \frac{1}{z^2 + c_1 z + c_0},$$

where $\tilde{g}(z)$ is the LAPLACE transform of $g(t)$. This is an algebraic equation in which $y(0)$ and $y'(0)$ are given constants.

LaplaceTransform[y''[t]+c1*y'[t]+c0*y[t]-g[t]==0,t,z]

gives

$-$LaplaceTransform$[g[t], t, z] + c0$ LaplaceTransform$[y[t], t, z] +$ z^2 LaplaceTransform$[y[t], t, z] + c1$ $(z$ LaplaceTransform$[y[t], t, z] - y'[0]) -$ z $y[0] - y'[0] == 0$

6. Solve $y''(t) + y(t) = t$, $y(0) = 1$, $y'(0) = -2$.

One gets $z^2 y - z + 2 + y = 1/z^2$ or $y(z) = 1/z^2 + z/(z^2 + 1) - 3/(z^2 + 1)$.

The inverse transform gives $y(t) = t + \cos t - 3 \sin t$, which solves the equation.

LaplaceTransform[y''[t]+y[t]-t==0,t,z]

gives

$-z^{(} - 2) +$ LaplaceTransform$[y[t], t, z] + z^2$ LaplaceTransform$[y[t], t, z] -$ z $y[0] -$ Derivative$[1][y][0] == 0$

7. Solve the *telegraph equation*

$$u_{xx} - a u_{tt} - b u_t - c u = 0 \tag{3.2.53}$$

together with initial conditions $u(x, 0) = 0$, $u_t(x, 0) = 0$ and the boundary conditions $u(0, t) = f_0(t)$, $u(l, t) = f_1(t)$. Use the LAPLACE transformation. You should obtain the ordinary differential equation

$$\frac{d^2 w}{dx^2} - (az^2 + bz + c) w = 0$$

with $w(0; z) = \bar{f}_0(z)$, $w(l; z) = \bar{f}_1(z)$, where $\bar{f}_0, \bar{f}_1$ are the transforms of f_0 and f_1 respectively. Try

DSolve[w''[x]-(a*z^2 +b*z+c)*w[x]==0,w[x],x]

$w[x] \rightarrow e^{x\sqrt{c+bz+az^2}} C[1] + e^{-x\sqrt{c+bz+az^2}} C[2]$.

It should be mentioned that a LIE *series* treatment of the telegraph equation is also known [3.29].

8. Apply the LAPLACE transformation on the equation of a string with friction between two rigid supports. Assume that the string undergoes a time dependent point force $f(t)\delta(x - x_0)$ at the point x_0 [1.2]. The transversal motion of this string is described by

$$\frac{1}{c^2}u_{tt} + \frac{2\kappa}{c^2}u_t - u_{xx} = \delta(x - x_0)f(t), \quad t > 0.$$

Here c is the velocity of the transverse waves on the string, κ describes the frictional resistance of the medium of the string and δ is the DIRAC *function (delta function* - Mathematicians prefer the expression "distribution," rather than "function"). This pathological function has the properties

$$\delta(x) = \lim_{\Delta \to 0} \begin{cases} 0 & \text{for} & x < -\Delta/2 \\ 1/\Delta & \text{for} & -\Delta/2 < x < \Delta/2 \text{ or } \delta(x) = 0, x \neq 0 \\ 0 & \text{for} & x > \Delta/2 \end{cases}$$

such that

$$\int\limits_{-\infty}^{+\infty} \delta(x)\mathrm{d}x = 1,$$

$$\int\limits_{-\infty}^{+\infty} f(\xi)\delta(\xi - x)\mathrm{d}\xi = f(x).$$

Multiplication by the kernel yields the ordinary differential equation

$$-c^2\frac{\mathrm{d}^2w}{\mathrm{d}x^2} + (z^2 + 2\kappa z)w(x, z)$$

$$= c^2\delta(x - x_0)F(z) + (z + 2\kappa)u(0, x) + u_t(0, x).$$

Here F is the transform of f and the last two rhs terms describe the initial conditions

DSolve[-c^2*w''[x]+(z^2+2*κ*z)*w[x]==c^2*
DiracDelta[x-x0]*F[z] +(z+2*κ)*u[0]+ut[0],w[x],x]

$w[x] \to C[1] \, \mathrm{Cos}[(x \, \sqrt{-z^2 - 2 \, z \, \kappa})/c] + C[2] \, \mathrm{Sin}[(x \, \sqrt{-z^2 - 2 \, z \, \kappa})/c] +$

$(c \, z \, \sqrt{-((z \, (z + 2 \, \kappa))/c^2)}] \, \mathrm{Cos}[(x \, \sqrt{-(z \, (z + 2 \, \kappa))})]]/c] \, \mathrm{Sign}[c]$

....

$(c \, z \, (1 + z^2) \, (z + 2 \, \kappa) \, \sqrt{-((z \, (z + 2 \, \kappa))/c^2)}] \, \mathrm{Sign}[c])$

9. Apply the similarity transformation $\eta = t/x$ on the nonlinear partial differential equation $u_{xx} = u_{tt} + u_t^2$. The calculation should yield $(\eta^2 - 1)f'' + \eta f' - f'^2 = 0$ for $f(\eta)$ [3.12].

3.3   The method of characteristics

In section 1.2 we found the asthonishing result that two arbitrary functions $f(x,t)$ and $g(x,t)$ were able to solve the partial differential equation (1.2.1) which represents the one-dimensional *wave equation* (D'ALEMBERT *equation*). Furthermore, the classification of partial differential equations by the conditions (1.2.9) to (1.2.11) seems to indicate that there is more behind this classification. It is easy to demonstrate that

$$u(x,y) = f(x + iy) + g(x - iy) \tag{3.3.1}$$

delivers a solution of the two-dimensional LAPLACE *equation*

$$u_{xx} + u_{yy} = 0, \tag{3.3.2}$$

a fact that will lead us to the method of conformal mapping in section 4.2. We call these aggregates $x \pm iy = const$ characteristics.

Let us again consider the most general *linear partial differential equation* of second order (1.2.8)

$$a(x,y)u_{xx} + 2b(x,y)u_{xy} + c(x,y)u_{yy} +$$
$$d(x,y)u_x + e(x,y)u_y + g(x,y)u = h(x,y). \tag{3.3.3}$$

If the coefficients a, b, c are constant and if $b = 0, a = c = 1$, then the medium described is isotropic and homogeneous. Anisotropy is exhibited by $a \neq c$ and if the coefficients are not constants, the medium is inhomogeneous.

Since in engineering problems partial differential equations of first order appear, see the continuity equation (3.2.43), we now will discuss the characteristics theory for partial differential equations of first order. LAGRANGE has shown that the general *quasilinear partial differential equation*

$$P(x,y,u)u_x(x,y) + Q(x,y,u)u_y(x,y) = R(x,y,u) \tag{3.3.4}$$

is solved implicitly by

$$F(\varphi(x,y,u), \psi(x,y,u)) = 0, \tag{3.3.5}$$

if F is differentiable and if

$$\varphi(x,y,u) = a = const, \quad \psi(x,y,u) = b = const \tag{3.3.6}$$

are two independent solutions of any combination of the differential equations of the *characteristics*

$$\frac{\mathrm{d}x}{P} = \frac{\mathrm{d}y}{Q} = \frac{\mathrm{d}u}{R} \quad \text{or} \quad \frac{\mathrm{d}x}{\mathrm{d}s} = P, \quad \frac{\mathrm{d}y}{\mathrm{d}s} = Q, \quad \frac{\mathrm{d}u}{\mathrm{d}s} = R. \tag{3.3.7}$$

Integration yields two surfaces (3.3.6). Their cut delivers space curves. s is the arc length along these curves. Let us consider an example:

$$u_t + u u_x = 0. \tag{3.3.8}$$

Comparing with (3.3.4) we find $P = 1, Q = u, R = 0$ and (3.3.7) reads

$$\frac{dt}{ds} = 1, \quad \frac{dx}{ds} = u, \quad \frac{du}{ds} = 0 \tag{3.3.9}$$

so that $t = s + c, dt = ds, u = a = const$. Due to

$$\frac{dx}{ds} = \frac{dx}{dt}\frac{dt}{ds} = \frac{dx}{dt} = u$$

one obtains $x = ut + b$ and the characteristics $\varphi = a = u, \psi = b = x = ut$. The quantities a, b are integration constants. The solution of (3.3.8) is then given by

$$F(a, b) = F(\varphi(x, y, u), \psi(x, y, u)) = F(u, x - ut) = 0. \tag{3.3.10}$$

This enables to solve a CAUCHY-problem, see Table 1.1. If

$$u(x, 0) = f(x) \tag{3.3.11}$$

is given, the argument replacement $x \to x - ut$ yields the solution $u(x, t) = f(x - ut)$. This expression satisfies (3.3.11) and (3.3.8): $u_t = f'(x - ut) \cdot (-u)$, $u_x = f'(x - ut) \cdot (1), u_t + u u_x = 0 \to -f'u + u f' = 0$. For an implicit nonlinear partial equation of first order

$$F(u(x, y), u_x(x, y), u_y(x, y), x, y) = 0, \tag{3.3.12}$$

the procedure has to be modified and the derivatives u_x and u_y as well as u itself have to be assumed as five independent variables: u, u_x, u_y, x, y. The characteristics are then determined by

$$\frac{dx}{F_{u_x}} = \frac{dy}{F_{u_y}} = \frac{du}{u_x F_{u_x} + u_y F_{u_y}}$$

$$= \frac{du_x}{-u_x F_u - F_x} = \frac{du_y}{-u_y F_u - F_y}. \tag{3.3.13}$$

These five ordinary differential equations have to be solved. Let us consider an example. For the nonlinear partial differential equation

$$F \equiv 16 u_x^2 u^2 + 9 u_y^2 u^2 + 4 u^2 - 4 = 0 \tag{3.3.14}$$

one gets

$$F_x = 0, \quad F_y = 0, \quad F_u = (32 u_x^2 + 18 u_y^2) u + 8u,$$
$$F_{u_x} = 32 u_x u^2, \quad F_{u_y} = 18 u_y u^2 \tag{3.3.15}$$

and thus from (3.3.13),

$$\frac{dx}{32u_xu^2} = \frac{dy}{18u_yu^2} = \frac{du}{32u_x^2u^2 + 18u_y^2u^2}$$

$$= \frac{du_x}{-32u_x^3u - 18u_y^2uu_x - 8uu_x} = \frac{du_y}{-32u_x^2uu_y - 18u_y^3u - 8uu_y}. \quad (3.3.16)$$

Multiplication of the five fractions by $1, 0, 4u_x, 4u$ and 0, respectively, collects the five fractions over a common denominator

$$N = 1(+32u_xu^2) + 4u_x(+32u_x^2u^2 + 18u_y^2u^2)$$
$$-4u(32u_x^3u + 18u_xu_y^2u + 8u_xu) = 0$$

and hence

$$dx + 4u_x du + 4u du_x = 0 = dx + 4d(u_xu). \quad (3.3.17)$$

Integration and inserting of u_x yields

$$(x-a)^2 + 9u_y^2u + 4u^2 - 4 = 0 \quad (3.3.18)$$

and the solution (after elimination of u_y)

$$\frac{x-a}{4} + \frac{(y-b)^2}{9/4} + u^2 = 1. \quad (3.3.19)$$

It is interesting that the JACOBI *equation* (3.1.92) has differential equations for characteristics that are the HAMILTON *equations* of motion of the objects described by (3.1.92) [3.35].

The basic equations of hydrodynamics have the form of quasilinear partial differential equations of first order

$$a_{11}u_x + a_{12}v_x + b_{11}u_y + b_{12}v_y = h_1(x,y),$$
$$a_{21}u_x + a_{22}v_x + b_{21}u_y + b_{22}v_y = h_2(x,y). \quad (3.3.20)$$

Here $u(x,y)$, $v(x,y)$ are the dependent variables and the coefficients $a_{ik}(x,y,u,v)$, $b_{ik}(x,y,u,v)$ are not constant. Introducing the two vectors $\vec{h} = \{h_1, h_2\}$ and $\vec{u} = \{u, v\}$ and the matrices A and B one may rewrite (3.3.20) in the form

$$A\vec{u}_x + B\vec{u}_y = \vec{h}. \quad (3.3.21)$$

The one-dimensional *continuity equation* taking into account a source $g(x,t)$ of gas by combustion of a liquid reads [3.36]

$$\rho_t + u\rho_x + \rho u_x = g(x,t) \quad (3.3.22)$$

and the one-dimensional EULER *equation* of motion takes the form

$$\rho u_t + \rho u u_x + p_x = f(x,t), \quad (3.3.23)$$

where $f(x,t)$ describes a momentum source from combustion of fuel etc., [3.36]. Assuming *adiabatic* behavior

$$p(\rho) = const \cdot \rho^\gamma, \tag{3.3.24}$$

where γ (or κ) is the adiabatic coefficient expressed by the ratio c_p/c_V of the specific heats of the fluid for constant pressure p and constant volume V respectively. Introducing the *velocity a of sound* by

$$a^2 = \frac{\mathrm{d}p}{\mathrm{d}\rho} = \frac{\gamma p}{\rho} \tag{3.3.25}$$

the equation of motion can be written as

$$\rho u_t + \rho u u_x + a^2 \rho_x = f. \tag{3.3.26}$$

Now we look for the characteristics of such a system like (3.3.20). They would probably be of the form $\psi(x,y) = const$ or possibly $y = k(x) + const$, $\mathrm{d}y/\mathrm{d}x = k'(x)$. Then the following two equations are valid along characteristic curves

$$\frac{\mathrm{d}v}{\mathrm{d}x} = \frac{\partial v}{\partial x} + \frac{\partial v}{\partial y} \cdot \frac{\mathrm{d}y}{\mathrm{d}x} = v_x + v_y k',$$

$$\frac{\mathrm{d}u}{\mathrm{d}x} = \frac{\partial u}{\partial x} + \frac{\partial u}{\partial y} \cdot \frac{\mathrm{d}y}{\mathrm{d}x} = u_x + u_y k'. \tag{3.3.27}$$

Mathematica helps to obtain u_x and v_x from (3.3.27). Introducing a notation $u_x \rightarrow ux, v_x \rightarrow vx, \mathrm{d}v/\mathrm{d}x \rightarrow \mathrm{d}vx, \mathrm{d}ux = \mathrm{d}u_x/\mathrm{d}x, ks = k'$ we may type

Solve[{vx+vy*ks==dvx,ux+uy*ks==dux},{ux,vx}]

yielding $v_x = \mathrm{d}v_x/\mathrm{d}x - v_y k'$ and $u_x = \mathrm{d}u_x/\mathrm{d}x - u_y k'$.

Inserting u_x, v_x into the system (3.3.20) one obtains

$$u_y(-a_{11}k' + b_{11}) + v_y(-a_{12}k' + b_{12}) = h_1 - a_{11}\frac{\mathrm{d}u}{\mathrm{d}x} - a_{12}\frac{\mathrm{d}v}{\mathrm{d}x}$$

$$u_y(-a_{21}k' + b_{21}) + v_y(-a_{22}k' + b_{22}) = h_2 - a_{21}\frac{\mathrm{d}u}{\mathrm{d}x} - a_{22}\frac{\mathrm{d}v}{\mathrm{d}x}. \tag{3.3.28}$$

If the values of u and v are given along the characteristic curves $y(x)$, then (3.3.28) allows the calculation of u_x and v_y according to the CRAMER *rule*. We define

$$R = \begin{vmatrix} a_{11}k' - b_{11} & a_{12}k' - b_{12} \\ a_{21}k' - b_{21} & a_{22}k' - b_{22} \end{vmatrix} = |Ay' - B|, \tag{3.3.29}$$

$$V_1 = \begin{vmatrix} a_{11}k' - b_{11} & h_1 - a_{11}\mathrm{d}u/\mathrm{d}x - a_{12}\mathrm{d}v/\mathrm{d}x \\ a_{21}k' - b_{21} & h_2 - a_{21}\mathrm{d}u/\mathrm{d}x - a_{22}\mathrm{d}v/\mathrm{d}x \end{vmatrix}, \tag{3.3.30}$$

$$V_2 = \begin{vmatrix} a_{12}k' - b_{12} & h_1 - a_{11}\mathrm{d}u/\mathrm{d}x - a_{12}\mathrm{d}v/\mathrm{d}x \\ a_{22}k' - b_{22} & h_2 - a_{21}\mathrm{d}u/\mathrm{d}x - a_{22}\mathrm{d}v/\mathrm{d}x \end{vmatrix}. \tag{3.3.31}$$

Three cases are possible:

1. $R \neq 0$: one can calculate u_y, v_y and all first derivatives u_k, v_k are determined along the characteristic curves,

2. $R = 0, V_1$ or $V_2 = 0$, the linear equations for u_y, v_y are linearly depending and an infinite manifold of solutions u_y, v_y exists,

3. $R = 0, V_1 \neq 0$ or $V_2 \neq 0$: no solutions u_y, v_y exist.

For our system (3.3.22), (3.3.24) one finds

$$\begin{array}{cccc} a_{11} = 0 & a_{12} = 1 & a_{21} = \rho & a_{22} = 0 \\ b_{11} = \rho & b_{12} = u & b_{21} = \rho u & b_{22} = a^2. \end{array}$$

Using again $k'(x) = y'(x), R = 0$ results in

$$y'^2 - 2y'u + u^2 - a^2 = 0 \tag{3.3.32}$$

so that

$$y' = u \pm a. \tag{3.3.33}$$

The propagation of small waves in a fluid follows (3.3.33) downstream (+) or upstream (−).

Characteristics may also be derived for a system of quasilinear partial differential equations of second order. For m independent variables $x_k, k = 1 \ldots m, l = 1 \ldots m$ and n depending variables $u_j, j = 1 \ldots n$ such a system may be written as

$$\sum_{j=1}^{n} \sum_{k,l=1}^{m} A_{ij}^{(kl)} (x_k, u_j) \frac{\partial^2 u_j}{\partial x_k \partial x_l} + H\left(x_k, u_j, \frac{\partial u_j}{\partial x_l}\right) = 0, \quad i = 1 \ldots m. \tag{3.3.34}$$

The characteristics of "second order" of this system obey a partial differential equation of first order and of degree n^2, which reads

$$\left| \sum_{k,l=1}^{m} A_{ij}^{(kl)} \frac{\partial \varphi}{\partial x_k} \frac{\partial \varphi}{\partial x_l} \right| = 0. \tag{3.3.35}$$

This system has again to be solved using the characteristics of partial differential equations of first order.

In modern aerodynamics, nonlinear partial differential equations of second order appear. If the equation is implicit and of the form

$$F(x, y, z, p, q, r, s, t) = 0, \tag{3.3.36}$$

where we used the notation $p = u_x, q = u_y, r = u_{xx}, s = u_{xy}, t = u_{yy}$, then one can define

$$a = \frac{\partial F}{\partial z_{xx}}, \quad 2b = -\frac{\partial F}{\partial z_{xy}}, \quad c = \frac{\partial F}{\partial z_{yy}} \tag{3.3.37}$$

and inserting into(1.2.9), (1.2.10) and (1.2.11) one can classify the equation.

Let us consider a problem of gasdynamics [3.30]. The continuity equation for a three-dimensional *potential flow*

$$\vec{v} = \nabla\varphi \tag{3.3.38}$$

of a compressible nonviscous medium reads

$$\frac{\partial\rho}{\partial t} + \sum_j \frac{\partial}{\partial x_j}(\rho v_j) = 0, \quad j = 1,2,3. \tag{3.3.39}$$

The equation of motion is

$$\frac{\partial v_i}{\partial t} + \sum_j \frac{\partial v_i}{\partial x_j} = -\frac{1}{\rho}\frac{\partial p}{\partial x_j} \tag{3.3.40}$$

and the energy theorem reads

$$\frac{\partial}{\partial t}\left(U + \frac{v^2}{2}\right) + \sum_j \frac{\partial}{\partial x_j}\left(U + \frac{v^2}{2}\right) = -\frac{1}{\rho}\sum_i \frac{\partial}{\partial x_i}(p v_i). \tag{3.3.41}$$

If (3.3.24) is satisfied, one has potential flow and the CROCCO *theorem* is satisfied [3.31], [3.32], [3.33]. $U = c_V T$ is the internal energy of the ideal gas and T temperature. Using (3.3.38) one may write (3.3.39) in the form

$$\frac{1}{\rho}\frac{\partial\rho}{\partial t} + \sum_i \frac{\partial^2\varphi}{\partial x_i \partial x_i} + \sum_j \frac{1}{\rho}\frac{\partial\varphi}{\partial x_j}\frac{\partial\rho}{\partial x_j} = 0. \tag{3.3.42}$$

The equation of motion now reads

$$\frac{\partial}{\partial t}\frac{\partial\varphi}{\partial x_i} + \sum_j \frac{\partial\varphi}{\partial x_j}\frac{\partial^2\varphi}{\partial x_j \partial x_i} =$$

$$-\frac{1}{\rho}\frac{\partial p}{\partial x_i} = -\frac{1}{\rho}\frac{dp(\rho)}{d\rho}\frac{\partial\rho}{\partial x_i} = -\frac{1}{\rho}a^2\frac{\partial\rho}{\partial x_i}, \tag{3.3.43}$$

where (3.3.25) has been used. On the other hand, integration of (3.3.40) over space yields the BERNOULLI *integral*

$$\frac{\partial\varphi}{\partial t} + \frac{v^2}{2} + \int\frac{dp(\rho)}{\rho} = const; \quad p(\rho) = const \cdot \rho^\gamma \tag{3.3.44}$$

or

$$\frac{\partial^2\varphi}{\partial t^2} + v_j\frac{\partial v_j}{\partial t} + \frac{a^2}{\rho}\frac{\partial\rho}{\partial t} = 0 = \frac{\partial^2\varphi}{\partial t^2} + \sum_j \frac{\partial\varphi}{\partial x_j}\frac{\partial^2\varphi}{\partial x_j\partial t} + \frac{a^2}{\rho}\frac{\partial\rho}{\partial t}. \tag{3.3.45}$$

Insertion of ρ_x from (3.3.43) multiplied by v_j and ρ_t from (3.3.45) into (3.3.42) results in the nonlinear *potential flow equation*

$$(a^2 - \varphi_x^2)\varphi_{xx} + (a^2 - \varphi_y^2)\varphi_{yy} + (a^2 - \varphi_z^2)\varphi_{zz} - 2\varphi_x\varphi_y\varphi_{xy}$$

$$-2\varphi_y\varphi_z\varphi_{yz} - 2y_z\varphi_x\varphi_{zx} = \varphi_{tt} + 2\varphi_x\varphi_{xt} + 2\varphi_y\varphi_{yt} + 2\varphi_z\varphi_{zt} \tag{3.3.46}$$

and (3.3.44) is now found to be

$$2a^2 + (\gamma - 1)(\varphi_x^2 + \varphi_y^2 + \varphi_z^2) + 2(\gamma - 1)\varphi_t = const = 2(\gamma - 1)a_0^2, \quad (3.3.47)$$

where a_0 is the velocity of sound in the gas at rest [3.30], [3.34]. A two-dimensional stationary (time-independent) compressible flow is then subject to

$$(a^2 - \varphi_x^2)\varphi_{xx} + (a^2 - \varphi_y^2)\varphi_{yy} - 2\varphi_x\varphi_y\varphi_{xy} = 0. \quad (3.3.48)$$

Using the characteristic theory for nonlinear partial differential equations of second order (3.3.24), (3.3.35) one obtains the differential equation

$$\frac{dy}{dx} = \frac{\varphi_x\varphi_y}{a^2 - \varphi_y^2} \pm \frac{1}{a^2 - \varphi_y^2}\sqrt{4\varphi_x^2\varphi_y^2 - 4(a^2 - \varphi_x^2)(a^2 - \varphi_y^2)}. \quad (3.3.49)$$

Since φ_x, φ_y depend on x, y it is necessary to know these functions in order to be able to integrate (3.3.49) and to find the characteristics $y(x)$. Special methods will be necessary to solve (3.3.48), see sections 6.4 and 6.5 of this book.

For $u^2 = \varphi_x^2 + \varphi_y^2 > a^2$ one has a *suprasonic* (not *ultrasonic*) flow and the partial differential equation is *hyperbolic*. For $u \approx a$ the the flow is *transsonic* and (3.3.49) is *parabolic*. For $u < a$ subsonic flow occurs and the potential equation is *elliptic*. Flow with $u >> a$ is sometimes called hypersonic and for $u << a$ the sonic speed a becomes constant, compressibility and φ_x^2, φ_y^2 may be neglected and (3.3.48) becomes the LAPLACE equation $\varphi_{xx} + \varphi_{yy} = 0$. Ultrasonics is normal sound with a frequency higher than the 20.000 Hertz representing the threshold of hearing of the human ear. Thus the partial differential equation (3.3.48) is of mixed type, its type depends on the domain in the $\varphi_x = u, \varphi_y = v$ plane. Another equation of mixed type is the very simple equation $u_{xx} + xu_{yy} = 0$ (TRICOMI *equation*). For $x < 0$ the characteristics are $y \pm 2/3(-x)^{2/3} = const$ and for $x > 0$ they read $y + 2/3(ix)^{3/2} = const$. The EULER *equation* $au_{xx} + 2bu_{xy} + cu_{yy} = 0$ with constant coefficients a, b, c is of constant type.

Problems

1. Find characteristics and solution of $u_t + F(u)u_x = 0$ ($u = f(x - t.F(u))$).

2. Find characteristics and solution $u(x, y)$ of $2xu_x - yu_y + 2u_x - 4xy^2 = 0$. (3.3.7) yields $dx/2x = -dy/y = du/(4xy^2 - 2u)$. Integrations yield

$$xy^2 = a = \varphi(x, y), \quad u = 2xy^2 + y^2\varphi(x, y), \quad u/y^2 - 2x = b = \psi(x, y)$$

so that the solution reads $F(xy^2, u/y^2 - 2z) = F(\varphi, \psi)$.

3. $xu_x + yu_y = 0$, $u = a$, $y/x = b$, $u = \psi(b)$.

4. $u_x + u_y = 0$, $u = F(x - y)$.

5. $ux^2 - xu_x - u_y = 0$,

$$\frac{du_x}{-u_x} = \frac{du_y}{0} = \frac{dx}{-2u_x + x} = \frac{dy}{1} = \frac{du}{-2u_x^2 + xu_x + u_y},$$

which are solved by $\ln u_x = -y + \ln a$,

$u_y = -xu_x + u_x^2 + u_y^2$, $u = ax \exp(-y) - (1/2)a^2 \exp(-2y) + b$.

6. $u - u_x^2 - u_y^2 = 0$, $u = (x + ay + b)^2/4(1 + a^2)$.

7. $u_{xy} - \dfrac{n}{x-y}u_x + \dfrac{m}{x-y}u_y = 0$, $u(x, y) = \dfrac{\partial^{m+n-2}}{\partial x^{m-1} \partial y^{n-1}} \left(\dfrac{X(x) - Y(y)}{x - y} \right)$.

8. $u_{tt} = x^2 u_{xx} + u/4$, $u(x, t) = \sqrt{x}\, F(\ln x - t)$.

9. Derive (3.3.46).

10. Show that the heat conduction equation (3.2.3) written with all material constants and including a source term, namely

$$\frac{\partial \tilde{T}}{\partial t} = \frac{\lambda}{\rho c} \frac{\partial^2 \tilde{T}}{\partial x^2} - \frac{\alpha \tilde{T}}{\rho c} \tag{3.3.50}$$

(ρ density, λ heat conductivity, c specific heat, α heat transfer coefficient) can be transformed by $\tilde{T} = T(x, t) \exp(-\alpha t / \rho c)$ into

$$\frac{\partial T}{\partial t} = \frac{\lambda}{\rho c} \frac{\partial^2 T}{\partial x^2}. \tag{3.3.51}$$

Solve this using the *similarity transformation*
$\eta = x/2\sqrt{\tau}$, $\tau = \lambda t/\rho c$, $T_t = T'(\eta) \cdot \partial\eta/\partial t$; $T_x = T'(\eta) \cdot \partial\eta/\partial x$
etc., so that

$$T''(\eta) + 2\eta T'(\eta) = 0, \tag{3.3.52}$$

$$T(x, t) = T(\eta) = A \int_0^{-\eta} \exp(-\eta^2)d\eta = A\frac{\sqrt{\pi}}{2} \Phi \left(\frac{x}{2}\sqrt{\frac{\rho c}{\lambda t}} \right) + B. \tag{3.3.53}$$

Φ is the GAUSS *error integral*. Determine the integration constants from the boundary and initial conditions

$$T(0, t) = T_1 = 0 \rightarrow B = 0,$$
$$T(x, 0) = T_0 = A\sqrt{\pi}/2 \quad x > 0$$

since $\Phi(0) = 0$, $\Phi(\infty) = 1$.

11. Use $\eta = x^2/4\tau$ to solve (3.3.51) and use also a separation setup $T(x,t) = U(x)V(t)$ giving $T(x,t) = A\exp(-ak^2t)\cos[k(x-\beta)]$.
    Verify the solution $T(x,t) = T_0\delta \exp(-x^2/4\tau)/(2\sqrt{a\pi t})$ and plot $T(x_it_i)$ for some arbitrary fixed values of $t(t_1,t_2,t_3)$
    and $\lambda/\rho c = a = 2 \cdot 10^{-3}\text{cm}^{-2}\text{s}^{-1}$ for earth.
    We choose $T_0\delta/2 = 1, a = 2$ and give the commands:

```
a=2.;
t=0.05;
Plot[(1/(Sqrt[a Pi t]))*Exp[-x^2/(4*a*t)],{x,-5,5},
PlotRange->{0.,1.8}]
```

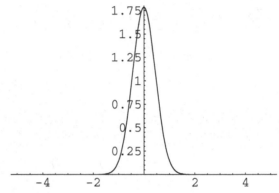

Figure 3.1
$T(x,t)$ at $t = 0.05$

```
a=2.;
t=0.3;
Plot[(1/(Sqrt[a Pi t]))*Exp[-x^2/(4*a*t)],{x,-5,5},
PlotRange->{0.,1.8}]
```

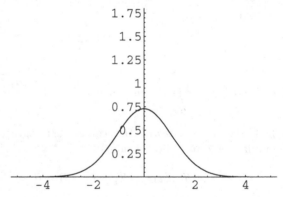

Figure 3.2
$T(x,t)$ at $t = 0.03$

```
a=2.;
t=1.;
Plot[(1/(Sqrt[a Pi t]))*Exp[-x^2/(4*a*t)],{x,-5,5},
PlotRange->{0.,1.8}]
```

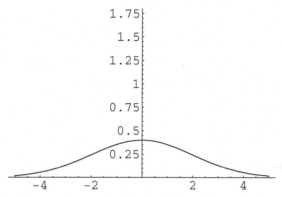

Figure 3.3
$T(x,t)$ at $t = 1.0$

12. Let us see what *Mathematica* can do for us in connection with problems 10 and 11.

```
Clear[T,TW]; TW[x_,t_]==T[x_,t_]*Exp[-α*t/(ρ*c)];
```

Now insert this separation setup into (3.3.51).

13. `D[TW[x,t],t]-λ*D[TW[x,t],{x,2}]/(c*ρ)+ α*TW[x,t]/(ρ*c)`

This does not help! We obtain

$$\frac{\alpha\, TW[x,t]}{c\,\rho} + TW^{(0,1)}[x,t] - \frac{\lambda\, TW^{(2,0)}[x,t]}{c\,\rho}.$$

Apparently, *Mathematica* is not able to execute a partial derivative of an unknown function. We will investigate the situation. But remember the result for problem 4 in section 3.1

```
Clear[T,η,U];τ=λ*t/(ρ*c);
```

this yields $\left(\dfrac{x}{2\sqrt{\frac{t\lambda}{c\rho}}}\right)^{(1,0)}\left[x, \dfrac{t\,\lambda}{c\,\rho}\right].$

O.K., but let's see now if a derivative of an unknown function depending on one single variable works: `U[x]=f[x]*Sin[x];`   `D[U[x],x]`

This results in $\mathrm{Cos}[x]\ f[x] + \mathrm{Sin}[x]\ f'[x]$

This works! Now we try again:

```
Clear[T];T[x_,t_]=T[η[x,t]];D[T[η],η]
```

This gives $T'[\eta]$.

Another way:

D[T[η[x,t]],x] yields $T'[\eta[x,t]]\,\eta^{(1,0)}[x,t]$

D[T[η[x,t]],{x,2}]

yields an ordinary differential equation for $T(\eta)$

$T''[\eta[x,t]]\,\eta^{(1,0)}[x,t]^2 + T'[\eta[x,t]]\,\eta^{(2,0)}[x,t].$

Now let us solve the ordinary differential equation (3.3.52)

DSolve[T''[η]+2*η*D[T[η],η] ==0,T[η],η]

which gives exactly the solution (3.3.53).

3.4   Nonlinear partial differential equations

It is fortunate that there exist transformations making it possible to transform nonlinear partial differential equations into linear partial equations or into ordinary differential equations. These transformations can be classified as follows:

1. transforming only the dependent variable (transformation according to KIRCHHOFF, HOPF, RIEMANN, functional transformation),

2. transforming only the independent variables (VON MISES, BOLTZMANN, similarity transformations),

3. both types of variables are transformed (LEGENDRE, MOLENBROEK, LAGRANGE, hodograph transformation).

Some industrial and engineering problems will be discussed. The *perfume industry* has the interest to create products satisfying two contrary conditions: on the one side, the perfume should evaporate and diffuse and on the other side the perfume should adhere to the ladies as long as possible. This has the physical consequence that the diffusion coefficient D becomes a function of the local perfume concentration $c(x,t)$, so that the *diffusion equation* (FICK *equation*) reads

$$\frac{\partial c}{\partial t} = \frac{\partial}{\partial x}D(c(x,t))\frac{\partial c}{\partial x}. \tag{3.4.1}$$

Now one may try

$$D(c) = c^n \tag{3.4.2}$$

yielding a nonlinear partial differential equation

$$\frac{\partial c}{\partial t} = \frac{\mathrm{d}D}{\mathrm{d}c}\left(\frac{\partial c}{\partial x}\right)^2 + D\frac{\partial^2 c}{\partial x^2} = nc^{n-1}\left(\frac{\partial c}{\partial x}\right)^2 + D\frac{\partial^2 c}{\partial x^2}. \tag{3.4.3}$$

Apparently it is useful to consider equations of a more general type than (3.4.1). We thus introduce the *pseudo* Laplacian $\mathrm{div}(F(U)\mathrm{grad}U)$ and use the KIRCHHOFF *transformation* $\psi = \psi(U(x,y,z))$. Now define $F(U)$ and

$$F(U) = \frac{\mathrm{d}\psi}{\mathrm{d}U}, \quad \mathrm{grad}\,\psi = F(U)\,\mathrm{grad}\,U, \quad \psi(U) = \int F(U)\mathrm{d}U. \qquad (3.4.4)$$

This manipulation transforms the nonlinear pseudo-Laplacian into a linear Laplacian

$$\mathrm{div}\,(F(U)\,\mathrm{grad}\,U) = \mathrm{div}\,\mathrm{grad}\,\psi = \Delta\psi. \qquad (3.4.5)$$

This looks nice, but, according to the "conservation theorem for the mathematical difficulty," there are now problems with $\partial c/\partial t$. If $\Delta\psi = 0$ would be valid, (3.4.4) would offer an advantage, but let us see:

$$D(c) \to F(c) = c^n, \quad c^n = \mathrm{d}\psi/\mathrm{d}c, \quad \psi = \int c^n \mathrm{d}c,$$

$$\psi = c^{n+1}/(n+1), \quad c = (\psi(n+1))^{1/n+1}.$$

So we come back to (3.4.3) for $c(x,t)$ or to

$$\frac{\partial c}{\partial t} = \frac{\mathrm{d}c}{\mathrm{d}\psi}\cdot\frac{\partial\psi}{\partial t} = \frac{1}{(n+1)}\cdot(\psi(n+1))^{-n/(n+1)}\cdot\frac{\partial\psi}{\partial t} = \frac{\partial^2\psi}{\partial x^2}.$$

The KIRCHHOFF transformation is a *functional transformation* used to transform nonlinear partial differential equations into linear partial differential equations. What is the inverse transformation? Can we transform linear partial differential equations into nonlinear equations? Let us start with

$$\Delta\psi = \lambda\frac{\partial\psi}{\partial t} \quad \text{or} \quad \lambda\frac{\partial^2\psi}{\partial t^2}, \quad \psi(x,y,z,t). \qquad (3.4.6)$$

Insertion of $\psi = F(U)$ results in

$$\Delta U + \frac{F''}{F'}(\mathrm{grad}\,U)^2 = \lambda U_t,$$

$$\Delta U + \frac{F''}{F'}[(\mathrm{grad}\,U)^2 - U_t^2] = \lambda U_{tt}. \qquad (3.4.7)$$

Let us now consider a good example of exact linearization. We take the one-dimensional NAVIER-STOKES *equation* (BURGERS *equation*)

$$u_t + uu_x - \nu u_{xx} = 0. \qquad (3.4.8)$$

$u(x,t)$ is the fluid velovity, ν a parameter describing the viscosity of the fluid, $u_t = \partial u/\partial t$ etc. Now we make the transformation $u = v_x$ resulting in

$$v_{xt} + v_x v_{xx} - \nu v_{xxx} = 0 \quad \text{(or } D(t)\text{)}, \qquad (3.4.9)$$

where D is a momentum source [3.36], compare (3.3.23). For constant ν a partial integration yields ($v_{xx} = \partial^2 v/\partial x^2$)

$$v_t + \frac{1}{2}v_x^2 - \nu v_{xx} = 0 \quad \text{(or } D(t)\text{)}. \tag{3.4.10}$$

For a rocket nozzle the periodicity condition

$$D(x,t) = D(x + 2\pi, t) \tag{3.4.11}$$

has to be taken into account [3.9] (page 23). Then the periodic initial condition reads:

$$v(x,0) = f(x) = f(x + 2\pi), \; v(x + 2\pi, t) = v(x,t). \tag{3.4.12}$$

Making the setup

$$v(x,t) = \varphi(x,t) + \int\limits_0^t D(t)\mathrm{d}t, \tag{3.4.13}$$

one can eliminate the inhomogeneous rhs term. With $\lambda = 1/\nu$ and regarding (3.4.7) one can put

$$\frac{F''}{F'} = -\frac{1}{2}\lambda, \quad F = A\exp\left(\frac{-U\lambda}{2}\right) + B. \tag{3.4.14}$$

Then (3.4.7) takes the homogeneous form of (3.4.10). Now we solve this equation using the HOPF *transformation* putting

$$u = -2\nu\frac{\partial}{\partial x}\ln\psi = -2\nu\frac{\psi_x}{\psi}, \quad v = -2\nu\ln\psi. \tag{3.4.15}$$

We then obtain the simple linear equation

$$\nu\psi_{xx} = \psi_t. \tag{3.4.16}$$

Thus the linearization succeeded.

More general functional transformations are also in use like

$$v(x,t) = F(G(x,t)). \tag{3.4.17}$$

Applying on (3.4.10) yields

$$F'(G_t - \nu G_{xx}) - G_x^2\left(\nu F'' - \frac{1}{2}F'^2\right) = 0. \tag{3.4.18}$$

The two "degrees of freedom" $F(G)$ and $G(x,t)$ may be used for exact linearization of complicated nonlinear partial differential equations. One might demand $G_t - \nu G_{xx} = 0$ so that $\nu F'' - F'^2/2 = 0$, so that

$$F(G) = -2\nu\ln(G - A) + B, \tag{3.4.19}$$

where A and B are integration constants (HOPF-COLE *transformation*). Another possibility could be (HOPF-AMES *transformation*)

$$v(x,t) = -2\nu \operatorname{grad} \ln F. \tag{3.4.20}$$

Some of the functional transformations can be combined together with similarity transformations [3.37]. Occasionally, the method of the unknown function $g(x)$ may help: in the *boundary layer* theory of a fluid, the equation [3.9]

$$\psi_{xy}\psi_y - \psi_{yy}\psi_x = \nu\psi_{yyy} \tag{3.4.21}$$

appears. $\psi(x,y)$ is the *stream function*. The setup

$$\psi = f(y + g(x)) \tag{3.4.22}$$

results in the ordinary differential equation

$$f'f''g' - f'g'f'' \equiv 0 = \nu f''', \tag{3.4.23}$$

so that

$$f = a(y + g(x))^2 + b(y + g(x)) + c. \tag{3.4.24}$$

a, b, c are integration constants, $g(x)$ is still arbitrary.

Another important transformation of the dependent variable is given by the RIEMANN *transformation*. As an example, we consider

$$u_{tt} = [F(u_x)]^2 u_{xx}, \tag{3.4.25}$$

compare (3.3.46) for instance. We define

$$v = u_x, \quad w = u_t, \tag{3.4.26}$$

so that

$$v_t - w_x = 0, \quad w_t - F^2(v)v_x = 0. \tag{3.4.27}$$

Multiplication of the first equation by $F(v)$ and addition deliver

$$F(v)v_t - F(v)w_x + w_t - F^2(v)v_x = 0 \tag{3.4.28}$$

and multiplication and subtraction yield

$$F(v)v_t - F(v)w_x - w_t + F^2(v)v_x = 0. \tag{3.4.29}$$

According to RIEMANN new dependent variables $r(x,t)$ and $s(x,t)$, called RIEMANN *invariants*, are introduced:

$$\begin{aligned} r_t &= w_t + F(v)v_t, \quad r_x = w_x + F(v)v_x, \\ s_t &= -w_t + F(v)v_t, \quad s_x = -w_x + F(v)v_x, \end{aligned} \tag{3.4.30}$$

where $r_t = \partial r/\partial t$, $r_x = \partial r/\partial x$ etc. Using these new variables, (3.4.28) may be written as

$$r_t - F(v)r_x = 0, \quad s_t + F(v)s_x = 0. \tag{3.4.31}$$

The new variables are called invariants, because for $dx/dt = -F$ the variable r and for $dx/dt = +F$ the variable s does not vary. Integration of (3.4.30) results in

$$r = w + \int F(v)dv = w + B(v),$$

$$s = -w + \int F(v)dv = -w + B(v). \tag{3.4.32}$$

Let $\tilde{B} = v(r,s)$ the inverse function of $B(v)$, then (3.4.32) delivers

$$r + s = 2B(v), \quad v = \tilde{B}\left(\frac{r+s}{2}\right), r - s = 2w. \tag{3.4.33}$$

As a consequence (3.4.31) results in two nonlinear partial differential equations

$$r_t - F\left[\tilde{B}\left(\frac{r+s}{2}\right)\right]r_x = 0, \quad s_t + F\left[\tilde{B}\left(\frac{r+s}{2}\right)\right]s_x = 0. \tag{3.4.34}$$

It depends on the type of $F(u_x)$ in (3.4.35) if a formal integration of (3.4.34) is possible or not. In a special case covered later on, even a linear partial differential equation of second order will be derived from (3.4.34).

Up to now we had discussed transformations of the dependent variables. Now we shall transform the independent variables. For a time-dependent, two-dimensional, incompressible viscous boundary flow along a plate in the x direction, the continuity equation and the equation of motion read respectively

$$u_x + v_y = 0, \quad uu_x + vu_y = \nu u_{yy}, \tag{3.4.35}$$

where $u(x,y), v(x,y)$ are the fluid velocity components in the x and y direction respectively. $\nu = \eta/\rho$ is the kinematic viscosity. Using the stream function $\psi(x,y)$

$$u = \psi_y, \quad v = -\psi_x \tag{3.4.36}$$

and the VON MISES *transformation* $u = u(x,\psi)$, $v = v(x,\psi)$ one obtains

$$\left(\frac{\partial u}{\partial x}\right)_y = \left(\frac{\partial u}{\partial x}\right)_\psi + \left(\frac{\partial u}{\partial \psi}\right)_x\left(\frac{\partial \psi}{\partial x}\right)_y = \left(\frac{\partial u}{\partial x}\right)_\psi - v\left(\frac{\partial u}{\partial \psi}\right)_x,$$

$$\left(\frac{\partial u}{\partial y}\right)_x = \left(\frac{\partial u}{\partial \psi}\right)_x\left(\frac{\partial \psi}{\partial y}\right)_x = u\left(\frac{\partial u}{\partial \psi}\right)_x,$$

$$\left(\frac{\partial^2 u}{\partial y^2}\right)_x = \frac{\partial}{\partial y}\left[u\left(\frac{\partial u}{\partial \psi}\right)_x\right]_x = u\frac{\partial}{\partial \psi}\left[u\left(\frac{\partial u}{\partial \psi}\right)_x\right]_x.$$

One can write for the equation of motion

$$\frac{\partial u}{\partial x} = \frac{\partial}{\partial \psi}\left(\nu u\frac{\partial u}{\partial \psi}\right), \tag{3.4.37}$$

which now exactly has the form (3.4.1). Similarity transformations and the BOLTZMANN transformation also belong in the class of transformations that transform the independent variables.

The class of transformations that transform both types of variables is for instance represented by the *hodograph transformation*. The problem with this transformation is however that it works only in two independent variables. Consider the plane steady potential flow of a nonviscous compressible fluid. We may use (3.3.47) and (3.3.48) in the form

$$a^2 = a_0^2 - (\kappa - 1)(u^2 + v^2)/2 \qquad (3.4.38)$$

and

$$\left(a^2 - u^2\right) u_x - uv\left(u_y + v_x\right) + \left(a^2 - v^2\right) v_y = 0. \qquad (3.4.39)$$

The potential condition $\text{curl}\vec{v} = 0$ reads

$$u_y - v_x = 0. \qquad (3.4.40)$$

Now the solution of the equations (3.4.39) and (3.4.40) shall not be searched for as $u(x, y), v(x, y)$, but in hodographic form $x(u, v), y(u, v)$ in the *velocity plane*. Such a hodographic solution informs on these locations (x, y) where the velocity components u, v assume given values. Thus (3.4.39) and (3.4.40) are replaced by

$$\left(a^2 - u^2\right) y_v + uv\left(x_v + v_y\right) + \left(a^2 - v^2\right) x_u = 0, \qquad (3.4.41)$$

$$x_v - y_u = 0. \qquad (3.4.42)$$

With $x = x(v, w)$, $y = y(v, w)$, $dx/dx = 1 = x_v v_x + x_w w_x$, $dy/dx = 0 = y_v v_x + y_w w_x$, $v_x = y_w/(x_v y_w - x_w y_v)$, etc., one may linearize (3.4.27) to build

$$x_w - y_v = 0, \qquad x_v - F^2(v)y_w = 0. \qquad (3.4.43)$$

This enables now to linearize (3.4.34), which we obtained by the RIEMANN transformation. The replacement r, s versus x, t gives

$$r_x = Jt_s, \qquad r_t = -Jx_s, \qquad s_x = -Jt_r, \qquad s_t = Jx_r, \qquad (3.4.44)$$

where the JACOBI determinant J is given by

$$J = r_x s_t - s_x r_t, \qquad (3.4.45)$$

so that

$$x_s + F\left[\tilde{B}\left(\frac{r + s}{2}\right)\right] t_s = 0, \qquad x_r - F\left[\tilde{B}\left(\frac{r + s}{2}\right)\right] t_r = 0 \qquad (3.4.46)$$

results. Assuming $F^2(u_x)$ in (3.4.25) to be $(1 + u_x)^\alpha$ or $(1 + v)^\alpha$, $F(v) = (1 + v)^{\alpha/2}$, equation (3.4.25) takes the form

$$v_t - w_x = 0, \qquad w_t - (1 + v)^\alpha v_x = 0. \qquad (3.4.47)$$

These equations are nonlinear for $v(x,t)$. But one may use (3.4.32) to obtain

$$r = w + \frac{2(1+v)^{1+\alpha/2}}{\alpha+2}, \quad s = -w + \frac{2(1+v)^{1+\alpha/2}}{\alpha+2}. \tag{3.4.48}$$

Addition and inversion result in

$$v = -1 + \left[\frac{1}{4}(r+s)(\alpha+2)\right]^{2/(\alpha+2)}, \quad w = \frac{1}{2}(r-s). \tag{3.4.49}$$

Using $F = [(r+s)(\alpha+2)/4]^n$, $n = \alpha/(2+\alpha)$ the equations (3.4.46) take the form

$$x_s = F \cdot t_s = 0, \quad x_r - F \cdot t_r = 0. \tag{3.4.50}$$

These equations are linear partial differential equations of first order for $x(s,r)$ and $t(s,r)$. Eliminating x_s and x_r by differentiation ($x_{sr} = x_{rs}$) results in a linear partial differential equation of second order

$$\frac{\partial^2 t}{\partial r \partial s} + \frac{n}{2} \frac{(\partial t)/(\partial r) + (\partial t/\partial s)}{r+s} = 0. \tag{3.4.51}$$

This is the DARBOUX *equation*. Analogously,

$$x_{rs} - \frac{n}{2} \cdot \frac{x_r + x_s}{r+s} = 0. \tag{3.4.52}$$

Finally we consider transformations of the type where both variables were transformed. The LEGENDRE *transformation*, which is used in thermodynamics and gasdynamics, is able to transform the quasilinear two-dimensional, nonlinear potential equation (3.3.48) for $\varphi(x,y)$ into a linear partial differential equation for $\psi(u,v)$. For this purpose we make the ansatz (setup)

$$\psi(u,v) = ux + yv - \varphi(x,y) \tag{3.4.53}$$

in the u,v velocity plane. Then

$$\psi_u = x, \quad \psi_v = y, \quad \psi_{uu} = x_u, \quad \psi_{vv} = y_u$$

(remember that now $x(u,v)$, $y(u,v)$ are dependent variables and that the independent variables are now u,v).

$$\varphi_x = u, \quad \varphi_y = v, \quad \varphi_{xx} = u_x, \quad \varphi_{yy} = v_y,$$

$$d\varphi = \varphi_x dx + \varphi_y dy = u dx + v dy,$$

$$d\psi = \psi_u du + \psi_v dv = x du + y dv.$$

Integration of $d\phi + d\psi = d(ux) + d(vy)$ results in (3.4.53). Insertion into (3.3.48) gives the linear equation

$$\left(\bar{a}^2 - u^2\right)\psi_{vv} + 2uv\psi_{uv} + \left(\bar{a}^2 - v^2\right)\psi_{uu} = 0, \tag{3.4.54}$$

where $\bar{a}$ now depends on ψ according to (3.3.47). Although equation (3.4.54) is linear, it cannot be solved because the actual solution depends on the boundary conditions that depend on $\varphi(x,y)$. Combined graphic-numerical methods allowing an alternating use of ψ and φ are able to find solutions, see chapter 6.

Problems

1. Under what conditions is it possible to solve a CAUCHY problem for (3.4.54)? Remember Table 1.1 and the criteria (1.2.9) to (1.2.11) in chapter 1.

2. To describe large amplitude transversal oscillations of a string, the CARRIER *equations* must be used [3.38]. They can be transformed into a nonlinear *integro-differential equation* of the form

$$\frac{1}{c^2} v_{tt} - v_{xx} \left(1 + \frac{Eq}{2lp_0} \int\limits_0^l v_x^2 \mathrm{d}x \right) = 0. \qquad (3.4.55)$$

Here $v(x,t)$ is the oscillation amplitude, p_0 the tension of the string, q the string cross section, E the modulus of elasticity, l the string length, ρ the material density and the propagation speed c is given by $\sqrt{p_0/\rho q}$. Solve (3.4.55) by

$$v(x,t) = F(x)G(t), \quad \int\limits_0^l v_x^2 \mathrm{d}x = G^2(t) \int\limits_0^l \left(\frac{\mathrm{d}F}{\mathrm{d}x} \right)^2 \mathrm{d}x. \qquad (3.4.56)$$

This procedure results in two ordinary differential equations

$$F'' + \nu^2 F = 0,$$

$$G'' + \nu^2 c^2 \left[1 + \frac{Eq}{2lp_0} G^2 \int\limits_0^l F'^2(x) \mathrm{d}x \right] G = 0. \qquad (3.4.57)$$

ν is the separation constant.

Let *Mathematica* solve the two equations

DSolve[F''[x]+ν^2*F[x]==0,F[x],x]

which should yield

$F[x] \rightarrow C[1] \, \mathrm{Cos}[x \, \nu] + C[2] \, \mathrm{Sin}[x \, \nu]$

Then the second equation takes another form. First we compute the *definite integral* $\int_0^l F'(x)^2 \mathrm{d}x$. To be able to do this, we now use the InputForm and repeat the integration of the ordinary differential equation for $F[x]$ by giving the command

InputForm[DSolve[F''[x]+ν^2*F[x]==0,F[x],x]]

Interesting, now the star * appears for the multiplication.

F[x_]=C[1]*Cos[x*ν]+C[2]*Sin[x*ν]

To be able to apply operations on $F(x)$, we define $F'(x)$ as a function

```
F'[x_]=D[F[x],x]
```

which results in

$\nu\, C[2]\, \text{Cos}[x\,\nu] - \nu\, C[1]\, \text{Sin}[x\,\nu]$

Now let us play. We ask for $F'(x)^2$?

```
F'^2[x]
```
yields nonsense.

$\nu\, C[2]\, \text{Cos}[\nu\,\#1] - \nu\, C[1]\, \text{Sin}[\nu\,\#1]\,\&)^{2[x]}$.

Aha, we should write $F'(x)^2 \rightarrow$ **F'[x]^2**

yielding

$(\nu\, C[2]\, \text{Cos}[x\,\nu] - \nu\, C[1]\, \text{Sin}[x\,\nu])^2$,

but we want the result! So we give the command

```
Expand[F'[x]^2]
```

yielding

$\nu^2\, C[2]^2\, \text{Cos}[x\,\nu]^2 -$
$2\,\nu^2\, C[1]\, C[2]\, \text{Cos}[x\,\nu]\, \text{Sin}[x\,\nu] + \nu^2\, C[1]^2\, \text{Sin}[x\,\nu]^2$

and we now integrate from $0 \rightarrow l$

```
a=Integrate[F'[x]^2,{x,0,1}]
```

resulting in

$(2\,l\,\nu^2\, C[1]^2 - 2\,\nu\, C[1]\, C[2] + 2\,l\,\nu^2\, C[2]^2 +$
$2\,\nu\, C[1]\, C[2]\, \text{Cos}[2\,l\,\nu] - \nu\, C[1]^2\, \text{Sin}[2\,l\,\nu] + \nu\, C[2]^2\, \text{Sin}[2\,l\,\nu])/4$

Then the equation for $G(t)$ reads

$$G'' + \nu^2 c^2 (1 + aG^2)G = 0, \tag{3.4.58}$$

where a is given above and

```
DSolve[G''[t]+ν^2*c^2*(1+a*G[t]^2)*G[t]==0,G[t],t]
```

results in an horrible long expression in the form of an unsolved algebraic equation for $G(t)$. But from Figure 2.4 in section 2.6 we know that an equation of the type $y'' + y + y^3 = 0$ represents a stable nonlinear oscillator and (3.4.58) is of the same type. To avoid the study of the horrible expression, we try

```
DSolve[y''[x]+y[x]+y[x]^3==0,y[x],x]
```

and obtain an aggregate of the elliptic functions cn and sn.

3. We now consider the boundary and initial condition belonging to (3.4.25). If this equation describes a clamped nonlinear string, what would be the conditions? Apparently at the clamped ends $x = 0$ and $x = l$, $u(0,t) = 0$, $u(l,t) = 0$ is valid. When one plucks the string at the time $t = 0$, then the initial conditions would read $u(x,0) = h(x), u_t(x,0) = H(x)$, where h and H are given arbitrary functions like e.g., $h(x) =$

$a \sin(\pi x/l), H(x) = 0$. Could one transform the boundary value problem into an initial value problem? To do this we introduce a new variable $\xi, -\infty < \xi < \infty$ instead of $0 \leq x \leq l$ and we demand that $u(-\xi, t) = -u(\xi, t)$ (uneven solution) and that $u(\xi + 2, t) = u(\xi, t)$ with $\Delta \xi = 2$. Now the boundary conditions are satisfied: $u(0, t) = -u(0, t)$ so that $u(0, t) = 0$. Furthermore due to the periodicity over the interval $2l$ one has $u(l, t) = -u(-l, t) = -u(l, t)$ and thus $u(l, t) = 0$. The initial conditions are now $u(\xi, 0) = a \sin(\pi \xi/l)$ and $u_t(\xi, o) = 0$ for $-\infty < \xi < \infty$.

4. The LAGRANGE *transformation* is a transition from the fluid picture giving the two-dimensional local velocity field $u(x, y), v(x, y)$ to a particle picture that describes the trajectory of the particles by $x(t, x_0, y_0)$ and $y(t, x_0, y_0)$, where the actual initial location of particle number 0 $x(t = 0) = x_0, y(t = 0) = y_0$ gives the name to particle and its trajectory. Discuss this transition and transform continuity equation and equation of motion

$$\frac{d\rho}{dt} = \frac{\partial \rho}{\partial t} + \frac{\partial \rho}{\partial x}\frac{dx}{dt} + \frac{\partial \rho}{\partial y}\frac{dy}{dt} = \rho_t + u\rho_x + v\rho_y. \tag{3.4.59}$$

The JACOBI matrix becomes

$$J = \frac{\partial(x, y)}{\partial(x_0, y_0)} = x_{x_0} y_{y_0} - x_{y_0} y_{x_0}, \tag{3.4.60}$$

since $x(x_0, y_0), y(x_0, y_0)$ and the continuity equation results in

$$\frac{1}{\rho}\frac{d\rho}{dt} + \frac{1}{J}\frac{dJ}{dt} = 0, \quad \rho J = const \tag{3.4.61}$$

whereas the equation of motion is now linear

$$\frac{du}{dt} = 0, \quad \frac{dv}{dt} = 0, \quad \frac{\partial x}{\partial t} = u, \quad \frac{\partial y}{\partial t} = v. \tag{3.4.62}$$

5. Solve (3.4.16). Inserting $\psi = U(x) \cdot V(t)$ into $\psi_{xx} = \psi_t$ yields

$$\nu\frac{U''(x)}{U(x)} = \frac{V'(t)}{V} \quad \text{or} \quad \nu U'' + aU = 0, \; V' - a^2 V = 0,$$

where a is the separation constant. Seen the results for problem 4 in section 3.1 and problem 12 in section 3.3 we now apply the *Mathematica* commands

```
ψ[x_,t_]==U[x]*V[t];
Expand[ν*D[ψ[x,t],{x,2}]-D[ψ[x,t],t]/ψ[x,t]]
```

which results in $-\dfrac{\psi^{(0,1)}[x,t]}{\psi[x,t]} + \nu\psi^{(2,0)}[x,t]$.

Now we try

`D[U[x]*V[t],{x,2}]` resulting in `V[t]U''[x]`

This is O.K.. Hence, we try

`Expand[ν*D[U[x]*V[t],{x,2}]/V[t]-D[U[x]*V[t], t]/U[x]]` what results in $-V'[t] + \nu\, U''(x)$.

Learn by making errors!

6. Prove (3.4.5) in cartesian and spheroidal coordinates.

4

Boundary problems with one closed boundary

4.1 Laplace and Poisson equations

It is well known that a vector field $\vec{K}$ having a vanishing curl $\vec{K} = 0$ can be represented by a potential function U

$$K_x(x,y,z) = \pm\frac{\partial U}{\partial x}, \quad K_y(x,y,z) = \pm\frac{\partial U}{\partial y}, \quad K_z(x,y,z) = \pm\frac{\partial U}{\partial z}. \quad (4.1.1)$$

Setting up the divergence of such a field results in

$$\frac{\partial K_x}{\partial x} + \frac{\partial K_y}{\partial y} + \frac{\partial K_z}{\partial z} = \operatorname{div}\vec{K} = \frac{\partial}{\partial x}\frac{\partial U}{\partial x} + \frac{\partial}{\partial y}\frac{\partial U}{\partial y} + \frac{\partial}{\partial z}\frac{\partial U}{\partial z} = \Delta U. \quad (4.1.2)$$

There are many practical examples in physics and engineering for the Laplace *equation* $\Delta U = 0$. In hydrodynamics, in problems of heat conduction, in electromagnetism and in many other fields the Laplace equation or its inhomogeneous counterpart, the Poisson *equation*, appears as

$$\Delta U = \rho(x,y). \quad (4.1.3)$$

Let us now solve a simple two-dimensional boundary value problem for the Laplace equation. The closed domain should be a rectangle of dimensions a and b, see Figure 4.1.

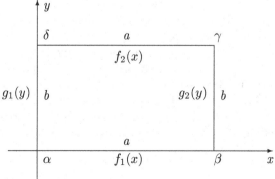

Figure 4.1
Rectangular boundary

These inhomogeneous boundary conditions are described by

$$U(x, y = 0) = f_1(x), \quad 0 \le x \le a, \tag{4.1.4}$$

$$U(x, y = b) = f_2(x), \quad 0 \le x \le a, \tag{4.1.5}$$

$$U(x = 0, y) = g_1(y), \quad 0 \le y \le b, \tag{4.1.6}$$

$$U(x = a, y) = g_2(y), \quad 0 \le y \le b. \tag{4.1.7}$$

So that this problem is well defined, the functions must be continuous at the corners, but it is not necessary that the boundary value functions satisfy $\Delta U = 0$. Let $\alpha, \beta, \gamma, \delta$ be the values of the boundary functions at the corner, then they have to satisfy

$$\begin{aligned} f_1(0) = g_1(0) = \alpha, \quad f_2(a) = g_2(b) = \gamma, \\ f_1(a) = g_2(0) = \beta, \quad f_2(0) = g_1(b) = \delta. \end{aligned} \tag{4.1.8}$$

Using the usual setup $U(x, y) = X(x) \cdot Y(y)$ for linear partial differential equations with constant coefficients one gets

$$X''(x) + p^2 X(x) = 0, \tag{4.1.9}$$

$$Y''(y) - p^2 Y(y) = 0, \tag{4.1.10}$$

where p is the separation constant. The general solutions of these equations read

$$X(x) = \sum_{m=1}^{\infty} A_m \sin\left(p_m x + C_m\right), \tag{4.1.11}$$

$$Y(y) = \sum_{n=1}^{\infty} B_n \sinh\left(p_n y + C_n\right), \tag{4.1.12}$$

but other solutions like harmonic polynomials (section 2.3) or functions of a complex variable like in (3.3.1) exist, too. The partial amplitudes A_m, B_n, the separation constants p_m and the constants C_m, C_n are still unknown. They have to be determined by the boundary conditions. Since $\Delta u = 0$ is linear, the sum over particular solutions is again a solution (*superposition principle*). Since the partial differential equation $\Delta u = 0$ is homogeneous each particular solution may be multiplied by a factor (partial amplitude). The phase constants C_m may be used to find a solution with two functions

$$A_m \sin\left(p_m x + C_m\right) = A_m \sin p_m x \cos C_m + A_m \cos p_m x \sin C_m \tag{4.1.13}$$

and analogously for sinh.

The solution of $\Delta U(x, y) = 0$ can now be written as

$$U(x,y) = \sum_{k=0}^{\infty} D_k \sin\left(p_k x + C_k\right) \sinh\left(p_k y + \tilde{C}_k\right). \qquad (4.1.14)$$

To simplify calculations we choose

$$U(0,y) = g_1(y) = f_1(0) = \alpha = 0, \qquad (4.1.15)$$

$$U(a,y) = g_2(y) = f_1(a) = \beta = 0, \qquad (4.1.16)$$

$$U(x,b) = f_2(x) = g_1(b) = \delta = 0, \qquad (4.1.17)$$

but

$$U(x,0) = f_1(x) \neq 0, \qquad (4.1.18)$$

and, for continuity, $f_2(a) = g_2(b) = \gamma = 0$ must be valid. To satisfy the boundary condition (4.1.15) by the solution (4.1.14), the expression

$$U(0,y) = \sum_{k=0}^{\infty} D_k \sin\left(C_k\right) \sinh\left(p_k y + \tilde{C}_k\right) = 0, \qquad (4.1.19)$$

must be valid. This can be satisfied by $C_k = 0$ for all k. The boundary condition (4.1.16) demands

$$U(a,y) = \sum_{k=0}^{\infty} D_k \sin\left(p_k a\right) \sinh\left(p_k y + \tilde{C}_k\right) = 0. \qquad (4.1.20)$$

From this condition the separation constants p_k will immediately be determined:

$$p_k a = k\pi, \quad p_k = k\pi/a. \quad k = 0,1,2\ldots. \qquad (4.1.21)$$

Like in many boundary value problems the separation constants are determined by the boundary conditions. Next we consider the boundary condition (4.1.17)

$$U(x,b) = \sum_{k=0}^{\infty} D_k \sin\frac{k\pi x}{a} \sinh\left(\frac{k\pi b}{a} + \tilde{C}_k\right) = 0, \qquad (4.1.22)$$

which can be satisfied by

$$\frac{k\pi b}{a} + \tilde{C}_k = 0, \quad \tilde{C}_k = -\frac{k\pi b}{a}. \qquad (4.1.23)$$

Finally condition (4.1.18) has to be taken into account.

$$U(x,0) = f_1(x) = \sum_{k=1}^{\infty} D_k \sin\frac{k\pi x}{a} \sinh\left(-\frac{k\pi b}{a}\right). \qquad (4.1.24)$$

Observe that now $k = 0$ is not allowed and that the continuity condition
(4.1.15) $f_1(0) = 0$ is satisfied automatically.

As one sees, (4.1.24) represents a FOURIER *series* expansion of the given
function $f_1(x)$. To find the FOURIER expansion coefficients we multiply
(4.1.24) by $\sin(m\pi x/a), m \neq k$ and integrate over x

$$\int_0^a f_1(x) \sin \frac{m\pi x}{a}\, dx = \int_0^a \sum_{k=1} D_k \sin \frac{k\pi x}{a} \sin \frac{m\pi x}{a} \sinh\left(-\frac{k\pi b}{a}\right) d_x. \quad (4.1.25)$$

In section 2.3 we discussed the orthogonality of some functions, compare
(2.3.17). The attribute of orthogonality is very important. It allows to expand given functions like $f_1(x)$ and to express them by a series according to
orthogonal functions. In (4.1.25) one has

$$\int_b^a \sin \frac{k\pi x}{a} \sin \frac{m\pi x}{a}\, dx = 0 \quad \text{for} \quad k \neq m. \quad (4.1.26)$$

This is easy to prove. The simple transformation $x = a\xi/\pi$ modifies (4.1.26)
into

$$\frac{a}{\pi} \int_0^\pi \sin k\xi \sin m\xi\, d\xi = \frac{a}{\pi} \begin{cases} 0 & \text{for} \quad k \neq m, \\ \pi/2 & \text{for} \quad k = m, \end{cases} \quad m, k = 1, 2 \ldots . \quad (4.1.27)$$

Thus we get

$$\int_0^a f_1(x) \sin \frac{k\pi x}{a}\, dx = \frac{a}{2} D_k \sinh\left(-\pi kb/a\right), \quad (4.1.28)$$

which gives the FOURIER expansion coefficients D_k. The solution of our
boundary value problem defined by (4.1.15) to (4.1.18) is then given by

$$U(x, y) = \sum_{k=1}^\infty \frac{2 \sinh(k\pi b/a - k\pi y/a)}{a \sinh(\pi kb/a)}$$

$$\cdot \int_0^a f_1(\xi) \sin(k\pi\xi/a)\, d\xi \cdot \sin(k\pi x/a). \quad (4.1.29)$$

But what about the more general problem (4.1.4) to (4.1.7)? We will come
back to it in problem 2 of this section.

The boundary problem we just solved is inhomogeneous because $f_1(x) \neq 0$.
Why could we not homogenize the problem as we learnt in section 1.2? There
we have shown that an inhomogeneous problem consisting of an homogeneous
equation together with an inhomogeneous condition may be converted into an
inhomogeneous equation of the type (4.1.3) with an homogeneous boundary
condition.

There are two methods to solve linear inhomogeneous partial differential
equations:

1. Search a function U satisfying the homogeneous boundary conditions and determine free parameters like expansion coefficients within U in such a way that U satisfies the inhomogeneous partial differential equations. This delivers a particular solution of the inhomogeneous equation that however does not satisfy the homogeneous equation.

2. First, solve the matching homogeneous partial differential equation and expand the inhomogeneous term ρ according to the solutions of the homogeneous equation.

We start with method 1. The differential equation (4.1.3) is inhomogeneous and the boundary conditions should be homogeneous. With regard to our experience we set up

$$U(x,y) = \sum_{\nu=1}^{\infty} c_\nu(y) \sin \frac{\nu \pi x}{a}, \qquad (4.1.30)$$

where $c_\nu(y)$ are the expansion coefficients. Now (4.1.30) should first satisfy the boundary conditions. We multiply (4.1.30) by $\sin(\mu \pi x/a)$ and integrate over x. Due to (4.1.27) we then obtain

$$\int_0^a \sin \frac{\nu \pi x}{a} \sin \frac{\mu \pi x}{a} \, dx = \frac{a}{2} \delta_{\nu \mu} \qquad (4.1.31)$$

and

$$c_\nu(y) = \frac{2}{a} \int_0^a U(x,y) \sin \frac{\nu \pi x}{a} \, dx. \qquad (4.1.32)$$

Due to the homogeneous boundary conditions

$$U(0,y) = U(a,y) = U(x,0) = U(x,b) = 0, \qquad (4.1.33)$$

the c_ν must satisfy $c_\nu(0) = c_\nu(b) = 0$ for all ν. We now set up

$$c_\nu(y) = \sum_{\nu=1}^{\infty} \gamma_{\nu \mu} \sin \frac{\mu \pi y}{b} \qquad (4.1.34)$$

so that

$$\gamma_{\nu \mu} = \frac{2}{b} \int_0^b c_\nu(y) \sin \frac{\mu \pi y}{b} \, dy, \qquad (4.1.35)$$

and the expression satisfying the homogeneous boundary conditions reads

$$U(x,y) = \sum_{\nu=1}^{\infty} \sum_{\mu=1}^{\infty} \gamma_{\nu \mu} \sin \frac{\nu \pi x}{a} \sin \frac{\mu \pi y}{b}, \qquad (4.1.36)$$

where now

$$\gamma_{\nu \mu} = \frac{4}{ab} \int_0^a \int_0^b U(x,y) \sin \frac{\nu \pi x}{a} \sin \frac{\mu \pi y}{b} \, dy dx, \qquad (4.1.37)$$

which are still unknown! (4.1.36) satisfies the boundary conditions, but U must satisfy the inhomogeneous equation too. To find a particular solution of the inhomogeneous partial differential equation expand the inhomogeneous term with respect to the solutions of the matching homogeneous equation. But wait, we do not yet have these solutions. So we just have to write down a setup compatible with the boundary conditions:

$$\rho(x,y) = \sum_{\nu=1}^{\infty} \sum_{\mu=1}^{\infty} d_{\nu\mu} \sin \frac{\nu\pi x}{a} \sin \frac{\mu\pi y}{b}, \tag{4.1.38}$$

where

$$d_{\nu\mu} = \frac{4}{ab} \int_0^a \int_0^b \rho(x,y) \sin \frac{\mu\pi y}{b} \sin \frac{\nu\pi x}{a} \, dy dx. \tag{4.1.39}$$

But from where do we get $\rho(x,y)$? We have to insert (4.1.36), which contains the homogeneous boundary conditions, into the inhomogeneous POISSON equation (4.1.3)

$$\rho(x,y) = U_{xx} + U_{yy}$$

$$= \sum_{\nu=1}^{\infty} \sum_{\mu=1}^{\infty} \sin \frac{\nu\pi x}{a} \sin \frac{\mu\pi y}{b} \cdot \gamma_{\nu\mu} \left[-\frac{\nu^2\pi^2}{a^2} - \frac{\mu^2\pi^2}{b^2} \right]$$

$$= \sum_{\nu=1}^{\infty} \sum_{\mu=1}^{\infty} d_{\nu\mu} \sin \frac{\nu\pi x}{a} \sin \frac{\mu\pi y}{b}. \tag{4.1.40}$$

Now we can read off the $\gamma_{\nu\mu}$ from (4.1.40)

$$\gamma_{\nu\mu} = \frac{-a^2 b^2}{\pi^2 \left(a^2\mu^2 + b^2\nu^2\right)} d_{\nu\mu} \tag{4.1.41}$$

and the boundary problem (4.1.3), (4.1.33) is solved and has the solution

$$U(x,y) = \sum_{\nu=1}^{\infty} \sum_{\mu=1}^{\infty} \frac{-a^2 b^2 d_{\nu\mu}}{\pi^2 \left(a^2\mu^2 + b^2 nu^2\right)} \sin \frac{\nu\pi x}{a} \sin \frac{\mu\pi y}{b}. \tag{4.1.42}$$

But remember, this only is a particular solution of (4.1.3) and does not solve the homogeneous equation.

Now let us look how method 2 works. We first have to solve the matching homogeneous equation $\Delta U = 0$ and then expand ρ with respect to the functions solving $\Delta U = 0$. We demonstrate the procedure on a three-dimensional spherical problem. We use (3.1.74) and the ordinary differential equations (3.1.76) - (3.1.78). The solutions to (3.1.77) had been given in section 2.3, $\Phi'' + m^2\Phi = 0$ is solved by $\cos m\varphi$ and $\sin m\varphi$ and *Mathematica* solves

$$R''(r) + \frac{2}{r} R'(r) - \frac{l(l+1)}{r^2} R(r) = 0, \tag{4.1.43}$$

which is NOT a BESSEL equation (2.2.48) but it is of the type (2.2.65) that is an EULER *equation* of type (2.2.33) with $x_0 = 0$.

`DSolve[R''[r]+2*R'[r]/r-(l*(l+1))*R[r]/r^2==0,R[r],r]`

yields

$R[r] \rightarrow r^{(-1-l)} C[1] + r^l C[2]$

Thus, the solution of the matching homogeneous equation $\Delta U = 0$ reads in spherical coordinates

$$U(r, \vartheta, \varphi) = \sum_{l=0}^{\infty} \sum_{m=0}^{\infty} (a_l^m r^l + b_l^m r^{-l-1}) \cdot N P_l^m (\cos \vartheta) e^{im\varphi}. \qquad (4.1.44)$$

N is a normalizing factor and a_l^m and b_l^m are the partial amplitudes. Expanding the inhomogeneous term ρ of the inhomogeneous equation $\Delta u = \rho$ gives

$$\rho(r, \vartheta, \varphi) = \sum_{l,m} \rho_{lm}(r) P_l^m (\cos \vartheta) e^{im\varphi}. \qquad (4.1.45)$$

Multiplication by $P_l^{-m}(\cos \vartheta) \exp(-im\varphi)$, integration and using (2.3.19) and $dx = -\sin \vartheta d\vartheta$, $\int \int \sin \vartheta d\vartheta d\varphi = 4\pi$ results in

$$\rho_{lm}(r) = \iint \rho(r, \vartheta, \varphi) P_l^{-m}(\cos \vartheta) e^{-im\varphi} \sin \vartheta d\vartheta d\varphi \frac{2l+1}{(-1)^m 8\pi}. \qquad (4.1.46)$$

If one inserts the expression

$$U = \sum_{l=0}^{\infty} \sum_{m=l}^{\infty} U_{lm}(r) P_l^m (\cos \vartheta) e^{im\varphi} \qquad (4.1.47)$$

into the inhomogeneous equation which we write now $\Delta U = -4\pi\rho$ as usual in electromagnetism and if one multiplies by the orthogonal functions and integrates, one gets

$$\frac{1}{r^2} \frac{d}{dr} r^2 \frac{dU_{lm}(r)}{dr} - \frac{l(l+1)}{r^2} U_{lm}(r) = -(-1)^m (2l+1)\rho_{lm}(r). \qquad (4.1.48)$$

This is an inhomogeneous ordinary differential equation. The matching homogeneous equation has a solution obtained by the analogous command as given above and reads like the solution **R[r]** given above. The *Wronskian determinant* (2.1.12) now takes the form

$$W = r^l(-l-1)r^{-l-2} - lr^{l-1}r^{-l-1} = -(2l+1)r^{-2}. \qquad (4.1.49)$$

Inserting W and (4.1.44) into (2.1.14) then a particular solution of (4.1.48) is obtained:

$$U_{lm}(r) = (-1)^m \left(r^l \int_r^{\infty} \rho_{lm}(r') r'^{-l+1} dr' + r^{-l-1} \int_0^r \rho_{lm}(r') r'^{l+2} dr' \right). \qquad (4.1.50)$$

Since the $\rho_{lm}(r)$ are unknown, *Mathematica* can give only a formal solution
of the inhomogeneous equation (4.1.48). The command

DSolve[R''[r]+2*R'[r]/r-(1*(1+1))*R[r]/r^2==B[r],R[r],r]

gives

$$R[r] \rightarrow r^{(-1-l)}\, C[1] + r^l\, C[2] + \frac{1}{1+2l}$$

$$\left(r^{-1-l} \left(-\int_{K\$118}^{R} [K\$117^{2+l}\, B[K\$117]dK\$117 + r^{1+2\,l} \right. \right.$$

$$\left. \left. \int_{K\$131}^{r} K\$130^{1-l}\, B[K\$130]dK\$130 \right) \right).$$

Using addition formulas [2.1], [2.9] for the spherical functions and the formula

$$\cos\Theta = \cos\vartheta \cos\vartheta' + \sin\vartheta \sin\vartheta' \cos(\varphi - \varphi'),$$

one can modify (4.1.50) to read

$$U(r,\vartheta,\varphi) = \int_0^\infty \int_0^\pi \int_0^{2\pi} \frac{r'^2 \rho(r',\vartheta',\varphi')}{\sqrt{r^2 + r'^2 - 2rr' \cos\Theta}} d\varphi' \sin\vartheta' d\vartheta' dr',$$

where ρ is given by (4.1.45). In series expansions of the type

$$p_{lm} = \iint r^l P_l^m(\cos\vartheta) e^{+im\varphi} \rho(r,\vartheta,\varphi) \sin\vartheta d\vartheta d\varphi,$$

the expansion coefficients p_{lm} are called *multipole moments*. There are many
physical and engineering problems in which such expansions appear: burden of
the organism by γ-rays due to the uranium content in bricks or the distribution
of electric charges on a body.

We treated an inhomogeneous problem that had been inhomogeneous because the differential equation (4.1.3) has been inhomogeneous, but the boundary conditions (4.1.33) were homogeneous. We presented two methods to solve
such a problem. But a problem of an homogeneous equation with inhomogeneous boundary conditions is inhomogeneous too! Due to the theorem that the
general solution U of an inhomogeneous linear equation $\Delta U = f(x,y)$ consists
of the superposition of the general solution W of the homogeneous equation
plus a particular solution V of the inhomogeneous equation we can execute
the following method. To solve $\Delta U = f(x,y)$ one may write $U = W + V$,
so that $\Delta U = \Delta W + \Delta V = f(x,y)$, where $\Delta V = f(x,y)$ and $\Delta W = 0$.
Since we demand homogeneous boundary conditions for U(boundary)$= 0 =$
V(boundary)$+W$(boundary), one has V(boundary)$= -W$(boundary). Now
the function $V(x,y)$ should satisfy the inhomogeneous boundary conditions

for W and at the same time it must be such that $\Delta V = f(x, y)$. It is quite difficult to create such a function V.

We start with the boundary conditions (4.1.15) to (4.1.18) for V for the rectangle as shown in Figure 4.1 in the specialized form

$$V(0, y) = V(a, y) = V(x, b) = 0, \quad V(x, 0) = 1. \tag{4.1.51}$$

A function satisfying these conditions has to be discovered. An example would be

$$
\begin{aligned}
V(x, y) = {} & \frac{x}{x + y} + \frac{x - a}{x - y - a} - 2 + \frac{x}{a} + \frac{(a + 2b)y}{b(a + b)} - \frac{2xy}{ab} \\
& + \frac{a - x}{a} \left[\frac{2y + a}{y + a} - \frac{(a + 2b)y}{b(a + b)} \right] \\
& + \left[\frac{y}{y + a} + \frac{ay}{b(a + b)} \right] \cdot \left[\frac{a}{a + b} + \frac{x}{a} - \frac{x}{x + b} - \frac{x - a}{x - a - b} \right]. \tag{4.1.52}
\end{aligned}
$$

If one calculates $f(x, y) = \Delta V$, then $\Delta U = f(x, y)$ is the inhomogeneous equation for U. Its solution satisfies the homogeneous boundary condition $U(\text{boundary}) = 0$.

Another method to solve the inhomogeneous POISSON equation is the use of the GREEN *function*, compare (2.4.29). The GREEN function for $\Delta U = \rho$ and for the rectangle may be derived as follows: inserting (4.1.30), which satisfies the boundary homogeneous conditions, $U(0, y) = U(a, y) = 0$ into the POISSON equation yields

$$\Delta U = \sum_{\nu=1}^{\infty} \sin \frac{\nu \pi x}{a} \left[c_\nu''(y) - \frac{\nu^2 \pi^2}{a^2} c_\nu(y) \right] = \rho(x, y). \tag{4.1.53}$$

This demonstrates that the expression in brackets plays the role of FOURIER expansion coefficients for $\rho(x, y)$ so that

$$c_\nu''(y) - \frac{\nu^2 \pi^2}{a^2} c_\nu(y) = \frac{2}{a} \int_0^a \rho(x, y) \sin \frac{\nu \pi x}{a} \, dx = g(y). \tag{4.1.54}$$

The new function $g(y)$ is the inhomogeneous term of the differential equation for $c_\nu(y)$

$$c_\nu''(y) - \frac{\nu^2 \pi^2}{a^2} c_\nu(y) = g_\nu(y). \tag{4.1.55}$$

Due to the boundary conditions (4.1.33) the c_ν have to satisfy

$$c_\nu(y) = U(0, y), \quad c_\nu(0) = c_\nu(b) = 0, \quad \nu = 1, 2, \dots. \tag{4.1.56}$$

Then, according to section 2.4 the solution has the form

$$c_\nu(y) = \int_0^b G_\nu(y,\eta) g_\nu(\eta) d\eta, \qquad (4.1.57)$$

where now the GREEN function for the rectangle can be read off

$$G_\nu(y,\eta) = \begin{cases} -\dfrac{a}{\nu\pi}\dfrac{\sinh(\nu\pi\eta/a)}{\sinh(\nu\pi b/a)}\sinh\dfrac{\nu\pi(b-y)}{a}, & 0 \le \eta \le y \le b \\[4mm] -\dfrac{a}{\nu\pi}\dfrac{\sinh(\nu\pi y/a)}{\sinh(\nu\pi b/a)}\sinh\dfrac{\nu\pi(b-\eta)}{a}, & 0 \le y \le \eta \le b \end{cases} . \qquad (4.1.58)$$

Please observe our notation: $\rho(x,y)$ is the inhomogeneous term in the POIS-SON equation determined by the actual physical problem, whereas $f(x,y)$ is the inhomogeneous term created by the homogenization of the inhomogeneous boundary conditions belonging to the homogeneous differential equation.

The GREEN function for a rectangle with homogeneous boundary conditions, namely

$$G(x,y;\xi,\eta) = \frac{2}{a}\sum_{\nu=1}^{\infty} G_\nu(y,\eta)\sin\frac{\nu\pi x}{a}\sin\frac{\nu\pi\xi}{a} \qquad (4.1.59)$$

enables one to write the solution of (4.1.3), (4.1.33) in the form

$$U(x,y) = \sum_{\nu=1}^{\infty} c_\nu(y)\sin\frac{\nu\pi x}{a} = \int_0^a\int_0^b G(x,y;\xi,\eta)f(\xi,\eta)d\xi d\eta. \qquad (4.1.60)$$

Up to now we have discussed simple regular rectangles. But how to solve problems like as shown in Figures 4.2 and 4.3?

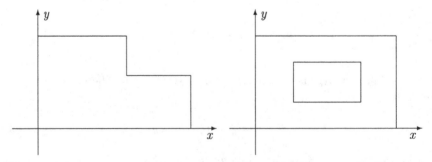

Figure 4.2
Rectangle with incision

Figure 4.3
Rectangle with excision

New methods however are necessary to solve such problems with corners or holes (admission of singularities, variational methods etc.).

Boundary problems for a circle are easier than the situations mentioned. According to problem 4 of section 3.1, the LAPLACE equation in cylindrical coordinates may be separated into two ordinary differential equations

$$R_\nu''(r) + \frac{1}{r}R_\nu'(r) - \frac{\lambda_\nu^2 R_\nu(r)}{r^2} = 0, \tag{4.1.61}$$

(we used $b = 0$, since our problem is independent on z and $a \to \lambda^2$) and ($\Theta \to \Phi$)

$$\Phi_\nu''(\varphi) + \lambda_\nu^2 \Phi_\nu(\varphi) = 0. \tag{4.1.62}$$

The index ν indicates that we expect a set of solutions since the original partial differential equation is linear. (4.1.61) is an EULER equation of type (2.2.33) having the solution $R(r) \sim r^\nu, r^{-\nu}$, so that $\lambda_\nu = \pm\nu, \nu = 1, 2, \ldots$ for a solution valid around the whole circle ($0 \le \varphi \le 2\pi$). Thus the general solution of the LAPLACE equation reads

$$U(r, \varphi) = \frac{a_0}{2} + \sum_{\nu=1}^{\infty} c_\nu r^\nu (a_\nu \cos\nu\varphi + b_\nu \sin\nu\varphi). \tag{4.1.63}$$

The solution $r^{-\nu}$ is singular at $r = 0$ and had been excluded. We demand $U(r = 0, \varphi) = $ regular, not singular. If the boundary condition along the circular line does not cover the whole circumference then the λ_ν are no longer integers.

Now we need boundary conditions on the circular circumference like

$$U(r = R, \varphi) = f(\varphi). \tag{4.1.64}$$

Using this condition, (4.1.63) reads

$$U(R, \varphi) = \frac{a_0}{2} + \sum_{\nu=0}^{\infty} c_\nu R^\nu (a_\nu \cos\nu\varphi + b_\nu \sin\nu\varphi) = f(\varphi) \tag{4.1.65}$$

and the FOURIER-coefficients are

$$\genfrac{}{}{0pt}{}{a_\nu}{b_\nu} = \frac{1}{\pi}\int_0^{2\pi} f(\varphi)\genfrac{}{}{0pt}{}{\cos\nu\varphi}{\sin\nu\varphi}\,d\varphi, \quad a_0 = \frac{1}{\pi}\int_0^{2\pi} f(\varphi)d\varphi. \tag{4.1.66}$$

To see the convergence of the sum $\sum_\nu$ we can demand $c_\nu R^\nu = 1, c_\nu = R^{-\nu}$, so that

$$U(r, \varphi) = \frac{a_0}{2} + \sum_{\nu=1}^{\infty} \left(\frac{r}{R}\right)^\nu (a_\nu \cos\nu\varphi + b_\nu \sin\nu\varphi). \tag{4.1.67}$$

Using the addition theorem for sin and cos and (4.1.66) we can modify (4.1.67) to read

$$U(r,\varphi) = \frac{1}{2\pi}\left[\int_0^{2\pi}f(\psi)\mathrm{d}\psi + 2\sum_{\nu=1}^{\infty}\left(\frac{r}{R}\right)^{\nu}\int_0^{2\pi}f(\psi)(\cos\nu\varphi\cos\nu\psi + \sin\nu\varphi\sin\nu\psi)\mathrm{d}\psi\right]$$

$$= \frac{1}{2\pi}\int_0^{2\pi}f(\psi)\left[1 + 2\sum_{\nu=1}^{\infty}\left(\frac{r}{R}\right)^{\nu}\cos\nu(\varphi-\psi)\right]\mathrm{d}\psi. \qquad (4.1.68)$$

Taking advantage of the formula for the infinite geometric series we write

$$\frac{1}{1-\rho e^{i\varphi}} = \sum_{\nu=0}^{\infty}\rho^{\nu}e^{i\nu\varphi} = \sum_{\nu=0}^{\infty}\rho^{\nu}(\cos\nu\varphi + i\sin\nu\varphi), \qquad (4.1.69)$$

where $\rho = r/R$ and

$$\sum_{\nu=0}^{\infty}\rho^{\nu}\cos\nu\varphi = \mathrm{Re}\frac{1}{1-\rho e^{i\varphi}} = \frac{1-\rho\cos\varphi}{1-2\rho\cos\varphi + \rho^2}. \qquad (4.1.70)$$

Comparing with (2.3.9) one recognizes the last rhs term as the generating function of the polynomials ρ^{ν}. Using

$$1 + 2\sum_{\nu=1}^{\infty}\rho^{\nu}\cos\nu\varphi = \frac{1-\rho^2}{1-2\rho\cos\varphi + \rho^2} \qquad (4.1.71)$$

one gets the so-called POISSON *integral* for a circle

$$U(r,\varphi) = \frac{1}{2\pi}\int_0^{2\pi}\frac{R^2 - r^2}{R^2 - 2Rr\cos(\varphi-\psi) + r^2}f(\psi)\mathrm{d}\psi \qquad (4.1.72)$$

from which the GREEN function for the circle can be read off.

POISSON integrals also exist in three dimensions for electrostatic or gravitational problems, since both potentials satisfy the POISSON equation. Let ξ, η, ζ or $r_0, \vartheta_0, \varphi_0$ be the coordinates of the *source point*, where a small charge element $\rho\mathrm{d}\tau$ is located and x, y, z or r, ϑ, φ the coordinates of the *field point*, then their relative distance d is given by

$$\mathrm{d} = \sqrt{(x-\xi)^2 + (y-\eta)^2 + (z-\zeta)^2} \text{ or } \sqrt{r^2 + R^2 - 2rR\cos(\Theta)}, \qquad (4.1.73)$$

if the source point is located on the surface of a sphere with radius $r = R$. *Mathematica* did the work for us. Let us go slowly step by step. According to (3.1.24) we have

```
x[r_,φ_,ϑ_]==r*Cos[φ]*Sin[ϑ];y[r_,φ_,ϑ_]==
r*Sin[φ]*Sin[ϑ];z[r_,φ_, ϑ_]==r*Cos[ϑ];
ξ[R_,φ0_,ϑ0_]== R*Cos[φ0]*Sin[ϑ0];η[R_,φ0_, ϑ0_]==
R*Sin[φ0]*Sin[ϑ0]; ζ[R_,φ0_,ϑ0_]== R*Cos[ϑ0];          (4.1.74)
```

```
d=Simplify[Sqrt[(r*Cos[φ]*Sin[ϑ]-
R*Cos[φ0]*Sin[ϑ0])^2+
(r*Sin[φ]*Sin[ϑ]-R*Sin[φ0]*Sin[ϑ 0])^2+
(r*Cos[ϑ]-R*Cos[ϑ0])^2]]                                (4.1.75)
```

yielding

$$\left(r^2 + R^2 - 2\,r\,R\,\text{Cos}[\vartheta]\,\text{Cos}[\vartheta 0] - 2\,r\,R\,\text{Cos}[\varphi]\,\text{Cos}[\varphi 0]\,\text{Sin}[\vartheta]\,\text{Sin}[\vartheta 0] - \right.$$
$$\left. 2\,r\,R\,\text{Sin}[\vartheta]\,\text{Sin}[\vartheta 0]\,\text{Sin}[\varphi]\,\text{Sin}[\varphi 0]\right)^{1/2}. \qquad (4.1.76)$$

To be able to continue using the commands copy and paste we form

```
InputForm[Simplify[%]]                                  (4.1.77)
```

which results in

Sqrt$\{r^2 + R^2 - 2\,r\,R\,\text{Cos}[\vartheta]\,\text{Cos}[\vartheta 0] - 2\,R\,\text{Cos}[\varphi]\,\text{Cos}[\varphi 0]\,Sin[\vartheta]\,\text{Sin}[\vartheta 0] - 2\,r\,R\,\text{Sin}[\vartheta]\,\text{Sin}[\vartheta 0]\,\text{Sin}[\varphi]\,\text{Sin}[\varphi 0]\}.$ \qquad (4.1.78)

Using the abbreviation (which is actually a formula from spherical trigonometry)

```
Cos[Θ]==Cos[ϑ]*Cos[ϑ0]+ Sin[ϑ]*Sin[ϑ0]*(Cos[φ]*Cos[φ 0]
+Sin[φ]*Sin[φ0])                                        (4.1.79)
```

we rewrite d as

```
Clear[d];d=InputForm[
Sqrt[r^2+R^2-2*r*R*(Cos[ϑ]*Cos[ϑ0]-
Cos[φ]*Cos[φ0]*Sin[ϑ]
*Sin[ϑ0]-Sin[ϑ]*Sin[ϑ0]
*Sin[φ]*Sin[φ0])]];                                     (4.1.80)
```

so that we receive

```
d=Sqrt[r^2+R^2-2*r*R*Cos[Θ]]                            (4.1.81)
```

$$\sqrt{r^2 + R^2 - 2\,r\,R\,\text{Cos}[\Theta]}$$

If the center of the sphere with radius R lies in the origin and both the source point and the field point are located in a plane then simply $\Theta = \vartheta$.

Now the potential U in the field point originating from a charge distribution $\rho d\tau$ is apparently given by

$$U(x, y, z) = \int \frac{\rho}{d} d\tau = \iiint \frac{\rho(\xi, \eta, \zeta)}{\sqrt{(x-\xi)^2 + (y-\eta)^2 + (z-\zeta)^2}} d\xi d\eta d\zeta, \qquad (4.1.82)$$

which satisfies $\Delta U = \rho$.

By the way, one can show that $1/d$, where d is given by (4.1.81) is nothing else than the GREEN function of a point charge.

If the surface potential $U(F)$ is given on the surface F of a sphere then the potential in the distance r from the origin is given by the POISSON *integral* for the sphere

$$U(r) = R^3 \left(1 - \frac{r^2}{R^2}\right) \int\limits_0^{2\pi} \int\limits_0^{\pi} \frac{U_F(\vartheta, \varphi) \sin \vartheta}{(R^2 + r^2 - 2rR\cos\Theta)^{3/2}} d\vartheta d\varphi. \qquad (4.1.83)$$

This is the solution of a DIRICHLET *problem*. In many other electrostatic problems, as well as in stationary thermal problems, the methods described here are used.

Up to now, we have treated boundary value problems of a rectangle, of a sphere and of a circle. Now, to finish this section, we discuss cylindrical problems. For an infinitely long cylinder the potential $U(r, \varphi)$ depends on polar coordinates in the x, y plane. Such problems will be treated in section 4.2. If the cylindrical problem of the LAPLACE equation is axially symmetric, then the two-dimensional problem is described by a potential $U(r, z)$ satisfying

$$\frac{\partial^2 U}{\partial r^2} + \frac{1}{r} \frac{\partial U}{\partial r} + \frac{\partial^2 U}{\partial z^2} = 0. \qquad (4.1.84)$$

Separation into two ordinary differential equations and the boundary conditions

$$U(R, 0) = U_0, \quad U(R, z) = U_0 \text{ for } 0 < z < l, \quad U(R, l) = 0, \qquad (4.1.85)$$

where R and l are radius and length of the cylinder respectively, result in

$$U(r, z) = U_0 \left[1 - 2 \sum_{n=1}^{\infty} \frac{J_0(\gamma_n r/R) \cdot \sinh(\gamma_n z/R)}{\gamma_n J_1(\gamma_n) \cdot \sinh(\gamma_n l/R)}\right], \qquad (4.1.86)$$

where γ_n are the positive roots of $J_0(\gamma) = 0$ and where

$$\int\limits_0^R r J_0(\gamma_n r/a) = R^2 J_1(\gamma_n)/\gamma_n.$$

J_0 and J_1 are BESSEL functions.

Problems

1. Calculate $\operatorname{div} \operatorname{grad} U$ in cartesian and spherical coordinates.

2. Solve the boundary problem defined by $\Delta U = 0$ and the inhomogeneous conditions (4.1.4) to (4.1.7). The result should be

$$
U_1(x,y) = \frac{2}{a} \sum_{k=1}^{\infty} \left\{ \frac{\sinh(k\pi y/a)}{\sinh(k\pi b/a)} \int_0^a f_2(\xi) \sin(k\pi\xi/a)\, \mathrm{d}\xi \cdot \sin(k\pi x/a) \right.
$$

$$
\left. + \frac{\sinh((b-y)k\pi/a)}{\sinh(k\pi b/a)} \int_0^a f_1(\xi) \sin(k\pi\xi/a)\, \mathrm{d}\xi \cdot \sin(k\pi x/a) \right\}
$$

$$
+ \frac{2}{b} \sum_{k=1}^{\infty} \left\{ \frac{\sinh(k\pi x/b)}{\sinh(k\pi a/b)} \int_0^b g_2(\eta) \sin(k\pi\eta/b)\, \mathrm{d}\eta \cdot \sin(k\pi y/b) \right.
$$

$$
\left. + \frac{\sinh((a-x)k\pi/b)}{\sinh(k\pi a/b)} \int_0^b g_1(\eta) \sin(k\pi\eta/b)\, \mathrm{d}\eta \cdot \sin(k\pi y/b) \right\}.
$$

Is this solution continuous on a corner where $U \neq 0$? (No) If the boundary conditions are simplified to (4.1.8) then the solution will be $U_2(x,y) = U_1(x,y) + U_0(x,y)$, where

$$
U_0(x,y) = \cos\frac{\pi y}{2b} \left(\frac{\beta}{\sinh(\pi a/2b)} \sinh\frac{\pi x}{2b} + \frac{\alpha}{\sinh(\pi a/2b)} \sinh\frac{\pi}{2b}(a-x) \right)
$$

$$
+ \sin\frac{\pi y}{2b} \left(\frac{\gamma}{\sinh(\pi a/2b)} \sinh\frac{\pi x}{2b} + \frac{\delta}{\sinh(\pi a/2b)} \sinh\frac{\pi}{2b}(a-x) \right)
$$

and where the expansion coefficients given by the integrals now read

$$
\int_0^a (f_2(\xi) - U_0(\xi,b)) \sin(k\pi\xi/a)\mathrm{d}\xi, \qquad \int_0^a (f_1(\xi) - U_0(\xi,0)) \sin(k\pi\xi/a)\mathrm{d}\xi,
$$

$$
\int_0^b (g_2(\eta) - U_0(a,\eta)) \sin(k\pi\eta/b)\mathrm{d}\eta, \qquad \int_0^b (g_1(\eta) - U_0(0,\eta)) \sin(k\pi\eta/b)\mathrm{d}\eta.
$$

3. Solve $U_{xx} + U_{yy} = -2$ for the rectangle defined by $0 \leq x \leq a$, $-b/2 \leq y \leq b/2$ with the homogeneous boundary condition $U(\text{boundary}) = 0$. Use $U = V + W$, $V(0,y) = 0$, $V(a,y) = 0$, $V(x,y) = Ax^2 + Bx + C$. The function W will be defined by $\Delta W = 0$ and the boundary conditions

$W(0) = 0$, $W(a, y) = 0$, $W(x, -b/2) = -V(x)$, $W(x, b/2) = -V(x)$.
The solution should be

$$U(x, y) = x(a - x) - \frac{8a^2}{\pi^3} \sum_n \frac{\cosh\left[(2n + 1)\pi y/a\right] \cdot \sin\left[(2n + 1)\pi x/a\right]}{(2n + 1)^3 \cosh\left[(2n + 1)\pi b/2a\right]}.$$

By the way, the similar equation $U_{xx} + U_{yy} = -4$ has the simple solution
$U(x, y) = a^2 - (x^2 + y^2)$.

4. Solve $U_{xx} + U_{yy} = -xy$ for a circular domain. The circle has the center at
$x = 0, y = 0$, the radius is a and the homogeneous boundary condition
is given by $U(r = a, \varphi) = 0$. Again one should use $U = V + W = -(1/12)xy(x^2 + y^2) + W$, $\Delta W = 0$, $W(r = a, \varphi) = -V(r = a, \varphi)$ and
the solution is

$$U(r, \varphi) = -\frac{r^3}{24} \sin 2\varphi + \frac{a^4}{48\pi} \int_{-\pi}^{+\pi} \sin 2t \frac{a^2 - r^2}{a^2 - 2ar\cos(t - \varphi) + r^2} \, dt.$$

5. Solve the boundary value problem $\Delta U = 0$, $U(0, y) = A$, $U(a, y) = Ay$
for $0 \le x \le a$, $0 \le y \le b$, $\partial u(x, y = 0)/\partial y = 0$, $\partial u(x, y = b)/\partial y = 0$.
The solution should be

$$U(x, y) = A + A(b - 2)x/2a$$

$$- \frac{4Ab}{\pi^2} \cdot \sum_{k=0}^{\infty} \frac{1}{(2k + 1)^2} \cdot \frac{\sinh[(2k + 1)\pi x/b]}{\sinh[(2k + 1)\pi a/b]} \cos \frac{(2k + 1)\pi y}{b}.$$

6. Assume a cylinder of height l defined by its radius R, $0 \le r \le R$, $0 \le z \le l$. Solve the steady heat conduction problem. Due to assumed
symmetry, $\partial/\partial\varphi = 0$ and the boundary conditions for the temperature
distribution $T(r, z)$

$$T(r, 0) = f_0(r), \quad T(r, l) = f_1(r), \quad T(R, z) = \varphi(z), \qquad (4.1.87)$$

the solution reads

$$T(r, z) = \frac{2}{l} \sum_{n=1}^{\infty} \frac{1}{I_0(n\pi R/l)} \left\{ I_0\left(\frac{n\pi r}{l}\right) \int_0^l \varphi(\xi) \sin\left(\frac{n\pi\xi}{l}\right) d\xi \right.$$

$$\left. + \frac{n\pi}{l} \int_0^R [(-1)^n f_1(\rho) - f_0(\rho)] \, \rho G_n(r, \rho) d\rho \right\} \sin \frac{n\pi z}{l},$$

where the GREEN function G is defined by

$$G_n(r,\rho) = \begin{cases} [\mathrm{K}_0(n\pi a/R)\mathrm{I}_0(n\pi r/l) \\ \quad -\mathrm{I}_0(n\pi a/l)\mathrm{K}_o(n\pi r/l)]\,\mathrm{I}_0(n\pi \rho/l), \quad \rho \leq r, \\ [\mathrm{K}_0(n\pi a/l)\mathrm{I}_0(n\pi \rho/l) \\ \quad -\mathrm{I}_0(n\pi a/l)\mathrm{K}_0(n\pi \rho/l)]\,\mathrm{I}_0(n\pi r/l), \quad \rho \geq r. \end{cases}$$

$\mathrm{I}_0(x) = \mathrm{J}(ix)$, $\mathrm{K}_0(x) = \mathrm{Y}_0(ix)$ are modified BESSEL functions, see (2.2.51), and problem 5 in section 2.2. Specialization of the boundary conditions (4.1.87) $f_0(r) = 0$, $f_1(r) = 0$, $\varphi(z) = T_0 = const$ yields

$$T(r,z) = \frac{4T_0}{\pi} \sum_{n=0}^{\infty} \frac{\mathrm{I}_0([2n+1]\pi r/l)\sin([2n+1]\pi z/l)}{\mathrm{I}_0([2n+1]\pi R/l)(2n+1)}$$

or

$$T(r,z) = 2T_0 \sum_{n=1}^{\infty} \frac{\mathrm{J}_0(\gamma_n r/R)\sinh(\gamma_n z/R)}{\gamma_n \mathrm{J}_1(\gamma_n)\sinh(\gamma_n l/R)}$$

if $f_0(r) = 0$, $f_1(r) = T_0$, $\varphi(z) = 0$.

7. A plane circular disk of radius R carries a surface charge Q that is distributed according to $\sigma = e/(4\pi R\sqrt{R^2 - r^2})$. Calculate the potential $U(P)$ in the field point P, which is located vertically above the center of the disk in a distance p. Solution:

$$U(P) = -\frac{Q}{4R}\arcsin\frac{-r^2 + R^2 - p^2}{\sqrt{R^4 + p^4 - R^2 p^2}} \quad \text{for} \quad 0 \leq r \leq R. \quad (4.1.88)$$

8. A diode consisting of two electrodes situated on the x axis in the distance d from each other emits an electron current $j = \rho\sqrt{2eU(x)/m}$ at the first electrode situated at $x = 0$. The potential distribution between the electrodes is then described by $\mathrm{d}^2 U/\mathrm{d}x^2 = -4\pi\rho$ and the boundary condition $U(a) = U_a$. Solve the one-dimensional homogeneous equation $U'' = const\, U^{-1/2}$ with the setup $U = const\, x^n$ resulting in $U(x) = const\, j^{2/3}x^{4/3}$. After insertion of the boundary condition one obtains the LANGMUIR *law* of plasma physics

$$j = \frac{U_a^{3/2}\sqrt{2e}}{9\pi a^2\sqrt{m}}. \quad (4.1.89)$$

Here m and e are mass and charge of an electron, respectively.

9. Solve

$$\frac{1}{r}\frac{\partial}{\partial r}r\frac{\partial U}{\partial r} + \frac{\partial^2 U}{\partial z^2} + \frac{1}{r^2}\frac{\partial^2 U}{\partial \varphi^2} = 0. \quad (4.1.90)$$

Solution: (p and q are separation constants)

$$U(r, z, \varphi) = (AJ_n(qr) + BN_n(qr)) \cdot \left(Ce^{qz} + De^{-qz}\right)$$
$$\cdot \; (E \sin p\varphi + F \cos p\varphi), \qquad (4.1.91)$$

$$U(r, z, \varphi) = (AJ_n(iqr) + BN_n(iqr)) \cdot (C \sin qz + D \cos qz)$$
$$\cdot \; (E \sin p\varphi + F \cos p\varphi). \qquad (4.1.92)$$

10. Assume a given potential $U(x, y, z)$. The field $\vec{K}$ belonging to this potential is given by

$$\vec{K} = \nabla U. \qquad (4.1.93)$$

Calculate the differential equations for the field lines attached to the field vector $\vec{K}$. They read

$$\frac{dx}{\partial U/\partial x} = \frac{dy}{\partial U/\partial y} = \frac{dz}{\partial U/\partial z}. \qquad (4.1.94)$$

Mathematica helps to visualize the field lines [2]

```
PlotVectorField3D[
{y/(3*z),-x/(2*z),1.},
{x,-1.,1.},{y, -1.,1.},{z,1.,3.}]                    (4.1.95)
```

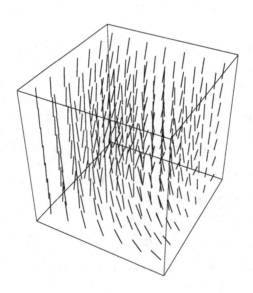

Figure 4.4
Visualization of the field given by (4.1.95)

11. A NEUMANN boundary condition may read in spherical coordinates

$$\left(\frac{\partial U}{\partial r}\right)_{r=R} = f(\vartheta_0, \varphi_0), \qquad (4.1.96)$$

where now the boundary values are expanded

$$f(\vartheta_0, \varphi_0) = \sum_{n=1}^{\infty} \frac{2n+1}{4\pi} \int_0^{2\pi}\int_0^{\pi} f(\vartheta, \varphi) P_n(\cos\Theta) \sin\vartheta d\vartheta d\varphi, \qquad (4.1.97)$$

where $\cos\Theta$ is given by (4.1.79). For $r < R$ the solution is

$$U(r, \vartheta, \varphi) = \sum_{n=1}^{\infty} \frac{R}{n} \left(\frac{r}{R}\right)^n Y_n(\vartheta, \varphi) \qquad (4.1.98)$$

and for $r > R$

$$U(r, \vartheta, \varphi) = -R\sum_{n=0}^{\infty} \frac{1}{n+1} \left(\frac{R}{r}\right)^{n+1} Y_n(\vartheta, \varphi). \qquad (4.1.99)$$

Here

$$Y_n(\vartheta, \varphi) = P_n(\cos\vartheta) \exp(in\varphi) \qquad (4.1.100)$$

are called *spherical functions*.

12. Calculate the surface temperature distribution $T(r, \vartheta)$ of a sphere of radius R in which there is a continuous heat production $q = const$ and which loses the equal amount of energy by conduction [4.8]. This induces a boundary condition

$$\lambda \frac{\partial T(R, \vartheta)}{\partial r} + \lambda T(R, \vartheta) = f(\vartheta). \qquad (4.1.101)$$

$f(\vartheta)$ is arbitrary, α is the heat transfer coefficient which appears in the NEWTON *cooling down law*

$$\lambda\nabla T + \alpha(T - T_0) = 0. \qquad (4.1.102)$$

The solution of $\lambda\Delta T = -q$ is given by

$$T(r, \vartheta) = \frac{q}{6\lambda}(R^2 - r^2) + \frac{qR}{3\alpha} + \frac{\lambda}{2\alpha}\int_0^{\pi} f(\vartheta) \sin\vartheta d\vartheta$$

$$+\frac{R}{2}\sum_{n=1}^{\infty} \frac{2n+1}{R\alpha/\lambda+n} \left(\frac{r}{R}\right)^n P_n(\cos\vartheta)\int_0^{\pi} f(\vartheta) P_n(\cos\vartheta) \sin\vartheta d\vartheta. \qquad (4.1.103)$$

This solution can also be found by homogenization of the boundary condition.

4.2   Conformal mapping in two and three dimensions

The method of conformal mapping is a consequence of the CAUCHY-RIEMANN *equations*. These equations are basic in the theory of functions of a complex variable $z = x + iy$. A function $f(z) = u(x,y) + iv(x,y)$ is called *analytic* at the point $z_0 = x_0 + iy_0$ in the complex GAUSS *plane*, if it is possible to find a derivative, so that the function may be developed into a TAYLOR *series*. If $f'(z_0)$ does not exist then the point z_0 is called singular (*singularity*). The differentiability at z_0 must be independent from the direction from which the point z_0 is approached, see Figure 4.5.

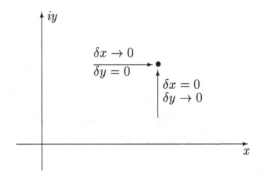

Figure 4.5
Approaching a point z_0 in the complex plane

Complex derivation is defined by

$$\lim_{\Delta z \to 0} \frac{\Delta f}{\Delta z} = \lim_{\Delta x \to 0} \left(\frac{\Delta u}{\Delta x} + i \frac{\Delta v}{\Delta x} \right) = \frac{\partial u}{\partial x} + i \frac{\partial v}{\partial x} \qquad (4.2.1)$$

and

$$\lim_{\Delta z \to 0} \frac{\Delta f}{\Delta z} = \lim_{\Delta y \to 0} \left(-i \frac{\Delta u}{\Delta y} + \frac{\Delta v}{\Delta y} \right) = -i \frac{\partial u}{\partial y} + \frac{\partial v}{\partial y}. \qquad (4.2.2)$$

For independence from direction the CAUCHY-RIEMANN equations must be valid:

$$\frac{\partial u}{\partial x} = \frac{\partial v}{\partial y}, \quad \frac{\partial u}{\partial y} = -\frac{\partial v}{\partial x}. \qquad (4.2.3)$$

As an example $\zeta = f(z) = z^2 = u(x,y) + iv(x,y), u(x,y) = x^2 - y^2, v = 2xy$ satisfies (4.2.3) and is therefore analytic, but $f(z) = z^* = x - iv, u = x, v = -y$ is not analytic. Derivation of (4.2.3) results in

$$\frac{\partial^2 u}{\partial x^2} = \frac{\partial^2 v}{\partial x \partial y}, \quad \frac{\partial^2 u}{\partial y^2} = -\frac{\partial^2 v}{\partial x \partial y} \quad \text{or} \quad \Delta u = 0, \Delta v = 0. \qquad (4.2.4)$$

Solutions of the two LAPLACE equations (4.2.4) may be written $u(x,y) = const$, $v(x,y) = const$. Any complex number can be translated into a point in the (x,y) plane and vice versa. If this is done for a set of points, one speaks of mapping of the x, y plane onto the u, v plane. *Mathematica* is of great help in producing such plots [4]. It offers the commands

```
<<ComplexMapPlot.m;  ComplexMapPlot[Sin[z],z,
RectangularGrid[{-1.,1.},{-1.,1.}]]
```
(4.2.5)

When you start *Mathematica*, don't forget to press shift and enter together with your first command. In (4.2.5) the rectangular grid in the u, v plane will be plotted onto the z(x,y) plane. The mapping is done by the function $\zeta[z] = Sin[z]$. The coordinates in brackets determine the left upper and left lower corner of the grid. The number of horizontal lines from -1 to $+1$ and of vertical lines from -1 to $+1$ has the default value (14) of the packages. If the reader has no access to the package **ComplexMapPlot**, the command **ComplexMap** may help. It is contained in Addons/StandardPackages/Graphics.
The command **echo $Packages** (4.2.6)
informs you which packages have been loaded. One may also force the loading of a package by the *Mathematica* command

```
AppendTo[$Path, "/usr/local/mathematica..."]
```
(4.2.7)

down to the directory, where the package is located. Another way is given by the command   **<<Graphics`ComplexMap`**

```
CartesianMap[Identity,{0,1.},{0,1.}]
```

which creates the rectangular grid in the u, v plane, see Figure 4.6

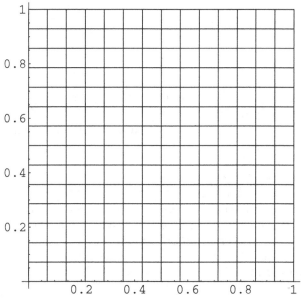

Figure 4.6
Rectangular grid in the u, v plane

To see the mapping of this grid into the x, y plane by the function z^2, we first
expand $z^2 = (x + iy)^2 = x^2 + 2ixy + y^2$, so that $u(x, y) = x^2 - y^2$, $v(x, y) = 2xy$. Now the command
```
RC=ContourPlot[u[x,y],{x,0,1.},{y,0,1.},
ContourShading->False]
```
yields

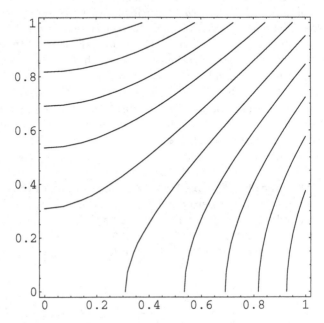

Figure 4.7
Plot of $u(x, y)$ in the x, y plane

Next we plot $v(x, y)$. The command   4.2.8)
```
RV=ContourPlot[v[x,y],{x,0,1.},{y,0,1.},
ContourShading->False]
```
(what happens if you drop, **ContourShading->False**?) and Figure 4.8
has been created.

Now we would like to put these two plots together. This is done very simply
by the command **Show[RC,RV]**   (4.2.9)
which results in Figure 4.9.

In (4.2.9) we learned a new *Mathematica* command how to combine two
Figures. If one wants to suppress the showing of a picture one can write
```
Plot[...DisplayFunction->Identity]
```
To combine several such plots and to show them together one uses
```
Show[Pl1,Pl2...
DisplayFunction->$DisplayFunction].
```

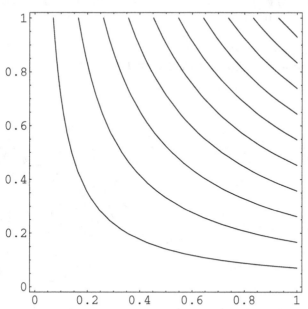

Figure 4.8
Plot of $u(x, y)$ in the x, y plane

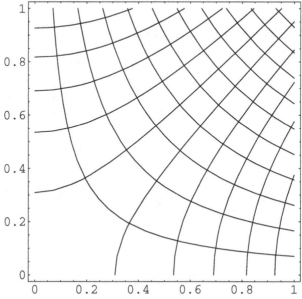

Figure 4.9
Combined plot $z(x, y)$

But would it be possible to create the last figure by a direct command?
CartesianMap has worked to produce the grid of Figure 4.6. The idea
to map z^2 into the cartesian plane u, v would not solve it, we have to use the
inverse function $\sqrt{z}$. We type

```
gun[z_]=Sqrt[z];
CartesianMap[gun,{0,1.},{0,1.}]
```
(4.2.10)

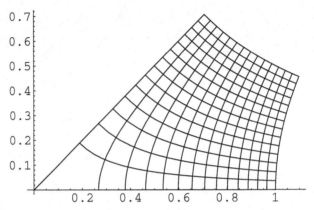

Figure 4.10
CartesianMap of $\sqrt{z}$

The disadvantage of conformal mapping is the fact, that the satisfaction of
given boundary conditions cannot be met: the function $\zeta(z)$ determines the
boundary condition. We need a method to find the mapping function as a
consequence of the given boundary conditions! For boundaries consisting of
straight lines the SCHWARZ-CHRISTOFFEL *transformation* is helpful. It maps
the real axis of the $\zeta(u, v)$ plane into any arbitrary polygon in the $z(x, y)$
plane, the upper half ($v > 0$) of the ζ plane into the interior of the polygon.
The inverse transformation maps the $z(x, y)$ plane polygon into the upper half
of the ζ plane. To understand the situation, we consider the expression $z'(\zeta)$

$$\frac{dz}{d\zeta} = A(\zeta - \zeta_1)^{-\alpha_1}. \qquad (4.2.11)$$

A is a complex constant and ζ_1 is real, the point a on the real u axis. A given
complex number $z = x+iy$ can be represented by $z = re^{i\vartheta}$ in polar coordinates
r, ϑ. Here $r = \sqrt{x^2 + y^2}$ is the magnitude of z and $\vartheta = \arctan(y/x)$ is called
argument arg. Now

$$\arg\left(\frac{dz}{d\zeta}\right) = \begin{cases} \arg A - \alpha_1 \pi \\ \arg A \end{cases} \quad \text{for} \quad \begin{matrix} \zeta < \zeta_1 \\ \zeta > \zeta_1. \end{matrix} \qquad (4.2.12)$$

Moving along the real u axis, $\arg(\zeta - \zeta_1) = \pi$ for $\zeta < \zeta_1$ and $\arg(\zeta - \zeta_1) = 0$
for $\zeta > \zeta_1$. Thus during moving along the u axis the argument of $dz/d\zeta$

discontinuously jumps by the amount $\alpha_1 \pi$ when ζ passes ζ_1. Based on these facts, the setup for the transformation is given by

$$\frac{dz}{d\zeta} = A(\zeta - \zeta_1)^{-\alpha_1}(\zeta - \zeta_2)^{-\alpha_2} \ldots (\zeta - \zeta_n)^{-\alpha_n}. \tag{4.2.13}$$

For a closed polygon the condition

$$\sum_{i=1}^{n} \alpha_i \pi = 2\pi \tag{4.2.14}$$

must be valid. Integration of (4.2.13) yields

$$z = A \int (\zeta - \zeta_1)^{-\alpha_1}(\zeta - \zeta_2)^{-\alpha_2} \ldots (\zeta - \zeta_n)^{-\alpha_n} d\zeta + B. \tag{4.2.15}$$

The complex integration constants allow to rotate (A) and to translate (B) the polygon in the z plane, respectively.

We have discussed boundary value problems on rectangles in earlier sections. What does now conformal mapping offer? Let us consider the mapping onto a rectangle in the z plane. We choose $\zeta_1 = -1/k$, $\zeta_2 = -1$, $\zeta_3 = 1$, $\zeta_4 = 1/k$, $0 \leq k \leq 1$. (4.2.15) yields

$$z = z_0 + A \int \frac{1}{\sqrt{(1 - \zeta^2)(1 - k^2\zeta^2)}} d\zeta = z_0 + A\mathrm{sn}^{-1}(\zeta, k). \tag{4.2.16}$$

The solution is given by the *Mathematica* command
`Integrate[((1-x^2)*(1-k^2*x^2))^(-1./2.),x]` (4.2.17)
This gives the hypergeometric function of two variables

$$\frac{(x \, (1 - x^2)^{0.5} \, (1 - k^2 \, x^2)^{0.5} \, \mathrm{AppellF1}[1/2, 0.5, 0.5, 3/2, x^2, k^2 \, x^2])}{((1 - x^2) \, (1 - k^2 \, x^2))^{0.5}}.$$

How? Remember: between 1 and 1., there is a difference. Repeat
`Integrate[((1-x^2)*(1-k^2*x^2))^(-1/2),x]` (4.2.18)

This yields

$$\frac{\sqrt{1 - x^2} \, \sqrt{1 - k^2 \, x^2} \, \mathrm{EllipticF}[\mathrm{ArcSin}[x], k^2]}{\sqrt{(1 - x^2)(1 - k^2x^2)}}. \tag{4.2.19}$$

But could we cancel the square roots?
`MM=%%`
`Cancel[MM]`
No, this does not work. The *Mathematica* book explains that cancel works for polynomials but not for roots.

What do we know about **`EllipticF`**? Let's ask *Mathematica*.
`?EllipticF` gives
EllipticF[phi, m]. This is the elliptic integral of the first kind that has been discussed in (2.7.15). Can we go to the

?JacobiSN[x,k]

Try

InverseFunction[ArcSin]   (4.2.20)

This results in Sin. O.K. Now:

InverseFunction[EllipticF[φ,m]] gives

InverseFunction[EllipticF[φ,m]]

Apparently this is too difficult. Test to plot these results using

CartesianMap

seems to fail too.

If the boundaries are no longer describable by straight lines, the choice of singularities may help. Let us investigate

$$\zeta(z) = \frac{q}{2\pi} \ln \frac{z+a}{z-a}.$$ (4.2.21)

This function describes a double source (electric charges etc.) of equal intensity q, which are situated at a and $-a$. *Mathematica* helps to clarify the situation. Let us learn by trial and error. We type

<<Graphics`ComplexMap`

and do not forget to hit the two keys Shift and Enter together, if this is the first command given in the *Mathematica* window. We continue by

echo $Packages
fz[z_]=Log[(z+1)/(z-1)];
CartesianMap[fz,{-1.,1.},{-1.,1.}]

Do not write **fz[z]** but write only **fz** in the last command. Otherwise the cartesian mapping does not work. The command given above results in a plot showing (4.2.21). We remember: when we want to plot (4.2.21) we have to use the inverse function of (4.2.21). Perhaps *Mathematica* may help? Type

InverseFunction[fz[z]]

but this does not help.We try

InverseFunction[fz] and we get fz^{-1}

So we find the inverse function ourselves and we type

gz[z_]=(1 + Exp[z])/(Exp[z]-1);
CartesianMap[gz,{-Pi,Pi},{-Pi,Pi}]

which produces Figure 4.11. In this picture we see that the lines $u(x,y) = const$ and $v(x,y) = const$ cross each other in an angle of $90°$: the mapping is *orthogonal* and it preserves the angles. Such analytic transformations are called *conformal*. Taking the limit $a \to 0, q \to \infty, aq \to R^2$ one gets the potential of a dipole

$$\zeta(z) = \frac{R^2}{z}.$$ (4.2.22)

This expression realizes a *transformation* by *reciprocal radii* in the two-dimensional z plane: the domain outside the circle with radius R is mapped onto the interior domain within the circle representing the cut of an infinite long cylinder standing vertically on the z plane. Adding the term $V_0 z$ representing a flow coming from left ($x < 0$) parallel to the x axis, one obtains

$$\zeta(z) = V_0 z + V_0 R^2 / z = V_0 (x + iy) + \frac{V_0 R^2 (x - iy)}{x^2 + y^2}. \qquad (4.2.23)$$

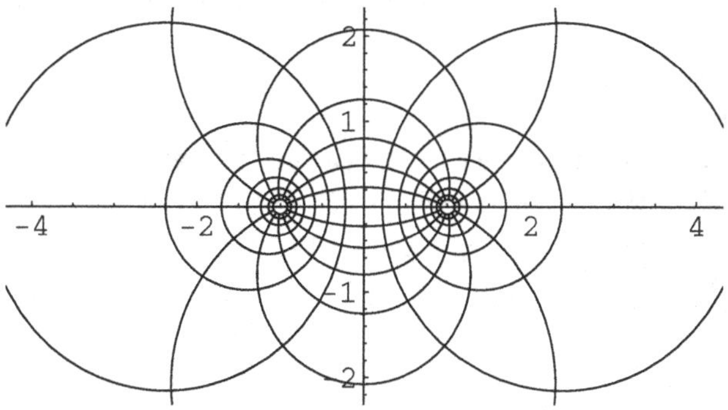

Figure 4.11
Conformal mapping of (4.2.21)

Equation (4.2.23) describes the field of flow around a circular cylinder. In the two points $x = \pm R, y = 0$ we find *stagnation points*, where the streaming velocity is zero, see Figure 4.12. The boundary condition that the velocity component vertical to a rigid wall $x^2 + y^2 = R^2$ must vanish, is satisfied. To see this we pay attention to the fact that $u(x, y) = const$ describes the potential of the flow and the *streamlines* are given by $v(x, y) = const$. Reading off from (4.2.23) we get

$$v(x, y) = V_0 y - \frac{V_0 R^2 y}{x^2 + y^2} = 0, \; u(x, y) = V_0 x + \frac{V_0 R^2 x}{x^2 + y^2}.$$

These functions may be plotted using the command (4.2.8)
ContourPlot

or by conformal mapping using

CartesianMap

For this goal we calculate the inverse function to (4.2.23).

It reads

```
giza1[z_]=z/(2*V0)+Sqrt[z^2/(4*V0^2)-R^2];
```
                                                                    (4.2.24)

In order to plot it, we type

```
<<Graphics`ComplexMap`
echo $Packages
```

to receive

Graphics`ComplexMap`echo,

Utilities`FilterOptions`echo,Global`echo,System`echo

and then

```
Clear[R,V0,P1];
R=1.;V0=1.;
P1=CartesianMap[giza1, {-6.,6.},{-6.,6.}]
```

Executing the last command one receives a strange picture, just the right side $(x > 0)$ of Figure 4.12. This is understandable, because root has two signs. We thus type

```
Clear[R,V0,giza2,P2];
R=1.;V0=1.;
giza2[z_]=z/(2*V0)-Sqrt[z^2/(4*V0^2)-R^2];
P2=CartesianMap[giza2,{-6.,6.},{-6.,6.}]
```

to receive the left side $(x < 0)$ of Figure 4.12. This is now the time to combine the two plots.

We type

```
Clear[PP];
PP=Show[P1,P2,PlotRange->All]
```

and finally we get Figure 4.12.

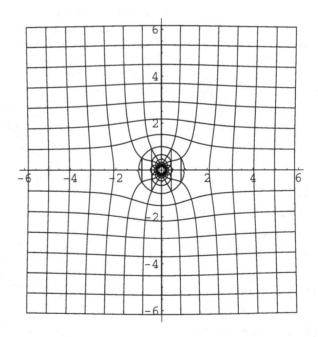

Figure 4.12
Flow around a cylinder. Map of (4.2.23)

Three-dimensional problems (x, y, z) exhibiting axial symmetry around the z axis, see Figure 4.13, may also be solved using conformal mapping. To do this a complex plane $E(\xi, z)$ is introduced that is located perpendicular to the x, y plane of a three-dimensional coordinate system x, y, z. Let α be the angle between the x axis and the new ξ axis. To create Figure 4.13 we give the commands

```
Clear[P1, G1]; P1 = {Text["x", {8., 0}],
    Text["z", {0., 8.}], Text["xi", {6.5, 2.0}];
    Text["E", {2., 2.5}], Text["alpha", {2.5, 0.45}],
    Text["y", {4.8, 3.9}], Line[{{0, 0}, {8.0, 0.},
    {0, 0}, {0, 8.0}, {0, 0}, {6.5, 2.0}}],
    Line[{{0., 4.0}, {4.0, 5.4}, {4.0, 5.4},
    {4.0, 1.2}}], Line[{{0., 0.}, {4.8, 3.9}}]};
G1 = Graphics[P1]; Show[G1]
```

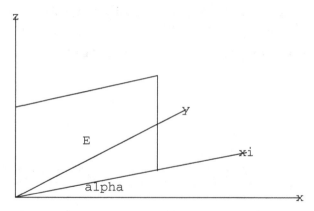

Figure 4.13
Conformal mapping in three-dimensional space

Let us assume that $\varphi(z \pm i\xi)$ is a solution of $\varphi_{zz} + \varphi_{\xi\xi} = 0$ within the plane $E(\xi, z)$. Then rotation around the z axis, described by the angle α may be effectuated by the transformation $\xi = x \cos \alpha + y \sin \alpha$. This delivers a three-dimensional solution $\varphi(z \pm i[x \cos \alpha + y \sin \alpha]) = \varphi(x, y, z; \alpha)$, which solves $\varphi_{xx} + \varphi_{yy} + \varphi_{zz} = 0$; for instance $\varphi(x, y, z; \alpha) = (z + i\xi)^2 = z^2 - x^2 \cos^2 \alpha - y^2 \sin^2 \alpha - 2xy \cos \alpha \sin \alpha + i2z\xi$. Since this expression is valid for arbitrary α, we may integrate over α resulting in

$$\varphi(x, y, z) = \int_0^{2\pi} F(z \pm i\xi) f(\alpha) \mathrm{d}\alpha = \int_0^{2\pi} F(z \pm i[x \cos \alpha + y \sin \alpha]) f(\alpha) \mathrm{d}\alpha. \quad (4.2.25)$$

The three-dimensional boundary conditions may be satisfied by the appropriate choice of $f(\alpha)$ [1.2].

Problems

1. The function (4.2.23) should receive an additional term reading

$$\zeta(z) = V_0 z + V_0 R^2/z + (i\Gamma/2\pi) \cdot \ln z. \qquad (4.2.26)$$

The new term describes *circulation* around the cylinder that means $\Gamma = \oint \vec{w} d\vec{s} \neq 0$, but $\mathrm{curl} \vec{w} = 0$ for the velocity field $\vec{w}$ of the flow parallel to the x axis. It has been shown that this circulation produces a force perpendicular to the direction of the free stream velocity $\vec{w}$. This force is called *lift*. According to the KUTTA-JOUKOWSKY *formula* the lift force in the y direction is given by

$$K_y = \rho \Gamma w_x (x = \infty, y = 0). \qquad (4.2.27)$$

Here ρ is the density of air. Investigate and plot the equation (4.2.26). A JOUKOWSKI *aerofoil* without circulation ($\Gamma = 0$) may be described by

ComplexMapPlot[z+1/z,z,{Circle[{0.25,0},1.25]}]

compare [4].

2. Investigate the so-called KARMAN-TREFFZ *profiles* described by

$$\zeta = \frac{\zeta - la}{\zeta + la} = \left(\frac{z-a}{z+a}\right)^k.$$

Here a and l are geometric parameters and $(2-k)\pi$ describes the angle between the upper and lower part of the airfoil at its end.

3. Two-dimensional, time-independent heat conduction problems satisfy the equation $\Delta T(x,y) = 0$, so that conformal mapping can be used. Let us consider a tube of radius R and temperature T_0 that is buried in a depth h below the surface of the earth ($x = 0$) exhibiting a temperature $T(x = 0, y) = 0$ [4.8]. This is a practical problem for a plumber to avoid the *freezing of water tubes* in cold winter times. Show that the problem can be solved by

$$\zeta = \frac{z-c}{z+c}, \quad c = \sqrt{h^2 - R^2}, \qquad (4.2.28)$$

which maps the upper half plane on-to the circular tube cross section. The solution reads

$$T(x,y) = \frac{T_0}{\ln\left[(h+c)/R\right]} \ln\left\{\frac{\sqrt{(x^2 - c^2 + y^2)^2 + 4c^2 y^2}}{(x-c)^2 + y^2}\right\}. \qquad (4.2.29)$$

4. Solve problem 3 by using a separation setup $T(x,y) = X(x) \cdot Y(y)$.

5. If you have the package [4] investigate and plot

```
ComplexMapPlot[Table[z+z^n,{n,1,20}],z,
{Circle[{0,0},1]},PlotPoints->100];
```

6. Investigate and plot [2]

```
PolarMap[Identity,{0,1.},{0, 2. Pi}]
```

7. Type **<<Graphics`PlotField`**

```
PlotVectorField[{Sin[x],Cos[y]},{x,0,Pi},{y,0,Pi}]
```

see [2]

8. Type the following commands:

```
Clear[psi];v=1.;

psi[x_,y_]=-v*y+v*y/(x^2+y^2)

C1=ContourPlot[psi[x,y],{x,-4.,4.},{y,-4.,4.},
ContourShading->False,
ContourSmoothing->2,
PlotPoints->80,Contours->{0},
DisplayFunction->Identity];

C2=ContourPlot[psi[x,y],{x,-4.,4.},{y,-4.,4.},
ContourShading->False,
ContourSmoothing->2,
PlotPoints->80,Contours->30,
DisplayFunction->Identity];

Show[ContourGraphics[C1],ContourGraphics[C2],
DisplayFunction->$DisplayFunction]
```

Try to understand the commands and what happens. Compare the plot with Figure 4.12.

9. The temperature of hot water produced by a solar collector depends on the boundary conditions taking into account irradiation, heat transfer and heat conduction. A *solar heat collector* consisting of two sheets of thickness $d = 0.005$ m, extended $2a$ in the x direction and infinite in the y direction, is subjected to a solar irradiation $I = 700$ W/m^2 and should deliver $T_R = 30°$C. For a heat transfer coefficient $\alpha = 4$ W/m^{2}K and a heat conductivity $\lambda = 400$ W/mK calculate the temperature $T_{\max}$ of the water heated by the collector for an environment temperature $T_0 = 10°$C. Assume
$\lambda T''(x) - \alpha(T-T_0)/d + I/d = 0$, $T(x=0) = T(2a) = T_R$, $(dT/dx)_{x=0} = 0$,
$T(x) = T_0 + I/\alpha + (T_R - T_0) - I/(\alpha) \cosh \sqrt{\alpha/\lambda d}(a-x)/ \cosh \sqrt{\alpha/\lambda d}$.

4.3   D'ALEMBERT **wave equation and string vibrations**

Elastic waves in bounded bodies are reflected at the body surface. The superposition of the incident wave and the wave reflected by the wall creates a *standing wave*. Its wave length depends on the dimensions of the body and thus on the boundary conditions. A string is a one-dimensional cylindrical body of very small diameter, so that the flexural stiffness may be neglected. The theory of elasticity [1.2] derives for longitudinal oscillations $u(x, t)$ (in the x direction along the cylinder axis) the one-dimensional D'ALEMBERT *wave equation*

$$\frac{\partial^2 u(x,t)}{\partial t^2} = \frac{E}{\rho} \frac{\partial^2 u(x,t)}{\partial x^2} \tag{4.3.1}$$

and for the lateral (transverse) oscillations $v(x, t)$ the equation

$$\frac{\partial^2 v(x,t)}{\partial t^2} = \frac{p}{\rho} \frac{\partial^2 v(x,t)}{\partial x^2}. \tag{4.3.2}$$

Here ρ is the (constant) line density, E the modulus of elasticity and p is the tension applied on the string. These equations are based on the assumption of the validity of the linear HOOKE *law* for small deflections. For large deflections the nonlinear CARRIER *equation* (3.4.55) has to be used. Usually the wave phase speeds $c_l = \sqrt{E/\rho}$ and $c_t = \sqrt{p/\rho}$ are introduced. Using c for c_l and c_t as well as w instead of u and v respectively, we may write

$$\frac{1}{c^2} \frac{\partial^2 w}{\partial t^2} = \frac{\partial^2 w}{\partial x^2}. \tag{4.3.3}$$

This equation has the solution (1.2.2) or

$$w(x,t) = f(x + ct) + g(x - ct). \tag{4.3.4}$$

This general solution of the partial differential equation of *second* order in x has to be adapted to the *two* boundary conditions of a string clamped on both ends ($x = 0, x = l$)

$$w(0,t) = 0, \qquad w(l,t) = 0. \tag{4.3.5}$$

Since there is also the second independent variable t in the partial equation of *second* order with respect to t, we have to choose *two* initial conditions:

$$w(x,0) = h(x), \quad w_t(x,0) = H(x), \tag{4.3.6}$$

where h and H are arbitrary but differentiable functions. Inserting $t = 0$ into (4.3.4) one obtains

$$w(x,0) = f(x) + g(x) = h(x) \tag{4.3.7}$$

and

$$w_t(x,0) = c\left(f'(x) - g'(x)\right) = H(x). \tag{4.3.8}$$

Integration of this equation and adding (4.3.7) results in

$$f(x) = \frac{1}{2}h(x) + \frac{1}{2c}\int H(x)\mathrm{d}x, \tag{4.3.9}$$

while subtraction gives

$$g(x) = \frac{1}{2}h(x) - \frac{1}{2c}\int H(x)\mathrm{d}x. \tag{4.3.10}$$

Now the argument replacement (see (3.3.11)) $x \to x \pm ct$ delivers the solution

$$w(x,t) = \frac{1}{2}\left[h(x + ct) + h(x - ct) + \frac{1}{c}\int_{x-ct}^{x+ct} H(x)\mathrm{d}x\right], \tag{4.3.11}$$

satisfying the initial conditions (4.3.6). To take into account the boundary conditions (4.3.5) we insert $x = 0$ giving

$$2w(0,t) = h(ct) + h(-ct) + \frac{1}{c}\int_{-ct}^{+ct} H(x)\mathrm{d}x = 0. \tag{4.3.12}$$

To satisfy (4.3.12) we now have to make assumptions concerning the arbitrary functions h and H: they must be odd (asymmetric), $h(-ct) = -h(ct)$, $H(-x) = -H(x)$. But this satisfies only the first boundary condition (4.3.5). In order to satisfy $w(l,t) = 0$ we use the STEFAN *trick* to expand the solution function over the boundaries $x < 0$, $x > l$ and $h(l - x) = -h(l + x)$, $H(l - x) = -H(l + x)$.

Another method to solve (4.3.3) is separation using $w(x,t) = X(x) \cdot T(t)$, giving

$$T(t) = A\sin\omega t + B\cos\omega t, \tag{4.3.13}$$

$$X(x) = C\sin\frac{\omega}{c}x + D\cos\frac{\omega}{c}x, \tag{4.3.14}$$

where ω^2 is the separation constant. To satisfy (4.3.5) we obtain $D = 0$, and the asymmetric *eigenfunction*

$$X(x) = C\sin\frac{n\pi x}{l}, \quad n = 1, 2, \ldots \tag{4.3.15}$$

Then the *eigenvalue* is given by

$$\omega_n = \frac{n\pi c}{l}, \tag{4.3.16}$$

and with $C = 1$ the solution to (4.3.3) reads

$$w(x, t) = \sum_{n=0}^{\infty} (A_n \sin \omega_n t + B_n \cos \omega_n t) \sin \frac{n\pi x}{l}. \qquad (4.3.17)$$

To satisfy the initial conditions we have to expand $h(x), H(x)$ with respect to the eigenfunctions:

$$t = 0, \quad w(x, 0) = h(x) = \sum_{n=0}^{\infty} B_n \sin \frac{n\pi x}{l}, \qquad (4.3.18)$$

$$t = 0, \quad w_t(x, 0) = H(x) = \sum_{n=0}^{\infty} A_n \omega_n \sin \frac{n\pi x}{l}. \qquad (4.3.19)$$

For given initial deflection defined by $h(x)$ and $H(x)$, the string oscillations are described by (4.3.17). But how does the solution read if an exterior force $K(x, t)$ excites the string continuously? Apparently (4.3.3) has to be modified:

$$\frac{\partial^2 w}{\partial x^2} - \frac{1}{c^2} \frac{\partial^2 w}{\partial t^2} = K(x, t) = K_0(x) \sin \omega t = \sum_n K_n \sin \frac{\pi n x}{l} \sin \omega t, \quad (4.3.20)$$

where

$$K_n = \frac{2}{l} \int_0^l K_0(x) \sin \frac{\pi n x}{l} dx. \qquad (4.3.21)$$

Since we know that the solution of (4.3.20) is given as a sum of the general solutions of the homogenous equation (4.3.3) and a particular solution of the inhomogeneous equation (4.3.20), we may write for the latter $U(x) \sin \omega t$, where ω is the frequency of the exterior force. We thus have

$$U''(x) + \frac{\omega^2}{c^2} U(x) = K_0(x), \qquad (4.3.22)$$

$$\sum_n \sin \frac{\pi n x}{l} \left(-A_n \frac{\pi^2 n^2}{l^2} + \frac{\omega^2}{c^2} A_n + K_n \right) = 0 \quad \text{or}$$

$$A_n = \frac{c^2 K_n}{\omega_n^2 - \omega^2}. \qquad (4.3.23)$$

Resonance and very large or infinite amplitudes A_n occur for $\omega \sim \omega_n$, if the frequency ω of the exterior force is nearly (or exactly) equal to one of the eigenfrequencies ω_n of the string. If a marching formation of soldiers crosses a bridge, one of the eigenfrequencies of the bridge may be excited and the bridge could collapse. To march in step when crossing a bridge is strictly forbidden for mathematical reasons. Even an army has to accept physical facts. But what about damping the oscillations? We will discuss this possibility in the problems attached to this section. On the other hand, are there time-independent solutions of (4.3.3)? Sure, in new denotation $w(x) \to G(x)$ one has $G''(x) = 0$ and

$$G(x) = ax + b. \tag{4.3.24}$$

In order to be able to satisfy the boundary conditions (4.3.5), we have to accept a discontinuous solution

$$G(x; \xi) = \begin{cases} G_0 x/\xi & \text{for} \quad 0 \le x \le \xi, \\ G_0(x - l)/(\xi - l) & \text{for} \quad \xi \le x \le l. \end{cases} \tag{4.3.25}$$

This strange solution describes a string that is continuously pulled out at $x = \xi$ by a force of strength G_o. This *point force* has all attributes of a GREEN *function*

1. symmetry $G(x, \xi) = G(\xi, x)$,

2. satisfies the pertinent homogeneous equation,

3. its first derivative satisfies the discontinuity condition (2.4.30) and $G'' = \delta(x - \xi)$ (DiracDelta).

But if $G(x, \xi)$ describes a local point force at ξ, then a distributed force of strength $f(\xi)$ and integration could construct another solution by

$$w(x) = \int_0^l G(x, \xi) f(\xi) \mathrm{d}\xi. \tag{4.3.26}$$

For $f(x) = \omega^2 w(x)/c^2$, (4.3.26) creates the *integral equation for the string*

$$w(x) = \frac{\omega^2}{c^2} \int_0^l G(x, \xi) w(\xi) \mathrm{d}\xi. \tag{4.3.27}$$

Here the GREEN function is the kernel of the integral equation, which contains the boundary conditions within the integration limits. The integral equation has the advantage that general theorems may be deduced quite easily. Let $\lambda_\nu = \omega_\nu^2/c^2$, $\lambda_\mu = \omega_\mu^2/c^2$ be the eigenvalues belonging to $w_\nu(x)$ and $w_\mu(x)$ respectively, then one may write down the identity

$$\lambda_\mu \int_0^l w_\mu(\xi) \cdot w_\nu(\xi) \mathrm{d}\xi = \lambda_\mu \int_0^l w_\mu(\xi) \mathrm{d}\xi \cdot \int_0^l \lambda_\nu G(x, \xi) w_\nu(x) \mathrm{d}x. \tag{4.3.28}$$

Exchanging ν and μ, writing a second identity and subtracting them one obtains

$$(\lambda_\mu - \lambda_\nu) \int_0^l w_\nu(\xi) w_\mu(\xi) \mathrm{d}\xi = 0, \tag{4.3.29}$$

which states that the eigenfunctions are *orthogonal.* It is also possible to prove that the eigenvalues are always real or that the kernel may be expanded as a bilinear form of the eigenfunctions:

$$G(x,\xi) = \sum_{\nu} \frac{w_\nu(x)w_\nu(\xi)}{\lambda_\nu} = \frac{2}{l}\sum_{\nu} \frac{\sin\frac{\nu\pi x}{l}\sin\frac{\nu\pi\xi}{l}}{\nu^2\pi^2/l^2}. \qquad (4.3.30)$$

Now the question arises if a similar solution like (4.3.26) exists for the time-dependent solution, too? We will discuss this question in problem 5.

What happens with our solutions in the limit $l \to \infty$? One may expect a continuous spectrum of eigenvalues as will be shown when treating membranes in the next section.

Problems

1. Investigate the damping of resonance oscillations. The pertinent equation is a modification of (4.3.3) and may read

$$\frac{1}{c^2}\frac{\partial^2 w^2}{\partial t^2} = \frac{\partial^2 w}{\partial x^2} - k\frac{\partial w}{\partial t}. \qquad (4.3.31)$$

A setup

$$w(x,t) = \exp(-at)\cos\omega t\sin(x\pi/l) \qquad (4.3.32)$$

satisfies the boundary conditions at $x = 0$ and $x = l$ and seems to be useful. Let us see if *Mathematica* could help.

Type **Clear[w];w[x,t]=Exp[-a*t]*Cos[ω*t]* Sin[Pi*x/l]**

Expand[(D[w[x,t],{t,2}]/c^2-D[w[x,t],{x,2}]+ k*D[w[x,t],t])/w[x,t]]

The result does not help very much. It reads

$$-k\,a + \frac{a^2}{c^2} - \frac{\omega^2}{c^2} - k\,\omega\,\mathrm{Tan}[t\,\omega] + \frac{2\,a\,\omega\,\mathrm{Tan}[t\,\omega]}{c^2} + \frac{\pi^2}{l^2}$$

Now try another way

S1=k*D[w[x,t],t];
S2=D[w[x,t],{t,2}]/c^2;
S3=-D[w[x,t],{x,2}]

which represents exactly the partial differential equation (4.3.31). Then give

Simplify[S1+S2+S3]

to receive

$$\frac{e^{-at}\text{Sin}[\frac{\pi x}{l}]\left(\left(a^2 l^2 - a c^2 k l^2 + c^2 \pi^2 - l^2 \omega^2\right)\text{Cos}[t\,\omega] + \left(2\,a - c^2\,k\right) l^2\,\omega\text{Sin}[t\,\omega]\right)}{c^2\,l^2}$$

Find that expression for a that satisfies the equation (4.3.31).

2. The equation describing damped oscillations excited by an exterior periodic force reads

$$\frac{1}{c^2}\frac{\partial^2 w}{\partial t^2} = \frac{\partial^2 w}{\partial x^2} - k\frac{\partial w}{\partial t} + f(x)\cos\omega t. \tag{4.3.33}$$

Try a solution using the setup

```
wd[x,t]=g[x]*Cos[ω*t]+h[x]*Sin[ω*t];
Simplify[D[wd[x,t],{t,2}]/c^2-D[wd[x,t],{x,2}]+
k*D[wd[x,t],t]-f[x]*Cos[ω*t]]
```

Does the result offer two ordinary differential equations for $g(x)$ and $h(x)$?

3. Prove that (4.3.26) solves $w''(x) + f(x) = 0$. Recall: $G'' = \delta(x - \xi)$.

4. Assume an inhomogeneous string with locally varying tension $p(x)$ and line density $\rho(x)$ so that the wave equation reads

$$\frac{\partial}{\partial x}\left(p(x)\frac{\partial u}{\partial x}\right) = \rho(x)\frac{\partial^2 u}{\partial t^2}. \tag{4.3.34}$$

Try a separation setup [3.13] $u(x,t) = T(t) \cdot X(x)$ and use the boundary conditions:

$$\alpha u(0,t) + \beta\frac{\partial u(0,t)}{\partial x} = 0,$$

$$\gamma u(l,t) + \delta\frac{\partial u(l,t)}{\partial x} = 0, \tag{4.3.35}$$

together with the initial conditions

$$u(x,0) = \varphi_0(x), \qquad \frac{\partial u(x,0)}{\partial t} = \varphi_1(x), \qquad 0 \le x \le l. \tag{4.3.36}$$

The result should read

$$T_n(t) = A_n\cos\lambda t + B_n\sin\lambda t,$$

$$u(x,t) = \sum_n (A_n\cos\lambda_n t + B_n\sin\lambda_n t)X_n(x),$$

where λ_n are the separation constants

$$u(x,0) = \varphi_0(x) = \sum_n A_n X_n(x),$$

$$\partial u(x,0)/\partial t = \varphi_1(x) = \sum_n \lambda_n B_n X_n(x),$$

$$A_n = \int_0^l \rho(x)\varphi_0(x)X_n(x)\mathrm{d}x,$$

$$B_n = (1/\lambda_n)\int_0^l \rho(x)\varphi_1(x)X_n(x)\mathrm{d}x.$$

5. A string of length l is clamped on both ends and at time $t = 0$ is plucked at $x = \xi$ such that $u(\xi, 0) = G_0$. There is no initial velocity, $u_t(\xi, 0) = 0$. Show that the solution of the string equation is given by

$$u(x, t) = \frac{2G_0}{\pi^2}\frac{l^2}{\xi(l-\xi)}\sum_{n=1}^{\infty}\frac{1}{n^2}\sin\frac{n\pi\xi}{l}\sin\frac{n\pi x}{l}\cos\frac{n\pi c}{l}t. \qquad (4.3.37)$$

6. A string of length $2l$ is clamped at $x = -l$ and $x = +l$. At time $t = 0$ it is plucked at $x = 0$ so that $u(x, 0) = 0$, $u_t(x, 0) = v_0$, $|x| < \varepsilon$ and $u_t(x, 0) = 0$, $\varepsilon < |x| \le l$. After taking the limit $\varepsilon \to 0$ the solution should be [4.8]

$$u(x, t) = \frac{4v_0}{\pi c}\sum_{n=0}^{\infty}\frac{1}{2n+1}\cos\left(\frac{2n+1}{2l}\pi x\right)\sin\left(\frac{2n+1}{2l}\pi ct\right). \qquad (4.3.38)$$

4.4   HELMHOLTZ **equation and membrane vibrations**

In physics and engineering a *membrane* is defined as a two-dimensional plane, very thin body, nearly a skin, with negligible bending rigity. For such a body the theory of elasticity derives first a two-dimensional wave equation that becomes a two-dimensional HELMHOLTZ *equation* after separation of the time dependent term $\sim \sin\omega t$. It reads

$$\Delta w(x, y) + \omega^2/c^2 \cdot w(x, y) = 0. \qquad (4.4.1)$$

For a clamped rectangular membrane $a \times b$ the boundary conditions read

$$\begin{aligned} w(x = \pm a/2, y) &= 0, \\ w(x, y = \pm b/2) &= 0 \end{aligned} \qquad (4.4.2)$$

and the solution of the wave equation is then

$$u(x, y, t) = \sum_{m,n}A_{mn}\cos\frac{n\pi x}{a}\cos\frac{m\pi y}{b}\sin(\omega_{mn}t + \delta_{mn}), \qquad (4.4.3)$$

where $n, m = 1, 3, \ldots$. The phase speed c of the transversal oscillation $u(x, y, t)$ is given by $\sqrt{p/\rho}$ and the eigenfrequencies, determined by (4.4.2), are

$$\omega_{mn}^2 = c^2\left(\frac{n^2\pi^2}{a^2} + \frac{m^2\pi^2}{b^2}\right). \qquad (4.4.4)$$

The partial amplitudes A_{mn} have to be determined from the initial conditions $u(x, y, 0) = g(x, y)$ and $u_t(x, y, 0) = 0$. For a dislocated rectangle (left lower corner is situated in the origin $x = 0, y = 0$ with modified boundary conditions) one obtains [4.8]

$$u(x, y, t) = \sum_{m,n=0}^{\infty} A_{mn} \cos\left(\frac{2n+1}{2a}\pi x\right) \cos\left(\frac{2m+1}{2b}\pi y\right) \cdot \cos(\omega_{mn} t),$$

$$A_{mn} = \frac{4}{ab} \int_0^a \int_0^b g(x, y) \cos\left(\frac{2n+1}{2a}\pi x\right) \cos\left(\frac{2m+1}{2b}\pi y\right) dx\, dy,$$

$$\omega_{mn}^2 = c^2\pi^2 \left[\left(\frac{2n+1}{2a}\right)^2 + \left(\frac{2m+1}{2b}\right)^2\right].$$

In the case of inhomogeneous boundary conditions one may either homogenize the conditions or one first solves the boundary problems for (4.4.1) and a superposition of the particular solutions with unknown amplitudes is used to solve the initial value problem.

If the rectangle is specialized into a square ($a = b$) equation (4.4.4) takes the form

$$\omega_{mn}^2 = c^2\pi^2(n^2 + m^2)/a^2. \tag{4.4.5}$$

Apparently these eigenvalues are multiple; two eigenfunctions belong to the same eigenvalue (*degenerate eigenvalue problem*). Lines $\xi(x)$ characterized by $u(x, \xi) = 0$ are called *nodal lines*. Along such nodal lines a membrane may be cut in subdomains. These smaller subdomains are associated with higher frequencies. The lowest eigenvalue is associated with the undivided whole domain (COURANT'S *theorem*) .

Degenerate eigenvalue problems offer the possibility to find nodal lines that are not identical with coordinate curves. Thus the two degenerate eigenfunctions of a square membrane

$$w(x, y) = \sin\frac{2\pi x}{a} \sin\frac{\pi y}{a} + \sin\frac{\pi x}{a} \sin\frac{2\pi y}{a}$$

$$= 2\sin\frac{\pi x}{a} \sin\frac{\pi y}{a} \left(\cos\frac{\pi x}{a} + \cos\frac{\pi y}{a}\right) = 0 \tag{4.4.6}$$

have not only the nodal lines $\sin(\pi x/a) = 0$, $\sin(\pi y/a) = 0$, but also

$$2\cos\frac{\pi}{2a}(x+y) \cos\frac{\pi}{2a}(x-y) = 0, \quad \text{or} \quad x + y = a, \ x - y = a.$$

Thus $x + y = a$ and $x - y = a$ are nodal lines, too. These straight lines are the diagonals of the square, which is now cut into four triangles.

We have seen that the lowest eigenvalue is important, but which membrane has the lowest eigenvalue? Can you hear the shape of a membrane? There has been a 75-year-old discussion. First, in 1923, it was proven [4.1] that for membranes of various shapes, but the same area, surface density and tension, the circular membrane has the lowest eigenvalue (FABER *theorem*). In 1966 KAC [4.2] asked if one can hear the shape of a drum and found the answer NO, but showed that one can hear the connectivity of the drum. On the other hand, in 1989, GORDON [4.3] stated that there are isospectral manifolds and that it is not possible to "hear the shape".

Anyway, we now investigate the eigenfrequencies of a circular membrane. We use the results of problem 4 of section 3.1 for solving (4.4.1) in cylindrical coordinates:

$$w(r, \varphi) = \sum_p A_p J_p \left(\frac{\omega_p}{c} r \right) [a_p \sin p\varphi + b_p \cos p\varphi]. \tag{4.4.7}$$

Inclusion of the time-dependent term yields the solution of the wave equation

$$w(r, \varphi, t) = \sum_p A_p J_p \left(\frac{\omega_p}{c} r \right) \cos p\varphi \sin (\omega t + \delta_p). \tag{4.4.8}$$

Both solutions satisfy the homogeneous boundary conditions $w(r = R, \varphi) = 0$, where R is the radius of the circular membrane, see (2.4.45). The unknown partial amplitudes may be used to satisfy initial conditions of the time-dependent wave equation. If for instance $w(r, \varphi, t = 0) = f(r, \varphi)$ and $w_t(r, \varphi, 0) = h(r, \varphi)$ are given, then also the amplitudes a_p, b_p in the time-dependent term $\sum_p (a_p \cos \omega_p t + b_p \sin \omega_p t)$ can be used to satisfy the initial condition. If inhomogeneous boundary conditions like $g(r, \varphi)$ are given, then it may be useful to start with $w(r, \varphi, t) = v(r, \varphi) \exp(i\omega t)$. Let $v_p(r, \varphi)$ the eigenfunctions of the HELMHOLTZ equation, then $w(r, \varphi, t) = v_p(r, \varphi) + \sum_p (a_p \cos \omega_p t + b_p \sin \omega_p t) \cdot v_p(r, \varphi)$ has to satisfy the initial conditions and one has to determine a and b so that for $t = 0$, $w(r, \varphi, 0) \to v_p(r, \varphi)$ vanishes.

The behaviour of the solution (4.4.7) is very interesting for $r \to R$. In this case the equation (2.4.43) describing BESSEL function becomes $y'' + y = 0$. This means that the BESSEL functions behave like $\sim \sin x$ or $\cos x$ for extremely large x. This recalls the result of problem 3 of section 2.5. Let us consider the infinite circular membrane. In the BESSEL equation (2.4.43) we neglect the term $\sim r^{-2}$ by assuming $p = 0$, but we keep the r^{-1} term. Thus, the equation reads

$$u''(r) + \frac{1}{r} u'(r) + \frac{\omega^2}{c^2} u(r) = 0. \tag{4.4.9}$$

Using $\omega^2/c^2 = k^2$, *Mathematica* solves this equation:
DSolve[u''[r]+u'[r]/r+k^2*u[r]==0,u[r],r]   (4.4.10)
results in
$u \to \text{BesselJ}[0, k\ r]\ C[1] + \text{BesselY}[0, k\ r]\ C[2]$
The NEUMANN *function* (BESSEL function of the second kind, also denoted by N) has a singularity at $r = 0$ and has to be excluded here. For the circular membrane bounded by a finite radius R the eigenvalues k are determined by the zeros of $J_0(kR)$, see (2.4.45). Now for $R \to \infty$, the argument kR tends to

∞, and we obtain an infinite continuous spectrum of eigenvalues. We expect to replace a sum $\sum J_0(kR)$ over the eigenvalues by an integral. We thus set up

$$u(r) = \int_0^\infty A(k)J_0(kr)dk. \qquad (4.4.11)$$

To solve the full problem including the time dependence and assuming symmetry $\partial/\partial\varphi = 0$ we write

$$u(r,t) = \int_0^\infty A(k)J_0(kr)\cos kct\,dk \qquad (4.4.12)$$

and give initial conditions

$$u(r,t=0) = u_0(r), \quad \frac{\partial u(r,t=0)}{\partial t} = 0, \quad 0 < r \le \infty. \qquad (4.4.13)$$

This describes an initial plucking of the membrane with velocity zero distributed over the radius. For $t = 0$ the solution (4.4.12) satisfies (4.4.13) if

$$u_0(r) = \int_0^\infty A(k)J_0(rk)dk. \qquad (4.4.14)$$

In order to determine the expansion coefficients $A(k)$ we write $(k-> \kappa)$

$$\left(\frac{1}{r}\frac{d}{dr}r\frac{d}{dr} + \kappa^2\right)J_0(\kappa r) = 0 \qquad \cdot J_0(\lambda r)$$

$$\left(\frac{1}{r}\frac{d}{dr}r\frac{d}{dr} + \lambda^2\right)J_0(\lambda r) = 0 \qquad \cdot (-J_0(\kappa r)) \qquad (4.4.15)$$

$$\frac{d}{dr}r\left[J_0(\lambda r)\frac{dJ_0(\kappa r)}{dr} - J_0(\kappa r)\frac{dJ_0(\lambda r)}{dr}\right] = \left(\lambda^2 - \kappa^2\right)J_0(\kappa r)J_0(\lambda r)r.$$

This demonstrates that the rhs term may be written as a differential. Multiplication of (4.4.14) by $rJ_0(\lambda r)$ and integration results in

$$\int_0^\infty u_0(r)rJ_0(\lambda r)dr = \int_0^\infty\int_0^\infty A(\kappa)J_0(\kappa r)d\kappa \cdot rJ_0(\lambda r)dr. \qquad (4.4.16)$$

Using (4.4.15) we can write

$$\int_0^\infty u_0(r)rJ_0(\lambda r)dr =$$

$$\lim_{R\to\infty}\int_0^\infty A(\kappa)\frac{R}{\lambda^2 - \kappa^2}\left[J_0(\lambda R)\frac{dJ_0(\kappa R)}{dR} - J_0(\kappa R)\frac{dJ_0(\lambda R)}{dR}\right]d\kappa. \qquad (4.4.17)$$

To evaluate the BESSEL functions for large arguments κR, we use an asymptotic expansion [1.2], [2.1]

$$J_p(x) = \frac{1}{\pi}\int_0^\pi \cos(x\sin\varphi - p\varphi)\,d\varphi, \tag{4.4.18}$$

which we write in the form

$$J_p(\kappa R) = \sqrt{2}\,\frac{\cos[\kappa R - (p+1/2)\pi/2]}{\sqrt{\kappa\pi R}} \tag{4.4.19}$$

so that for $p = 0$ (4.4.17) now reads

$$\int_0^\infty u_0(r)rJ_0(\lambda r)\,dr = \lim_{R\to\infty}\frac{1}{\pi}\int_0^\infty \frac{A(\kappa)}{(\lambda^2 - \kappa^2)\sqrt{\kappa\lambda}}$$

$$\cdot\left[\frac{\sin(\kappa - \lambda)R}{\kappa - \lambda} - \frac{\cos(\kappa + \lambda)R}{\kappa + \lambda}\right]d\kappa \tag{4.4.20}$$

(we used the addition theorem of trigonometric functions). The rhs term describes a resonance at $\kappa = \lambda$. Using the DIRICHLET *integral*

$$\int_0^\infty \frac{\sin z}{z}\,dz = \frac{\pi}{2} \tag{4.4.21}$$

one obtains

$$A(\kappa) = \kappa\int_0^\infty u_0(r)rJ_0(\kappa r)\,dr, \tag{4.4.22}$$

compare (4.4.14) and use it:

$$u_0(r) = \int_0^\infty \kappa\int_0^\infty u_0(r')r'J_0(\kappa r')J_0(\kappa r)\,dr'\,d\kappa. \tag{4.4.23}$$

Inserting (4.4.22) in (4.4.12), one obtains the solution for the infinite membrane

$$u(r,t) = \int_0^\infty \kappa\int_0^\infty u_0(r')r'J_0(\kappa r')\,dr'\,J_0(\kappa r)\cos\kappa ct\,d\kappa. \tag{4.4.24}$$

Waves in a very large lake or surface waves due to the *surface tension* of a liquid layer can be described by (4.4.24).

Solutions of the HELMHOLTZ equation for a boundary not describable by coordinate systems in which the partial differential equation is not easily separable, can only be found by other methods like numerical integration, collocation methods or variational methods, compare the sections of chapter 5.

The HELMHOLTZ equation plays an important role in electromagnetism. Applying the operator curl on $\text{curl}\vec{E} = -\partial\vec{B}/\partial t$ and the operator $\mu\mu_0\partial/\partial t$ on $\text{curl}\vec{H} = \partial\varepsilon\varepsilon_0\vec{E}/\partial t + \vec{j}$ on the MAXWELL *equations* one obtains the *electromagnetic wave equations*:

$$\Delta\vec{E} - \frac{1}{c^2}\frac{\partial^2\vec{E}}{\partial t^2} = \text{grad}\,\frac{\rho}{\varepsilon\varepsilon_0} + \mu\mu_0\frac{\partial\vec{j}}{\partial t}, \qquad (4.4.25)$$

$$\Delta\vec{B} - \frac{1}{c^2}\frac{\partial^2\vec{B}}{\partial t^2} = -\mu\mu_0\,\text{curl}\,\vec{j}. \qquad (4.4.26)$$

Here $\varepsilon_0 = 0.885\,419\cdot10^{-11}$As/Vm, $\mu_0 = 1.256\,637\cdot10^{-6}$Vs/Am and $\text{div}(\varepsilon\varepsilon_0\vec{E}) = \rho$, $\text{div}\vec{H} = 0$ have been used. $c = \sqrt{1/\mu\mu_0\varepsilon\varepsilon_0}$ is the velocity of light within a medium possessing the material parameters ε and μ. For a time dependence $\sim \exp(i\omega t)$ and a domain free of charges ρ and currents $\vec{j}$, the equations result in a vector HELMHOLTZ *equation* (3.1.80). It is not difficult to solve the equations in cartesian coordinates, see (3.1.81) - (3.1.83), but all other coordinates present difficulties, compare (3.1.84). Using the trick $(\Delta\vec{K})_l = \Delta K_l$ helps. As an example we may use cartesian components as functions of spherical coordinates $\vec{K}(r,\vartheta,\varphi) = K_x(r,\vartheta,\varphi)\cdot\vec{e}_x + K_y(r,\vartheta,\varphi)\vec{e}_y + K_z(r,\vartheta,\varphi)\vec{e}_z$

$$\frac{\partial^2 K_x}{\partial r^2} + \frac{2}{r}\frac{\partial K_x}{\partial r} + \frac{\partial^2 K_x}{r^2\partial\vartheta^2} + \frac{\cot\vartheta}{r^2}\frac{\partial K_x}{\partial\vartheta} + \frac{1}{r^2\sin^2\vartheta}\frac{\partial^2 K_x}{\partial\varphi^2} + k^2 K_x = 0 \quad (4.4.27)$$

(and analogous equations for K_y, K_z). As usual, $k = \omega/c$. In engineering problems cylindrical geometry is more important. One is interested in the solutions of (4.4.25) and (4.4.26) either for closed hollow boxes (*resonators*) or in infinitely long hollow cylinders (*wave guides*). For a z-independence $\sim \exp(i\gamma z)$ the MAXWELL equations read in general cylindrical coordinates $q_1, q_2, (q_3 = z)$:

$$\frac{1}{\sqrt{g_{22}}}\frac{\partial B_z}{\partial q_2} - i\gamma B_2 = \frac{i\omega}{c^2}E_1, \qquad (4.4.28)$$

$$i\gamma B_1 - \frac{1}{\sqrt{g_{11}}}\frac{\partial B_z}{\partial q_1} = \frac{i\omega}{c^2}E_2, \qquad (4.4.29)$$

$$\frac{1}{\sqrt{g_{11}}}\frac{\partial B_2}{\partial q_1} - \frac{1}{\sqrt{g_{22}}}\frac{\partial B_1}{\partial q_2} = \frac{i\omega}{c^2}E_z, \qquad (4.4.30)$$

$$\frac{1}{\sqrt{g_{22}}}\frac{\partial E_z}{\partial q_2} - i\gamma E_2 = -i\omega B_1, \qquad (4.4.31)$$

$$i\gamma E_1 - \frac{1}{\sqrt{g_{11}}}\frac{\partial E_z}{\partial q_1} = -i\omega B_2, \qquad (4.4.32)$$

$$\frac{1}{\sqrt{g_{11}}}\frac{\partial E_2}{\partial q_1} - \frac{1}{\sqrt{g_{22}}}\frac{\partial E_1}{\partial q_2} = -i\omega B_z. \qquad (4.4.33)$$

Expressing $E_2(q_1, q_2)$ by $B_1(q_1, q_2)$ and $B_z(q_1, q_2)$ etc., and inserting into (4.4.31), etc., results in the FOCK *equations*:

$$B_1(q_1, q_2) = \frac{i}{\omega^2 - \gamma^2 c^2} \left(\frac{\omega}{\sqrt{g_{22}}} \frac{\partial E_z}{\partial q_2} + \frac{\gamma c^2}{\sqrt{g_{11}}} \frac{\partial B_z}{\partial q_1} \right),$$

$$B_2(q_1, q_2) = \frac{i}{\omega^2 - \gamma^2 c^2} \left(-\frac{\omega}{\sqrt{g_{11}}} \frac{\partial E_z}{\partial q_1} + \frac{\gamma c^2}{\sqrt{g_{22}}} \frac{\partial B_z}{\partial q_2} \right), \qquad (4.4.34)$$

$$E_1(q_1, q_2) = \frac{-ic^2}{\omega^2 - \gamma^2 c^2} \left(\frac{\omega}{\sqrt{g_{22}}} \frac{\partial B_z}{\partial q_2} - \frac{\gamma}{\sqrt{g_{11}}} \frac{\partial E_z}{\partial q_1} \right),$$

$$E_2(q_1, q_2) = \frac{ic^2}{\omega^2 - \gamma^2 c^2} \left(\frac{\omega}{\sqrt{g_{11}}} \frac{\partial B_z}{\partial q_1} + \frac{\gamma}{\sqrt{g_{22}}} \frac{\partial E_z}{\partial q_2} \right), \qquad (4.4.35)$$

where E_z and B_z may be calculated from $(\Delta E)_z = \Delta E_z$, $(\Delta B)_z = \Delta B_z$ and the boundary conditions on a metallic wall $E_t(r = R, \varphi) = E_x = 0, B_n = 0$. For a circular cylinder the tangential component in the z direction is then given by

$$E_z(r, z, \varphi, t) = E_z^0 \exp(i\omega t - i\gamma z) \cdot \cos m\varphi \cdot J_m \left(\sqrt{k^2 - \gamma^2}\, r \right). \qquad (4.4.36)$$

The vanishing of this expression for $r = R$ yields

$$\sqrt{k^2 - \gamma^2}\, R = j_{mn},$$

see (2.4.45). It thus follows from the lowest eigenvalue 2.4 048 that there are no propagating waves with frequencies

$$\omega < c \sqrt{\frac{2.4\,048^2}{R^2} + \gamma^2}. \qquad (4.4.37)$$

This is the so-called *cut-off frequency*.

According to the FOCK equations there are two types of waves in a wave-guide:
1. transverse electric wave (TE *wave*), $E_z(q_1, q_2, z) = 0$, B_n(boundary, z) = 0, B_1, B_2, E_1, and E_2 are described by B_z from the FOCK equations,
2. transverse magnetic wave (TM *wave*), $B_z(q_1, q_2, z) = 0$, E_t(boundary, z) = 0, E_1, E_2, B_1, B_2 are described by E_z . For a wave-guide with rectangular cross section ($a \times b$) the results are

TM   TE

$$E_x = -fi\gamma\frac{m\pi}{a}\cos\frac{m\pi x}{a}\sin\frac{n\pi y}{b}, \qquad E_x = fik\frac{n\pi}{b}\cos\frac{m\pi x}{a}\sin\frac{n\pi y}{b},$$

$$E_y = -fi\gamma\frac{n\pi}{b}\sin\frac{m\pi x}{a}\cos\frac{n\pi y}{b}, \qquad E_y = -fik\frac{m\pi}{a}\sin\frac{m\pi x}{a}\cos\frac{n\pi y}{b},$$

$$E_z = f\pi^2\left(\frac{m^2}{a^2}+\frac{n^2}{b^2}\right)\sin\frac{m\pi x}{a}\sin\frac{n\pi y}{b}, \qquad E_z = 0,$$

$$B_x = fik\frac{n\pi}{b}\sin\frac{m\pi x}{a}\cos\frac{n\pi y}{b}, \qquad B_x = fi\gamma\frac{m\pi}{a}\sin\frac{m\pi x}{a}\cos\frac{n\pi y}{b},$$

$$B_y = -fik\frac{m\pi}{a}\cos\frac{m\pi x}{a}\sin\frac{n\pi y}{b}, \qquad B_y = fi\gamma\frac{n\pi}{b}\cos\frac{m\pi x}{a}\sin\frac{n\pi y}{b},$$

$$B_z = 0, \qquad B_z = f\pi^2\left(\frac{m^2}{a^2}+\frac{n^2}{b^2}\right)\cos\frac{m\pi x}{a}\cos\frac{n\pi y}{b}, \qquad (4.4.38)$$

where

$$f = \exp(i\omega t - i\gamma z), \quad k = \omega/c, \quad \gamma^2 = k^2 - \frac{m^2\pi^2}{a^2} - \frac{n^2\pi^2}{b^2}.$$

The factor $i = \exp(i\pi/2)$ indicates a relative phase shift. For a circular cross section of the waveguide one gets for the TM wave

$$E_r = -i\nu f\frac{j_{mn}}{R}\cos m\varphi\, \mathrm{J}'_m\left(j_{mn}\frac{r}{R}\right), \quad B_r = -ikf\frac{m}{r}\sin m\varphi\, \mathrm{J}_m\left(j_{mn}\frac{r}{R}\right),$$

$$E_\varphi = i\nu f\frac{m}{r}\sin m\varphi\, \mathrm{J}_m\left(j_{mn}\frac{r}{R}\right), \quad B_\varphi = -ikf\frac{j_{mn}}{R}\cos m\varphi\, \mathrm{J}'_m\left(j_{mn}\frac{r}{R}\right),$$

$$E_z = \frac{j_{mn}^2}{R^2}f\cos m\varphi\, \mathrm{J}_m\left(j_{mn}\frac{r}{R}\right), \quad B_z = 0, \qquad (4.4.39)$$

where $\nu^2 = k^2 - j_{mn}^2/R^2$. Again j_{mn} represents the n-th root of the BESSEL function J_m. Using the HERTZ *vector* $\vec{Z}$ one may investigate the emission of radiation from *cellular phones*. It is easy to prove that the setup

$$\vec{E} = -\frac{1}{c^2}\frac{\partial^2\vec{Z}}{\partial t^2} + \mathrm{grad\ div}\,\vec{Z}, \quad \vec{B} = \frac{1}{c^2}\mathrm{curl}\frac{\partial\vec{Z}}{\partial t} \qquad (4.4.40)$$

satisfies $\mathrm{curl}\vec{E} = -\partial\vec{B}/\partial t$. The equation $\mathrm{curl}\vec{B}/\mu\mu_0 = \partial\varepsilon\varepsilon_0\vec{E}/\partial t$ and integration with respect to time yield

$$\Delta\vec{Z} = \frac{1}{c^2}\frac{\partial^2\vec{Z}}{\partial t^2} = -k^2\vec{Z}. \qquad (4.4.41)$$

In the derivation of this equation (3.1.80) has been used. In spherical coordinates (4.4.41) has the coupled form of (3.1.84). In cartesian coordinates it would have the form (4.4.27) and could be solved analogously to (3.1.74). For a HERTZ *dipole* aerial of a cellular phone we make simplifications

$$\frac{\partial}{\partial \varphi} = 0,\ Z_\varphi = 0,\ Z_\vartheta = 0,\ Z \sim \exp(i\omega t) \tag{4.4.42}$$

and the setup

$$Z_r = rw(r, \vartheta). \tag{4.4.43}$$

The solution of (4.4.27) for $K_x \to Z_r$ reads

$$Z_r(r, \varphi) = rw(r, \vartheta) = P_l(\cos\vartheta)R_l(r), \tag{4.4.44}$$

where

$$R_l'' + \frac{2}{r}R_l' + \left(k^2 - \frac{l(l+1)}{r^2}\right)R_l = 0, \tag{4.4.45}$$

compare (2.2.65) and (3.1.76). These functions R_l are called *spherical* BESSEL *functions* or BESSEL functions of fractional order

$$R_l(r) = \frac{1}{\sqrt{r}}J_{l+1/2}(kr). \tag{4.4.46}$$

They are solutions of (2.2.65) and of the type discussed in problem 3 of section 2.5. The J functions describe standing waves. In our problem we are interested in propagating waves. Besides the standing wave solutions of (2.2.65) and of (2.2.48) for BESSEL functions of integer order, there exist propagating solutions, the HANKEL *functions* (BESSEL functions of third kind). They are defined by

$$H_\nu^{(1)}(z) = J_\nu(z) + iY_\nu(z),$$

$$H_\nu^{(1)}(z) = J_\nu(z) - iY_\nu(z). \tag{4.4.47}$$

For fractional order they degenerate into the R_l whereas the $J_{l+1/2}$ are related to standing waves like sin and cos, the R_l are related to propagating waves $\exp(\pm ikr)$. The $R_l(r)$ can be described by the RODRIGUEZ *formulas*

$$R_l(r) = r^l \left(\frac{1}{r}\frac{d}{dr}\right)^l \frac{\exp(\pm ikr)}{r}, \tag{4.4.48}$$

so that

$$R_0(r) = \frac{\exp(-ikr)}{r},$$

$$R_1(r) = R_0'(r) = \frac{\exp(-ikr)}{r}\left(ik - \frac{1}{r}\right),$$

$$R_2(r) = R_1'(r) - \frac{1}{r}R_1(r) = \frac{\exp(-ikr)}{r}\left(-k^2 - \frac{3ik}{r} + \frac{3}{r^2}\right) \tag{4.4.49}$$

with the addition theorems

$$R_l'(r) = R_{l+1}(r) + \frac{l}{r}R_l(r).\tag{4.4.50}$$

Due to $m = 0, P_0 = 1, R_0$ describes a central symmetric spherical wave, $l = 1$ describes a dipole radiation emission and $l = 2$ a quadrupole aerial. Inserting

$$Z_r = rw = \exp(-ikr)(ik - 1/r)\cos\vartheta\tag{4.4.51}$$

into (4.4.40) obtains

$$E_r = fE_r^0\left(\frac{1}{r^3} + \frac{ik}{r^2}\right)\cos\vartheta,$$

$$E_\vartheta = f\frac{E_r^0}{2}\left(\frac{1}{r^3} + \frac{ik}{r^2} - \frac{k^2}{r}\right)\sin\vartheta,$$

$$B_\varphi = fB_\varphi^0\left(\frac{1}{r^3} + \frac{ik}{r}\right)\sin\vartheta,\tag{4.4.52}$$

where $f = \exp(i\omega t - ikr)$ and E_r^0, B_φ^0 are partial amplitudes (integration constants). From (4.4.52) one may find that the radiation energy in the GHz range deposited in the head of a user ($r \to 0$) increases with decreasing distance r from the *cellular telephone*.

For a dipole aerial situated at the surface of the earth $w(R, \vartheta)$ is given by [4.5]

$$w_{pr}(r, \vartheta) = \sum_l C_l\sqrt{\frac{\pi}{2kr}}H_{l+1/2}^{(2)}(kr)P_l(\cos\vartheta), \quad r > R.\tag{4.4.53}$$

For $\sim \exp(+i\omega t)$ $H^{(2)}$ describes an outgoing wave and has to satisfy the boundary condition in infinity

$$\lim_{r\to\infty}\left(\frac{\partial w}{\partial r} + ikw\right) = 0.\tag{4.4.54}$$

In the interior of the earth a standing wave is created so that at the surface $r = R$ the boundary condition

$$\sum_l C_l\sqrt{\frac{\pi}{2kR}}H_{l+1/2}^{(2)}(kR)P_l(\cos\vartheta) = \sum_l D_l\sqrt{\frac{\pi}{2kR}}J_{l+1/2}^{(2)}(kR)P_l(\cos\vartheta)$$

$$\tag{4.4.55}$$

must be satisfied due to continuity: some of the series expansions of type (4.4.53) have very slow convergence for $kR > 1000$ so that it is necessary to transform the series into a complex integral (WATSON *transformation*) [4.5]. This represents another example of an integral representation of the solution of a partial differential equation, see also (4.4.24).

Problems

1. Find the nodal lines and boundary shape by plotting $y(x)$ from the degenerate eigenfunctions

$$A_{mn} \sin \frac{m\pi x}{a} \sin \frac{n\pi y}{a} + A_{nm} \sin \frac{n\pi x}{a} \sin \frac{m\pi y}{a} = 0 \qquad (4.4.56)$$

for $m = 1, n = 3, A_{31} = \pm A_{13}, a = 1$. Use $\sin 3\alpha = \sin\alpha\cos 2\alpha + \cos\alpha\sin 2\alpha$, $\sin 2\alpha = 2\sin\alpha\cos\alpha$, $2\cos^2\alpha = 1 + \cos 2\alpha$ to find $y(x) = \arccos(f(x))$. Play with various values of A_{31}.

2. Find the time-dependent asymmetric solution for a circular membrane of radius R that is plucked at its center at time $t = 0$. The initial conditions will then be

$$u(r, \varphi, t = 0) = \frac{h(R - r)}{R}, \quad \frac{\partial u(r, \varphi, t = 0)}{\partial t} = 0. \qquad (4.4.57)$$

The eigenfrequencies $\omega_{\nu p} = j_{\nu p}^2 c/R$ and the $j_{\nu p}$ are determined by the boundary condition (2.4.45). Start with

$$u(r, \varphi, t) = \sum_{\nu, p} A_{\nu p} J_p \left(\frac{\omega_{\nu p}}{c} r\right) \cos p\varphi \sin (\omega_{\nu p} t + \delta_{\nu p}) . \qquad (4.4.58)$$

The initial conditions result in $\delta_{\nu p} = \pi/2$. Multiplication by the orthogonal $w_{\mu l}$ and integration yield

$$\int_0^{2\pi}\int_0^R \frac{h(R - r)}{R} w_{\mu l} r\,dr\,d\varphi = \sum_{\nu, p} A_{\nu p} \int_0^{2\pi}\int_0^R w_{\nu p} w_{\mu l} r\,dr\,d\varphi . \qquad (4.4.59)$$

Due to (2.4.46) and $\int_0^{2\pi} \cos^2 p\varphi\, d\varphi = \pi$ one obtains the expansion coefficients $A_{\nu p}$ from

$$\int_0^{2\pi}\int_0^R \frac{h(R - r)}{R} J_p \left(\omega_{\nu p} \frac{r}{c}\right) r \cos(p\varphi) dr\,d\varphi = A_{\nu p} \frac{\pi}{2} J_{p+1}^2 \left(\frac{\omega_{\nu p}}{c} R\right) . \qquad (4.4.60)$$

3. Exterior continuous and periodic excitations K can be described by

$$\Delta u = \frac{1}{c^2} \frac{\partial^2 u}{\partial t^2} + K(x, y) \sin \omega t. \qquad (4.4.61)$$

Show that the particular solution

$$u(x, y, t) = \sin \omega t \sum_{\nu, p} w_{\nu p}(x, y) A_{\nu p},$$

$$K(x, y) = \sum_{\nu, p} K_{\nu p} w_{\nu p}(x, y), \qquad (4.4.62)$$

where $w_{\nu p}$ satisfy the homogeneous equation, results in the resonance condition

$$A_{\nu p} = \frac{c^2 K_{\nu p}}{\omega^2 - \omega_{\nu p}^2}. \tag{4.4.63}$$

If the whole circular membrane is excited by $K(t) = K_0 \sin(\omega t + \varphi)$, the solution is given [4.8], [4.2] by

$$u(r,t) = -\frac{K_0}{\omega^2}\left[1 - \frac{J_0(\omega r/c)}{J_0(\omega R/c)}\right]\sin(\omega t + \delta). \tag{4.4.64}$$

4. An elliptic membrane of eccentricity e and semiaxes a, b will be described by the elliptic cylindrical coordinate system, see problem 2 of section 3.1. The HELMHOLTZ equation may be separated into two ordinary differential equations. Derive these equations using *Mathematica*. The result should be

$$X'' - \left(p^2 + e^2 k^2 \cosh^2 u\right) X = 0,$$

$$Y'' + \left(p^2 - e^2 k^2 \cos^2 v\right) Y = 0, \tag{4.4.65}$$

where $\omega^2/c^2 = k^2$ and p is the separation constant. Let *Mathematica* solve these equations. The result should be given by MATHIEU functions

$$X(u) = \text{ce}\left[\frac{1}{2}(e^2\, k^2 + 2\, p^2), -\frac{1}{4}e^2\, k^2, -iu\right] +$$

$$\text{se}\left[\frac{1}{2}(e^2\, k^2 + 2\, p^2), -\frac{1}{4}e^2\, k^2, -iu\right],$$

$$Y(v) = \text{ce}\left[\frac{1}{2}(e^2\, k^2 + 2\, p^2), -\frac{1}{4}e^2\, k^2, v\right] +$$

$$\text{se}\left[\frac{1}{2}(e^2\, k^2 + 2\, p^2), -\frac{1}{4}e^2\, k^2, -v\right]. \tag{4.4.66}$$

The stability of these solutions had been discussed in problem 3 in section 2.6. For more general ordinary differential equations like (2.4.1) with $f(x) = 0$ and periodic coefficients $p_1(x), p_2(x)$ the FLOQUET theorem guarantees the existence of a generalized periodic solution $y(x+\tau) = sy(x)$, where s is the FLOQUET exponent [1.2].

5. The equation (4.4.9) had been solved using *Mathematica*. Try to solve

$$F''(r) + \frac{1}{r}F'(r) - k^2 F(r) = 0, \tag{4.4.67}$$

using the LIE *series method* [4.4]. Since

$$\frac{\mathrm{d}F}{\mathrm{d}r} = Z, \quad \frac{\mathrm{d}Z}{\mathrm{d}r} = -\frac{1}{r} - Z + k^2 F$$

one finds

$$F(r) = Z_1(r), Z = Z_2(r) = \vartheta_1(Z_0, Z_1, Z_2), \ r = Z_0, \ \vartheta_2(r) = -\frac{1}{r}Z + k^2 F,$$

$$\frac{dZ_0}{dr} = \vartheta_0 = 1, \ \frac{dZ_1}{dr} = \vartheta_1(Z_0, Z_1, Z_2), \ \frac{dZ_2}{dr} = \vartheta_2(Z_0, Z_1, Z_2).$$

With the initial conditions $r = 0$, $Z_0(0) = 0$, $Z_1(0) = x_1$, $Z_2(0) = x_2$ the solution is

$$Z_i(r) = \sum_{\nu=0}^{\infty} \frac{r^\nu}{\nu!} (D^\nu Z_i)_{r=0}, \tag{4.4.68}$$

where

$$D = \sum_{\nu=0}^{\infty} \vartheta_\nu (Z_0, Z_1, \ldots Z_n) \frac{\partial}{\partial Z_\nu}. \tag{4.4.69}$$

The reader might ask why this complicated sorcery since the BESSEL functions as solutions for (4.4.9) are very well known and tabulated numerically. There are, however, nuclear engineering problems exhibiting very large arguments (~ 50) so that even large mainframes have problems with the numerical calculations. Now the LIE *series method* offers a way to avoid the BESSEL function by splitting the problem into

$$F_{\text{cyl}} = F_{\text{plane}} - F_{\text{correction}}. \tag{4.4.70}$$

The functions ϑ_ν must be holomorphic (no singularities). Now ϑ_2 has a pole at $r = 1$, so that a transformation is necessary. Let $d_\mu = r_\mu - r_{\mu-1}$ the thickness of the μ-th radial domain, one may put $\xi = r - r_{\mu-1}$, where ξ and r are variables and $r_{\mu-1}$ is the radial distance of the μ-th subdomain. Then

$$\frac{dZ_0}{d\xi} = 1 = \vartheta_0,$$

$$\frac{dZ_1}{d\xi} = Z_2 = \vartheta_1,$$

$$\frac{dZ_2}{d\xi} = \frac{Z_2}{r_{\mu-1} + \xi} + k^2 Z_1 = \vartheta_2 \tag{4.4.71}$$

and the new operator reads

$$D = \frac{\partial}{\partial Z_0} + Z_2 \frac{\partial}{\partial Z_1} + \left\{ k^2 Z_1 - \frac{Z_2}{r_{\mu-1} + Z_0} \right\} \frac{\partial}{\partial Z_2}. \tag{4.4.72}$$

The initial conditions are now $F_0 = F(\xi = 0), F'(0) = dF(\xi = 0)/d\xi$, so that the solution reads now

$$Z_1 \equiv F(\xi) = F_0 \cosh k\xi + \frac{F_0'}{k} \sinh k\xi - \sum_{\nu=0}^{\infty} \xi^\nu \frac{f_\nu}{\nu!}$$

$$= F_{\text{plane}} \qquad\qquad - F_{\text{corr}}. \tag{4.4.73}$$

Here $f_\nu(Z_0, Z_1, Z_2)$ is evaluated for $Z_0 = \xi = 0, Z_1 = F_0, Z_2 = F'_0$ and

$$f_\nu = k^2 f_{\nu-2} + D^{\nu-2}\left[\frac{Z_2}{r_{\mu-1} + Z_0}\right]. \tag{4.4.74}$$

The operations can be calculated by recursion

$$D^k\left[\frac{Z_2}{r_{\mu-1} + Z_0}\right] = \frac{(l-2)k^2}{r_{\mu-1} + Z_0}D^{l-3}\left[\frac{Z_2}{r_{\mu-1} + Z_2}\right] \tag{4.4.75}$$

$$+ k^2 D^{l-2}\left[\frac{Z_2}{r_{\mu-1} + Z_0}\right] + \frac{l+1}{r_{\mu-1} + Z_0}D^{l-1}\left[\frac{Z_2}{r_{\mu-1} + Z_0}\right].$$

Since the analytic solution is known, one can compare. For $F_0 = F'_o = 1, k = 0.1, r_{\mu-1} = 100, d = 20$ one has

$$F_{\text{correction}} = F_{\text{plane}} - F_{\text{analytic}} =$$

$$= \cosh 2 + \frac{\sinh 2}{0.1} - 10K_1(10)I_0(12) + I_1(10)K_0(12) +$$

$$+ 100(K_0(10)I_0(12) - I_0(10)K_0(12) = 3.3\,667,$$

whereas the LIE solution up to D^4 gives 3.3 664, I and K are the *modified* BESSEL *function*, see problem 5 in section 2.2.

6. Membranes with varying surface mass density $\rho(x,y)$ are described by [4.6]

$$u_{xx}(x,y) + u_{yy}(x,y) + \frac{\omega^2}{E}\rho(x,y)u(x,y) = 0. \tag{4.4.76}$$

For special density distributions like

$$\rho(x,y) = f(x) + g(x),$$

separation of (4.4.76) into ordinary differential equations is now possible. Using $\rho(x,y)$ we obtain

$$X'' + X\left(\frac{\omega^2}{E}f(x) - \beta^2\right) = 0, \tag{4.4.77}$$

$$Y'' + Y\left(\frac{\omega^2}{E}g(y) + \beta^2\right) = 0, \tag{4.4.78}$$

where β^2 is the separation constant. To simplify matters we may choose $g(y) = 1$ and $f(x) = bx + 1$. For a square membrane of length a the boundary conditions may read $X(0) = 0, X(a) = 0, Y(0) = 0, Y(a) = 0$. Then

$$Y(y) = A\sin\left(\sqrt{\frac{\omega_n^2}{E} + \beta^2}\,\frac{y}{a}\right) \quad \text{and} \quad \sqrt{\frac{\omega_n^2}{E} + \beta^2} = n\pi$$

determines the eigenfrequencies ω_n, if β were known. On the other hand, one can reason that ω is not yet known and that the boundary condition determines $\beta(\omega)$. Then $\beta = \sqrt{n^2\pi^2 - \omega^2/E}$. Insertion into (4.4.78) yields

$$X'' + X\left(\frac{\omega^2 b}{E}x - n^2\pi^2\right) = 0.$$

Solve this equation using *Mathematica*. You should get AIRY *functions* as a result. Determine the two integration constants C[1] and C[2] using the boundary conditions for X. You will get two homogeneous linear equations for C[1] and C[2] and the vanishing of the determinant yields ω. The transcendental equation for ω should read

$$\mathrm{AiryAi}\left[\frac{n^2\pi^2}{\left(-\dfrac{b\omega^2}{E}\right)^{2/3}}\right]\mathrm{AiryBi}\left[\frac{n^2\pi^2 - \dfrac{ab\omega^2}{E}}{\left(-\dfrac{b\omega^2}{E}\right)^{2/3}}\right]$$

$$-\mathrm{AiryBi}\left[\frac{n^2\pi^2}{\left(-\dfrac{b\omega^2}{E}\right)^{2/3}}\right]\mathrm{AiryBi}\left[\frac{n^2\pi^2 - \dfrac{ab\omega^2}{E}}{\left(-\dfrac{b\omega^2}{E}\right)^{2/3}}\right] = 0.$$

For a circular membrane of radius R one might choose

$$\rho(r) = \rho_0\left[(R^2 - r^2) + 1\right].$$

Solve the equation and find the eigenfrequency ω for the special case $E = 1, R = 1, \rho_0 = 1$.

4.5 Rods and the plate equation

Whereas strings and membranes are thin bodies with negligible rigidity, rods, beams and plates are rigid. Their elastic characteristics are described by three parameters: YOUNG'S *modulus* E of elasticity, POISSON'S *ratio* μ, and the *shear modulus* G is described by

$$G = E/2(1 + \mu). \tag{4.5.1}$$

In the SI-system of units one has E [kg m^{-1} s^{-2}]. Using the boundary conditions on the surfaces of the bodies, variational and other methods derive the equations of motions for various types of oscillations [1.2], [4.7], [4.9], [4.10]. These might be:

1. longitudinal (dilational) oscillations of a rod in the x direction

$$\frac{\partial^2 u(x,t)}{\partial t^2} = \frac{E}{\rho}\frac{\partial^2 u(x,t)}{\partial x^2} + \frac{1}{q}f(x,t), \qquad (4.5.2)$$

(q cross section, ρ density, f exterior force)

2. transversal (bending) vibrations of a rod

$$\frac{\partial^2 u(x,t)}{\partial t^2} = \frac{EI}{\rho q}\frac{\partial^4 u(x,t)}{\partial x^4} + \frac{1}{\rho q}f(x,t), \qquad (4.5.3)$$

where

$$I = I_y = \int\limits_q \int y^2 \mathrm{d}f \quad \text{or} \quad I_z = \int\limits_q \int z^2 \mathrm{d}f \qquad (4.5.4)$$

is the *bending moment* orthogonal to the x axis. For a beam of rectangular cross section $a \times b$, one has

$$I_y = \int\limits_{-a/2}^{+a/2} \int\limits_{-b/2}^{+b/2} y^2 \mathrm{d}z\mathrm{d}y = \frac{a^3 b}{12}, \qquad I_z = \frac{ab^3}{12}, \qquad (4.5.5)$$

3. torsional vibrations of a rod around the x axis

$$\frac{\partial^2 u(x,t)}{\partial t^2} = \frac{G}{\rho}\frac{\partial^2 u}{\partial x^2}, \qquad (4.5.6)$$

4. bending vibrations of a plate of thickness $2h$

$$\frac{\partial^2 u(x,y,t)}{\partial t^2} = -\frac{Eh^2}{3\rho(1-\mu^2)}\Delta\Delta u(x,y,t) \qquad (4.5.7)$$

(*plate equation*). Using a time dependence $\sim \exp(i\omega t)$, from case to case the parameter $k^4 = 3\rho(1-\mu^2)\omega^2/Eh^2$ is introduced, so that (4.5.7) may be written $\Delta\Delta u - k^4 u = 0$.

To be able to solve all these equations we need boundary conditions. The end of a massive body may be free of forces, may be supported by another body or may be clamped. For a rod we may have the *longitudinal* oscillations:

a. free end at $x = l$

$$u_x(l,t) = 0, \qquad (4.5.8)$$

b. clamped end at $x = 0$

$$u(0,t) = 0. \qquad (4.5.8a)$$

For *transversal* oscillations one assumes

a. clamped end $x = 0$

$$u(0,t) = 0, \qquad u_x(0,t) = 0, \qquad (4.5.9)$$

b. free end $x = l$

$$u_{xx}(l,t) = 0, \qquad u_{xxx}(l,t) = 0, \qquad (4.5.10)$$

c. supported end $x = l$

$$u(l,t) = 0, \qquad u_{xx}(l,t) = 0, \qquad (4.5.11)$$

d. free end with a fixed mass

$$mu_{tt}(l,t) = EIu_{xxx}(l,t). \qquad (4.5.12)$$

Since the equations contain derivatives with respect to time t, we need initial conditions too. They may be

$$u(x,0) = h(x), \qquad u_t(x,0) = g(x). \qquad (4.5.13)$$

Torsional oscillations of a rod with one end $x = 0$ clamped $u(0,t) = 0$ and free at the other end $x = l$ are described by

$$u_x(l,t) = 0. \qquad (4.5.14)$$

But if at the end $x = l$ a body with moment of inertia Θ is fixed, then

$$GIu_x(l,t) = -\Theta u_{tt}(l,t) \qquad (4.5.15)$$

is valid. The more complicated boundary conditions for a plate will be discussed later.

Let us now discuss *longitudinal oscillations* of a rod. Using $E/\rho = c^2$, equation (4.5.2) gets the same form as the equation for a string (4.3.20). If both ends of the rod are clamped, we may immediately take over all solutions of (4.3.20) for the longitudinal oscillations of the rod. The same is valid for *torsional vibrations* for $c^2 = G/\rho$, compare (4.5.6) and (4.3.3). However for the *free bending vibrations* of the rod described by

$$\frac{1}{c^2}\frac{\partial^2 u}{\partial t^2} + \frac{\partial^4 u}{\partial x^4} = 0, \qquad c^2 = \frac{EI_y}{\rho q}, \qquad (4.5.16)$$

where c has the dimension m^2/s and is *not* a phase velocity, but its second power. We now have for the first time a partial differential equation of fourth order. Separating time dependence by $u(x,t) = X(x) \cdot (A\sin\omega t + B\cos\omega t)$, see section 4.3, we obtain with $\omega/c = k$

$$X''''(x) - k^2 X(x) = 0. \qquad (4.5.17)$$

Since this equation is homogeneous and has constant coefficients we may use the ansatz $X(x) = \exp(\sqrt{\omega/c}\,x)$ to get the general solution

$$X(x) = \ A\left(\cos\sqrt{k}\,x + \cosh\sqrt{k}\,x\right)$$

$$+B\left(\cos\sqrt{k}\,x - \cosh\sqrt{k}\,x\right)$$

$$+C\left(\sin\sqrt{k}\,x + \sinh\sqrt{k}\,x\right)$$

$$+D\left(\sin\sqrt{k}\,x - \sinh\sqrt{k}\,x\right). \qquad (4.5.18)$$

Taking into account the two boundary conditions for the clamped end $x = 0$
one gets $A = 0, C = 0$ and for the clamped end $x = l$

$$X(l) = B \left(\cos \sqrt{k}\, l - \cosh \sqrt{k}\, l \right)$$

$$+ D \left(\sin \sqrt{k}\, l - \sinh \sqrt{k}\, l \right) = 0, \qquad (4.5.19)$$

$$X'(l) = - kB \left(\sin \sqrt{k}\, l + \sinh \sqrt{k}\, l \right)$$

$$+ kD \left(\cos \sqrt{k}\, l - \cosh \sqrt{k}\, l \right) = 0. \qquad (4.5.20)$$

These two homogeneous linear equations for the two unknowns B and D
possess only then a nontrivial solution if the determinant of the coefficients
vanishes. This condition delivers the *eigenvalue equation*

$$\cos \sqrt{k}\, l \cdot \cosh \sqrt{k}\, l - 1 = 0, \qquad (4.5.21)$$

which has the zeros at $\sqrt{k}\, l = 4.73, 7.85$. If the rod is clamped at only one
end, but the other being supported, then

$$\tanh \sqrt{k}\, l - \tan \sqrt{k}\, l = 0 \qquad (4.5.22)$$

yields the eigenvalue. If both ends are supported and no forces applied on both
ends, then $\omega = \pi^2 c/l^2 = \pi^2 \sqrt{EI/\rho q}\, / l^2$. In order to solve the *inhomogeneous
equation* (4.5.3) with the initial conditions (4.5.13), one expands the solution
with respect to the eigenfunctions of the homogeneous problem $u(x,t) = \sum_n T_n(t) \cdot X_n(x)$. Insertion into (4.5.3) and replacement of X'''' from (4.5.17)
results in the linear inhomogeneous ordinary differential equation

$$\sum_{n=1}^{\infty} \left(T_n''(t) + \omega_n^2 T_n(t) \right) X_n(x) = f(x,t)/\rho q. \qquad (4.5.23)$$

Multiplication by the orthogonal $X_m(x)$ and integration from $x = 0$ to $x = l$
gives

$$T_n''(t) + \omega_n^2 T_n(t) = \frac{\int_0^l f(x,t) X_n(x)\mathrm{d}x}{\rho q \int_0^l X_n^2(x)\mathrm{d}x} = b_n(t). \qquad (4.5.24)$$

The initial conditions (4.5.13) are then satisfied by

$$u(x,0) = h(x) = \sum_{n=1}^{\infty} T_n(0) \cdot X_n(x),$$

$$u_t(x,0) = g(x) = \sum_{n=1}^{\infty} T_n'(0) \cdot X_n(x), \qquad (4.5.25)$$

so that

$$T_n(0) = \frac{\int_0^l h(x)X_n(x)\mathrm{d}x}{\int_0^l X_n^2(x)\mathrm{d}x}, \qquad T_n'(0) = \frac{\int_0^l g(x)X_n(x)\mathrm{d}x}{\int_0^l X_n^2(x)\mathrm{d}x}. \qquad (4.5.26)$$

Then the general solution becomes

$$T_n(t) = T_n(0)\cos\omega_n t + \frac{T_n'(0)}{\omega_n}\sin\omega_n t + \frac{1}{\omega_n}\int_0^t b_n(\tau)\sin\omega_n(t-\tau)\mathrm{d}\tau \quad (4.5.27)$$

and the solution of the inhomogeneous partial differential equation is given by

$$u(x,t) = \sum_{n=1}^{\infty} \frac{X_n(x)}{\int_0^l X_n^2(x)\mathrm{d}x}\left[\cos\omega_n t\int_0^l h(\xi)X_n(\xi)\mathrm{d}\xi + \frac{1}{\omega_n}\sin\omega_n t\int_0^l g(\xi)X_n(\xi)\mathrm{d}\xi\right]$$

$$+\frac{1}{\rho q}\sum_{n=1}^{\infty}\frac{X_n(x)}{\omega_n\int_0^l X_n^2(x)\mathrm{d}x}\int_0^t\int_0^l f(\xi,\tau)X_n(\xi)\sin\omega_n(t-\tau)\mathrm{d}\xi\mathrm{d}\tau. \qquad (4.5.28)$$

The first term describes the free oscillations initiated by the initial conditions, the second term constitutes the oscillation excited by the exterior force f. The $X_n(x)$ are the solutions of (4.5.17), satisfying the boundary conditions. If there are no exterior forces so that equation (4.5.16) has to be solved, we can assume the boundary conditions $u(0,t) = 0$, $u(l,t) = 0$ (two clamped ends) and the initial conditions (4.5.13). Then the solution reads [4.8]

$$u(x,t) = \frac{2}{l}\sum_{n=1}^{\infty}\left[\cos\left(\frac{n^2\pi^2 ct}{l^2}\right)\int_0^l h(\xi)\sin\left(\frac{n\pi\xi}{l}\right)\mathrm{d}\xi\right.$$

$$\left. +\frac{l^2}{n^2\pi^2 c}\sin\left(\frac{n^2\pi^2 ct}{l^2}\right)\int_0^l g(\xi)\sin\left(\frac{n\pi\xi}{l}\right)\mathrm{d}\xi\right]\sin\frac{n\pi x}{l}. \qquad (4.5.29)$$

Longitudinal oscillations of plates (within the plane of the plate) are of interest if a plate rotates. In [4.7] the deformation of a circular plate of radius R and density ρ rotating with the angular frequency ω has been derived:

$$u(r) = \frac{\rho\omega^2(1-\mu^2)}{8E}r\left(\frac{3+\mu}{1+\mu}R^2 - r^2\right). \qquad (4.5.30)$$

Transversal vibrations of plates are described by the *plate equation* (4.5.7): the boundary conditions for a plate are quite complicated. Again three situations have to be considered.

1. The boundary of the plate is clamped. No motion whatever is possible. Then

$$u(\text{boundary}) = 0 \qquad (4.5.31)$$

and

$$\frac{\partial u}{\partial \vec{n}}(\text{boundary}) = 0 \qquad (4.5.32)$$

are valid. $\vec{n}$ is the normal unit vector on the boundary curve.

2. If the boundary is simply incumbent on a support, then one has

$$u(\text{boundary}) = 0, \qquad (4.5.33)$$

but (4.5.32) has to be replaced by

$$\mu \Delta u + (1 - \mu) \left[\frac{\partial^2 u}{\partial x^2} \cos^2 \vartheta + \frac{\partial^2 u}{\partial y^2} \sin^2 \vartheta + 2 \frac{\partial^2 u}{\partial x \partial y} \sin \vartheta \cos \vartheta \right] = 0, \qquad (4.5.34)$$

where ϑ is the local angle between the x axis and the normal vector $\vec{n}$ on the boundary curve. For a rectangular plate one has $\vartheta = \pi/2$ for a boundary parallel to the x axis and otherwise $\vartheta = 0$. Therefore one has

$$\Delta u + \frac{1 - \mu}{\mu} \frac{\partial^2 u}{\partial y^2} = 0 \qquad (4.5.35)$$

for the part of the boundary parallel to the x direction and otherwise

$$\Delta u + \frac{1 - \mu}{\mu} \frac{\partial^2 u}{\partial x^2} = 0. \qquad (4.5.36)$$

3. If the boundary is free, it means that neither forces nor torques are exerted, then (4.5.34) and

$$\frac{\partial}{\partial \vec{n}} \Delta u + (1 - \mu) \times$$

$$\frac{\partial}{\partial \vec{s}} \left[\left(\frac{\partial^2 u}{\partial y^2} - \frac{\partial^2 u}{\partial x^2} \right) \sin \vartheta \cos \vartheta + \frac{\partial^2 u}{\partial x \partial y} (\cos^2 \vartheta - \sin^2 \vartheta) \right] = 0 \qquad (4.5.37)$$

are valid. Here $\vec{s}$ is the tangential vector along the boundary curve.

A separation setup $u(x, y, t) = X(x) \cdot Y(y) \cdot T(t)$ and homogeneous boundary conditions $u(x = a, y, t) = 0$, $u(x, y = b, t) = 0$ yield the solution

$$u(x, y, t) = \sum_{n=1}^{\infty} \sum_{m=1}^{\infty} \sin \frac{n\pi x}{a} \sin \frac{m\pi y}{b}$$

$$\cdot (A_{nm} \cos \omega_{nm} t + B_{nm} \sin \omega_{nm} t) \qquad (4.5.38)$$

and the *eigenvalue equation*

$$\omega_{nm}^2 = \frac{Eh^2\pi^4}{3\rho\,(1-\mu^2)}\left(\frac{n^2}{a^2}+\frac{m^2}{b^2}\right)^2.$$

(4.5.39)

The solution (4.5.38) is called NAVIER *solution* and satisfies the boundary conditions $u = 0, \Delta u = 0$ at the rectangular boundary. It can be used to solve the problem of a plate under load $p(x, y)$.

For a circular plate the separation setup $u(r, \varphi, t) = \Phi(\varphi) \cdot T(t) \cdot R(r)$ ends up with the equation

$$\left(\frac{d^2}{dr^2}+\frac{1}{r}\frac{d}{dr}-\frac{n^2}{r^2}+\lambda k^2\right)\left(\frac{d^2}{dr^2}+\frac{1}{r}\frac{d}{dr}-\frac{n^2}{r^2}-\lambda k^2\right)R(r) = 0 \quad (4.5.40)$$

and $T(t) = C\cos\lambda t + D\sin\lambda t$, where λ is the separation parameter and $k^4 = 3\rho(1-\mu^2)/Eh^2$.

One sees that the problem for the plate is analytically more difficult than that for the membrane. It is not possible, for example, to treat the general case of the rectangular boundary in terms of functions known explicitly. The only plate boundary problem that has been explicitly treated is the circular plate.

For axial symmetry, the solution of this equation is given by BESSEL and modified BESSEL functions

$$u(r) = AJ_0(\sqrt{\lambda}\,kr) + BI_0(\sqrt{\lambda}\,kr),$$

(4.5.41)

where $I_0(kr) = J_0(ikr)$. We need two solutions since two boundary conditions have to be satisfied. For a clamped circular plate with radius R, the boundary conditions deliver $u(R) = 0, u'(R) = 0$ so that we have

$$AJ_0(\sqrt{\lambda}\,kR) + BI_0(\sqrt{\lambda}kR) = 0,$$

$$AJ_0'(\sqrt{\lambda}\,kR) + BI_0'(\sqrt{\lambda}kR) = 0.$$

(4.5.42)

To be able to solve these two homogeneous linear equations for the unknown partial amplitudes A and B, the determinant

$$J_0(\sqrt{\lambda}\,kR)I_0'(\sqrt{\lambda}\,kR) - J_0'(\sqrt{\lambda}\,kR)I_0(\sqrt{\lambda}\,kR) = 0$$

(4.5.43)

must vanish. Using

$$\frac{d}{dr}J_0(\sqrt{\lambda}\,kr) = -\sqrt{\lambda}\,kJ_1(\sqrt{\lambda}\,kr)$$

(4.5.44)

and analogously for I_0 the determinant determines the eigenvalue $k = 3.196\,220\,612$. For the initial values

$$u(r, 0) = h(r), \qquad u_t(r, 0) = g(r)$$

(4.5.45)

and the boundary value problem

$$u(r = R, t) = 0, \ u_r(r = R, t) = 0 \tag{4.5.46}$$

one obtains the solution [4.8]

$$
u(r, t) = \frac{1}{R^2} \sum_{n=1}^{\infty} \frac{Z_n(r)}{I_0^2(\gamma_n) J_0^2(\gamma_n)} \left[\cos\left(\frac{\gamma_n^2 ct}{R^2}\right) \int_0^R \rho h(\rho) Z_n(\rho) d\rho \right.
$$
$$
\left. + \frac{R^2}{c\gamma_n^2} \sin\left(\frac{\gamma_n^2 ct}{R^2}\right) \int_0^R \rho g(\rho) Z_n(\rho) d\rho \right], \tag{4.5.47}
$$

where

$$Z_\gamma(r) = J_0\left(\gamma \frac{r}{R}\right) I_0(\gamma) - J_0(\gamma) I_0\left(\gamma \frac{r}{R}\right) \tag{4.5.48}$$

and γ_n represents the n-th root of $Z'_\gamma(R) = 0$. To derive this expression

$$
\int_0^R Z_n^2(\rho) \rho d\rho = \frac{R^6}{4} \left\{ Z_n''^2 - Z_n' \frac{d}{dr} \left[\frac{1}{r} \frac{d}{dr} \left(r \frac{Z_n}{dr} \right) \right] - \frac{Z_n' Z_n''}{\gamma_n} - Z_n' Z_n''' \right\}_{r=R}
$$
$$
= R^2 I_0^2(\gamma_n) J_0^2(\gamma_n) \tag{4.5.49}
$$

has been used.

Up to now we treated only time-dependent problems. But there is a problem in the transition to $\partial/\partial t = 0$. Due to the surface boundary conditions of massive elastic bodies the simple transition $\partial/\partial t = 0$ is not possible in (4.5.3). The equation for the *bending of a girder* has to be derived from the basics. The usual assumptions for the bending of a cantilever beam of rectangular cross section $a \times b$ are

1. There exists a "neutral" centre layer within the beam that will not be expanded (*elastic axis* or *neutral filament*), which is described by $u(z)$, if the axis of the beam coincides with the z axis and the bending is into the negative x direction, $\partial/\partial y = 0$.

2. The cross sections in the x, y plane are assumed to cut the elastic axis along the z direction perpendicularly.

3. The cross sections will not be deformed.

Under these conditions the differential equation for the elastic line has been derived

$$\frac{d^2 u_x(z)}{dz^2} = \frac{P}{EI}(z - l). \tag{4.5.50}$$

P is the load acting at the end of the beam of length l. Integration yields

$$u_x(x, z) = \frac{P}{EI} \frac{z^3}{2 \cdot 3} - \frac{Pl}{EI} \frac{z^2}{2} + c_1(x)z + c_2(x). \qquad (4.5.51)$$

Due to the boundary condition (clamped at $z = 0$)

$$u(x, 0) = 0, \quad \partial u(x, 0)/\partial z = 0. \qquad (4.5.52)$$

This results in

$$u_x(l) = -\frac{Pl^3}{3EI} = -\frac{4Pl^3}{a^3 Eb}. \qquad (4.5.53)$$

In this derivation (4.5.5) had been used. (4.5.53) defines the *pitch of deflection sag*.

The bending of a girder is a transverse problem. We shall now investigate a longitudinal problem: the determination of the *buckling strength* and the *critical compressive strength* of a pillar. The longitudinal compression $u(x)$ of a vertical pillar of cross section q, density ρ, weight $\rho q g$ carrying a load P is described by

$$EIu''''(x) + \rho g q \frac{d}{dx}[(l - x)u'(x)] = 0 \qquad (4.5.54)$$

and the boundary conditions

$$u(0) = u'(0) = 0, \qquad (4.5.55)$$

which describe clamping of the lower end on earth's surface $x = 0$ and

$$u''(l) = 0, \qquad u'''(l) = 0 \qquad (4.5.56)$$

at the load free upper end. If there is a load P at $x = l$, one has

$$u''(l) = 0, \qquad EIu'''(l) = P. \qquad (4.5.57)$$

We now first neglect a load P and solve

$$EIu'''(x) + \rho g q(l - x)u'(x) = 0, \qquad (4.5.58)$$

which is obtained by integrating (4.5.54) and neglection of P and of integration constants depending on x. A transformation $u'(x) = v(x)$, $A = \rho g q/EI$ yields

$$v''(x) + A(l - x)v(x) = 0 \qquad (4.5.59)$$

and a substitution $\xi = 2\sqrt{A}\,(l-x)^{3/2}/3$, $v = w(\xi)\sqrt{\xi}$ results in a BESSEL equation

$$w''(\xi) + \frac{1}{\xi}w'(\xi) + \left(1 - \frac{1}{9\xi^2}\right)w(\xi) = 0 \qquad (4.5.60)$$

with the solution

$$w(\xi) = C_1 J_{1/3}(\xi) + C_2 J_{-1/3}(\xi). \qquad (4.5.61)$$

The boundary conditions $u''(l) = 0$ and $u(0) = 0$ result in $C_1 = 0, C_2 = 0$ so that $u = const$ is a stable equilibrium if

$$J_{-1/3}\left(\frac{2}{3}A^{1/2}l^{2/3}\right) = 0 \qquad (4.5.62)$$

is NOT satisfied. The first zero 1.87 of $J_{-1/3}$ gives the buckling length l_c of a pillar with no load

$$l_k = \left(\frac{1.87 \cdot 3}{2}\right)^{2/3} \cdot \sqrt[3]{\frac{EI}{\rho g q}}. \qquad (4.5.63)$$

If the weight is neglected ($g = 0$), one obtains

$$l_k = \pi \sqrt[2]{\frac{EI}{P}} \qquad (4.5.64)$$

(EULER's *critical load*). The full equation (4.5.54) may be solved by numerical or approximation methods, see later in section 4.6.

According to (4.5.7) plates under a distributed static load $p(x,y)$ are described by the biharmonic operator $\Delta\Delta$

$$\frac{Eh^2}{3\rho\,(1-\mu^2)}\Delta\Delta u(x,y) = p(x,y). \qquad (4.5.65)$$

Solutions of the homogeneous equation (4.5.65) are called *biharmonic*. Such functions are $x, x^2, x^3, y, y^2, y^3, xy, x^2y, \cosh\lambda x \cos\lambda y$ etc. If v, w are harmonic functions, then $xv + w$, $(x^2+y^2)v + w$, $(x^2+y^2)\ln\sqrt{x^2+y^2}$, etc., are also biharmonic. The biharmonic operator for the stream function appears also in the theory of the motion of small bodies within a viscous fluid like blood or mucus. The propagation of *bacteria* in blood or of *sperma cells* in the viscid vaginal mucus presents an interesting boundary problem of the biharmonic operator [4.11]. There seems to exist a statistical evidence that the probability of a male baby is the higher, the quicker the spermatozoon is and the shorter the time span between ovulation and fertilization (*sex determination* by a boundary condition).

The inhomogeneous static plate equation (4.5.65) determines the deflection of a plate under a distributed load $p(x, y)$. With exception of the NAVIER solution (4.5.38), no exact closed analytic solutions of (4.5.65) for general boundary conditions are known for rectangular plates [4.10]. Variational calculus has to be used to find solutions. A closed solution only exists for the circular plate. Using axial symmetry, the plate equation reads

$$u''''(r) + \frac{2}{r}u'''(r) - \frac{1}{r^2}u''(r) + \frac{1}{r^3}u'(r) = p(r) \cdot \frac{3\rho\left(1 - \mu^2\right)}{Eh^2}. \qquad (4.5.66)$$

To first solve the homogeneous equation we make the substitution $r = \exp\bar{\rho}$ so that

$$u''''(\bar{\rho}) - 4u'''(\bar{\rho}) + 4u''(\bar{\rho}) = 0 \qquad (4.5.67)$$

results. The solution of (4.5.61) is then

$$u(r) = c_1 + c_2 \ln r + c_3 r^2 + c_4 r^2 \ln r. \qquad (4.5.68)$$

Variation of parameters (see section 2.1) delivers a general solution of (4.5.66) if the load satisfies itself the homogeneous equation. For constant load $p = p_0$ a special solution is given by

$$u(r) = c_1 + c_2 \ln r + c_3 r^2 + c_4 r^2 \ln r + r^4 p_0 \frac{3\rho(1 - \mu^2)}{64Eh^2}. \qquad (4.5.69)$$

This solution may be specified by the following boundary conditions

$$u(0) = \text{finite}, \qquad u''(0) = \text{finite}, \qquad (4.5.70)$$

and boundary clamped all around

$$u(R) = 0, \qquad u'(R) = 0. \qquad (4.5.71)$$

These conditions result in $c_2 = 0$, $c_4 = 0$, so that the deflection is given by

$$u(r) = \frac{3\rho(1 - \mu^2)}{Eh^2}p_0\left(R^4 - 2R^2r^2 + r^4\right)/64. \qquad (4.5.72)$$

A warning seems to be necessary. In these days of *computer enthusiasm* many engineers calculate plates numerically. But several times it had been overseen that a pillar supporting a plate constitutes a singularity and that the normal grid used in numerical computations should have been extensively narrowed near the pillar. The consequence of using black box computer routines has been a breakdown of a ceiling, cases of death and lawsuits over millions of dollars.

If deflections are combined with warming, then the coefficient α of thermal expansion enters into the plate equation [4.7].

Problems

1. Solve the problem described by (4.5.17) to (4.5.21) using *Mathematica*:

   `Clear[X];DSolve[X''''[x]-k^2*X[x]==0,X[x],x]`

   yields (4.5.18) in the form

   $$X[x] \rightarrow e^{-\sqrt{k}\,x}\,C[2] + e^{\sqrt{k}\,x}\,C[4] + C[1]\,\text{Cos}[\sqrt{k}\,x] + C[3]\,\text{Sin}[\sqrt{k}\,x];$$

   To be able to handle the solution, we redefine by copy and paste

   `X[x_]=`$e^{-\sqrt{k}\,x}$` C[2]+`$e^{\sqrt{k}\,x}$` C[4]`
   `+C[1]Cos`$\sqrt{k}$` x+C[3]Sin[`$\sqrt{k}$` x];`

   and derive it

   `Y[x_]=D[X[x],x]`

   resulting in

   $$-e^{-\sqrt{k}\,x}\,\sqrt{k}\,C[2] + e^{\sqrt{k}\,x}\,\sqrt{k}\,C[4] + \sqrt{k}\,C[3]\,\text{Cos}[\sqrt{k}\,x]$$
   $$- \sqrt{k}\,C[1]\,\text{Sin}[\sqrt{k}\,x]$$

   This is equivalent to (4.5.18). The next step is the satisfaction of the boundary conditions at $x = 0$. We do this by typing

   `X[0]==0`

   and one obtains a condition for the integration constants

   $C[1] + C[2] + C[4] == 0$ or `C[4]==-C[1]-C[2]`

   And the vanishing of the derivative **Y** at $x = 0$ yields

   `Y[0]==0` or `C[3]==2*C[2]+C[1]`

   These expressions for `C[3]` and `C[4]` will now be inserted into `X[x]` (and later also into `Y[x]`). The result is a redefinition

   `Clear[X, a11, a12];`
   `X[x_]=`$e^{-\sqrt{k}\,x}$` C[2]+`$e^{\sqrt{k}\,x}$` *(-C[1]-C[2])+C[1]*`
   `Cos[`$\sqrt{k}$` x]+(2*C[2]+C[1])*Sin[`$\sqrt{k}$` x];`

   To be able to fill the determinant later, we have to factor out `C[1]` and `C[2]`. This is done by

   `Collect[X[1],C[1]]`   (4.5.73)

   yielding

   $$e^{-\sqrt{k}\,l}\,C[2] - e^{\sqrt{k}\,l}\,C[2] + 2\,C[2]\,\text{Sin}[\sqrt{k}\,l]$$
   $$+ C[1]\left(-e^{\sqrt{k}\,l} + \text{Cos}\left[\sqrt{k}\,l\right] + \text{Sin}[\sqrt{k}\,l]\right)$$

   Now one can read off and define

   `a11=(-e`$^{\sqrt{k}\,l}$`+Cos[`$\sqrt{k}$` l +Sin[`$\sqrt{k}$` l]);`

and analogously

a12=(e$^{-\sqrt{k}\,l}$-\E$^{\sqrt{k}\,l}$+2 Sin[$\sqrt{k}$ l]);

Redefinition of the derivative **Y** by inserting for **C[3]** and **C[4]** results in

$$e^{\sqrt{k}\,x}\,\sqrt{k}\,(-C[1]-C[2])-e^{-\sqrt{k}\,x}\,\sqrt{k}\,C[2]+$$
$$\sqrt{k}\,(C[1]+2\,C[2])\,\text{Cos}[\sqrt{k}\,x]-\sqrt{k}\,C[1]\,\text{Sin}[\sqrt{k}\,x];$$

Now one is able to satisfy the boundary condition (4.5.20)

Collect[Y[1],C[1]]

yields the input

a21=$\left(-e^{\sqrt{k}\,l}+\text{Cos}[\sqrt{k}\,l]-\text{Sin}[\sqrt{k}\,l]\right)$;

and analogously

a22=$\left(-e^{-\sqrt{k}\,l}-e^{\sqrt{k}\,l}+2\,\text{Cos}[\sqrt{k}\,l]\right)$;

Now we have all elements of the matrix M and define

M={{a11,a12},{a21,a22}}/. Sqrt[k]*1->y   (4.5.74)

and at the same time we are replacing the argument $\sqrt{k}\,l$ by y.

We can simplify the expression and define a new function $F(y)$ containing the determinant of the matrix M

Simplify[%];
Clear[F];F[y_]=Det[M];l=1.;   (4.5.75)

Now we are able to verify (4.5.21). The command

FindRoot[F[y]==0,{y,4.0}]

starts at $y = 4.0$ and finds $4.73\,004$.

2. A rod is clamped at $x = 0$ and free at $x = l$. Solve the homogeneous equation (4.5.2) for longitudinal oscillations using the boundary conditions (4.5.8) $u_x(l,t) = 0$ and (4.5.8a) $u(0,t) = 0$. Derive the eigenfrequencies $\omega_n = (2n - 1)\pi c/2l$ and the solution

$$u(x,t) = \sum_{n=1}^{\infty} \sin k_n x(A_n \cos ck_n t + B_n \sin ck_n t),$$

$$k_n = (2n - 1)\pi x/2l. \tag{4.5.76}$$

3. If a mass m is fixed at the end $x = l$ of a rod, then the boundary condition is modified and reads $mu_{tt}(l,t) = -Equ_x(l,t)$. If the initial conditions are $u(x,0) = h(x) = h_0 x/l$ and $u_t(x,0) = 0$ derive the solution

$$u(x,t) = \frac{4u_0}{l} \sum_{n=1}^{\infty} \cos(ck_n t) \frac{\sin k_n l \sin k_n x}{k_n (2k_n l + \sin 2k_n l)} \tag{4.5.77}$$

and the eigenvalue equation is

$$\cot kl = mk/\rho q. \tag{4.5.78}$$

4. Investigate the longitudinal oscillations of a rod of length l for the initial conditions $u(x,0) = h(x), u_t(x,0) = g(x)$, if the end $x = 0$ is clamped and the end $x = l$ is free [4.8]. The solution is

$$u(x,t) = \frac{2}{l} \sum_{n=0}^{\infty} \left[\cos\left(\frac{2n+1}{2l}\pi ct\right) \int_0^l h(\xi) \sin\left(\frac{2n+1}{2l}\pi\xi\right) d\xi + \frac{2l}{(2n+1)\pi c} \right.$$

$$\left. \sin\left(\frac{2n+1}{2l}\pi ct\right) \int_0^l g(\xi) \sin\left(\frac{2n+1}{2l}\pi\xi\right) d\xi \right] \sin\left(\frac{2n+1}{2l}\pi x\right). \tag{4.5.79}$$

5. A bridge supported at both ends $x = 0, x = l$ is loaded by a force P that crosses the bridge with constant speed v_0. The loading function f is assumed to act within the small interval ε

$$f(x,t) = \begin{cases} P/\varepsilon & v_0 t \le x \le v_0 t + \varepsilon \\ 0 & 0 \le x < v_0 t, \quad v_0 t + \varepsilon \le l, \end{cases}$$

and

$$\int_0^l f(x,t)\mathrm{d}t = \int_{v_0 t}^{v_0 t + \varepsilon} P/\varepsilon \cdot \mathrm{d}x = P.$$

The transverse deflection (vertical to the bridge in x direction) is then given by

$$u(x,t) = \frac{2P}{\rho q l} \sum_{n=1}^{\infty} \frac{\sin(n\pi x/l)}{\omega_n^2 - \bar{\omega}_n^2} (\sin \bar{\omega}_n t - \bar{\omega}_n \sin(\bar{\omega}_n t)/\omega_n)), \tag{4.5.80}$$

where

$$\omega_n = \frac{n^2 \pi^2}{l^2} \sqrt{\frac{EI}{\rho q}}. \tag{4.5.81}$$

If one of the exciting frequencies $\bar{\omega} = n\pi v_0/l$ is near or equal to the eigenfrequencies ω_n of the bridge, resonance and breakage will occur.

6. Derive (4.5.64) by solving (4.5.54) for $g = 0$ and calculate the buckling length l_c of a pillar according to (4.5.63), (4.5.64). Assume the dimensions $a = 0.1$ m, $b = 0.12$ m, use (4.5.5), $q = a \cdot b, \rho = 2.5\,\mathrm{kN/m^3}, E = 30$ $\mathrm{GN/m^2}$ (reinforced concrete) and vary P in (4.5.64).

7. Solve the plate equation (4.5.7) for a time dependence $\sim \exp(i\omega t)$. Use $k^4 = \omega^2 3(1 - \mu^2)\rho/Eh^2$. No boundary or initial conditions are given. Split the operator $\Delta\Delta - k^4$ into $(\Delta + k^2)(\Delta - k^2)$ and solve $\Delta u + k^2 u = 0, \Delta u - k^2 u = 0$. Verify the solution, which has been prepared to be

used by a collocation procedure (section 4.8). Recall equation (1.1.55) and find an analogous solution of (4.5.7)

$$u(x,y) = \sum_u \left[A_n \cos\left(\sqrt{k^2 - b_n^2}\, x\right) \cos(b_n y) \right.$$

$$\left. + B_n \cosh\left(\sqrt{k^2 - b_n^2}\, x\right) \cosh(b_n y) \right]. \qquad (4.5.82)$$

Verify this solution by inserting it into the plate equation.

8. According to (1.4.16) the free vibrations of a plate with varying thickness $h(x)$ are described by

$$\frac{Eh^2}{12(1-\mu^2)} \Delta\Delta u(x,y,t) + \rho_0 u_{tt} + \frac{E}{12(1-\mu^2)} \{6h[h_x(u_{xxx} + u_{xyy})]$$

$$+ 3u_{xx}(2h_x^2 + hh_{xx}) + 3u_{yy}(2\mu h_x^2 + \mu hh_{xx})\} = 0. \qquad (4.5.83)$$

Assume $h(x) = ax + b$ and a time dependence $\sim \cos(\omega t)$ and try to solve (4.5.83).

4.6   Approximation methods

In practical applications in physics and engineering, many problems arise that cannot be solved analytically. An immediate use of a black box numerical routine may give no satisfying results or may even bring about the breakdown of a construction. It is always advantageous to dispose of an approximate analytic expression. It can be obtained by:

1. collocation in section 1.3, equation (1.3.10),

2. successive approximation in section 2.7, equation (2.7.28),

3. averaging method in section 2.7, equation (2.7.31),

4. multiple time scales in section 2.7, equation (2.7.54),

5. WKB method in section 2.7, equation (2.7.64),

6. the moment method in section 3.2, equation (3.2.44),

7. expansion of the solution into a power series, equation (4.6.61).

In many practical problems described by partial differential equations, a small parameter $\varepsilon \ll 1$, the so-called *perturbation term*, appears. Let us consider an example. We assume $p(x) = 1$ and $\rho(x) = 1 + \varepsilon(x)$ in (4.3.34) and obtain for the inhomogeneous string

$$\frac{\partial^2 u(x,t)}{\partial x^2} = \frac{1 + \varepsilon(x)}{c^2} \frac{\partial^2 u(x,t)}{\partial t^2}. \tag{4.6.1}$$

If the string is clamped at both ends $x = 0$ and $x = l$ one may expect a discrete eigenvalue spectrum ω_n like in section 4.3, equation (4.3.16). To solve (4.6.1) we assume

$$u(x,t) = f(x)\cos(\omega t + \varphi) \tag{4.6.2}$$

and obtain

$$\frac{\mathrm{d}^2 f(x)}{\mathrm{d}x^2} = -\frac{\omega^2}{c^2}(1 + \varepsilon(x))f(x). \tag{4.6.3}$$

Since we know the solution (4.3.15) for $\varepsilon = 0$, we set up

$$f(x) = \sin\frac{n\pi x}{l} + g(x), \quad \omega = \omega_n(1 + \eta). \tag{4.6.4}$$

Since the perturbations ε, g, η are considered to be small, one may neglect $\varepsilon g, \eta^2, g\eta, \varepsilon\eta$ and higher orders. Insertion of (4.6.4) into (4.6.3) results in

$$g''(x) + \frac{\omega_n^2}{c^2} g = -\frac{n^2\pi^2}{l^2}[\varepsilon(x) + 2\eta]\sin\frac{n\pi x}{l}. \tag{4.6.5}$$

Expanding $g(x)$ into a series of the eigenfunctions of the unperturbed problem, (satisfying the boundary conditions)

$$g(x) = \sum_{\nu=1}^{\infty} a_\nu \sin\frac{\nu\pi x}{l} \tag{4.6.6}$$

and inserting into (4.6.5) yields

$$\sum_{\nu=1}^{\infty}\left(\omega_n^2 - \frac{\nu^2\pi^2 c^2}{l^2}\right) a_\nu \sin\frac{\nu\pi x}{l} = -\omega_n^2[\varepsilon(x) + 2\eta]\sin\frac{n\pi x}{l}. \tag{4.6.7}$$

If one multiplies by $\sin(n\pi x/l)$, integrates over 0 to l and takes orthogonality for $\nu \neq n$ into account, one obtains

$$\eta = -\frac{1}{l}\int_0^l \varepsilon(x)\sin^2\frac{n\pi x}{l}\,\mathrm{d}x. \tag{4.6.8}$$

In this derivation

$$\frac{1}{l}\int_0^l \sin^2\frac{n\pi x}{l}\,\mathrm{d}x = \frac{1}{2} \tag{4.6.9}$$

has been used. Multiplication of (4.6.7) by $\sin(m\pi x/l)$ and integration yields for $m \neq n$ the expansion coefficients a_m of (4.6.7)

$$a_m = \frac{\omega_n^2}{\omega_m^2 - \omega_n^2} \frac{2}{l} \int_0^l \varepsilon(x) \sin\frac{n\pi x}{l} \sin\frac{m\pi x}{l} \mathrm{d}x. \tag{4.6.10}$$

We thus have the solution of (4.6.1) for a discrete spectrum. For the investigation of the continuous spectrum we replace the standing waves (4.6.4) by travelling waves

$$u(x,t) = Ae^{ik(x-ct)} + u_1(x,t) = u_0 + u_1 \tag{4.6.11}$$

and receive by insertion and neglection of higher order terms $\varepsilon u_1 \approx 0$

$$\frac{\partial^2 u_1}{\partial x^2} - \frac{1}{c^2}\frac{\partial^2 u_1}{\partial t^2} = -\varepsilon Ak^2 e^{ik(x-ct)} \equiv \frac{\varepsilon}{c^2}\frac{\partial^2 u_0}{\partial t^2}. \tag{4.6.12}$$

This equation can be solved by

$$u_1(x,t) = v(x)e^{-ikct} \tag{4.6.13}$$

leading to

$$v'' + k^2 v = -\varepsilon Ak^2 e^{ikx}. \tag{4.6.14}$$

For higher accuracy one can use more terms: $u = u_0 + u_1 + u_2 \dots$. This works also for more dimensions in (x, y, t)

$$\frac{\partial^2 u_\nu}{\partial x^2} + \frac{\partial^2 u_\nu}{\partial y^2} - \frac{1}{c^2}\frac{\partial^2 u_\nu}{\partial t^2} = \frac{\varepsilon(x,y)}{c^2}\frac{\partial^2 u_{\nu-1}}{\partial t^2}. \tag{4.6.15}$$

For time-dependent perturbations one can use

$$u(x,t) = \sum_{\nu=1}^{\infty} a_\nu(t) \sin\frac{\nu\pi x}{l}. \tag{4.6.16}$$

What happens if we have a *degenerate eigenvalue problem* like in section 3.4? We shall see that the perturbation abolishes the degeneration. We investigate a square membrane of length a with the unperturbed solution

$$u_{mn}(x,y,t) = A_{mn} \sin\frac{m\pi x}{a} \sin\frac{n\pi y}{a} \cos(\omega_{mn}t - \varphi_{mn}). \tag{4.6.17}$$

To solve the perturbed membrane equation (4.6.15) we set up

$$u(x,y,t) = \left[\alpha \sin\frac{m\pi x}{a} \sin\frac{n\pi y}{a} + \beta \sin\frac{n\pi x}{a} \sin\frac{m\pi y}{a} + g(x,y)\right]$$
$$\cdot \cos(\omega t - \varphi). \tag{4.6.18}$$

Here

$$\omega = \omega_{mn}(1+\eta), \quad \omega_{mn} = \frac{c\pi}{a}\sqrt{m^2 + n^2} \tag{4.6.19}$$

is the perturbed eigenvalue. Insertion and neglection of higher orders yield

$$\Delta g + \frac{\omega_{mn}^2}{c^2} g = -\frac{\omega_{mn}^2}{c^2}(\varepsilon + 2\eta)$$
$$\cdot \left(\alpha \sin \frac{m\pi x}{a} \sin \frac{n\pi y}{a} + \beta \sin \frac{m\pi x}{a} \sin \frac{n\pi y}{a}\right). \quad (4.6.20)$$

To solve this equation we set up

$$g(x,y) = \sum_{\mu=1}^{\infty} \sum_{\nu=1}^{\infty} a_{\mu\nu} \sin \frac{\mu\pi x}{a} \sin \frac{\nu\pi y}{a}. \quad (4.6.21)$$

Multiplying (4.6.20) by

$$\sin \frac{m\pi x}{a} \sin \frac{n\pi y}{a} \quad \text{or} \quad \sin \frac{n\pi x}{a} \sin \frac{m\pi y}{a}$$

and integration result in

$$(\varepsilon_{mn} + 2\eta)\,\alpha + \zeta_{mn}\beta = 0,$$
$$\zeta_{mn}\alpha + (\varepsilon_{mn} + 2\eta)\,\beta = 0. \quad (4.6.22)$$

Here the abbreviations

$$\varepsilon_{mn} = \frac{4}{a^2} \int_0^a \int_0^a \varepsilon(x,y) \sin^2 \frac{m\pi x}{a} \sin^2 \frac{n\pi y}{a} dx dy,$$

$$\zeta_{mn} = \frac{4}{a^2} \int_0^a \int_0^a \varepsilon(x,y) \sin \frac{m\pi x}{a} \sin \frac{n\pi x}{a} \sin \frac{m\pi y}{a} \sin \frac{n\pi y}{a} dx dy$$

had been used. (4.6.22) is a homogeneous system of two linear equations for the unknown partial amplitudes α and β. To solve this system, the determinant of the coefficients must vanish. We thus obtain for the eigenvalue perturbation

$$\eta_{1,2} = \frac{1}{4}\left[-(\varepsilon_{mn} + \varepsilon_{nm}) \pm \sqrt{(\varepsilon_{mn} - \varepsilon_{nm})^2 + 4\zeta_{mn}^2}\right]. \quad (4.6.23)$$

This result is interesting; we have two solutions $\omega = \omega_{mn} + \eta_1$, $\omega = \omega_{mn} + \eta_2$, the degeneration (one eigenvalue belonging to two eigenfunctions) is suspended by the perturbation.

The procedure that we just discussed may be formalized. Let

$$\mathcal{L}\{u\} + \varepsilon\,\mathcal{S}\{u\} = \lambda\,\mathcal{N}\{u\} \quad (4.6.24)$$

be a linear self-adjoint differential *operator* equation subjected to an homogeneous boundary condition $\mathcal{R}\{u\} = 0$. Then

$$\lambda_k = \lambda_{0k} + \varepsilon\lambda_{1k} + \varepsilon^2\lambda_{2k} + \ldots, \quad \varphi_k = \varphi_{0k} + \varepsilon\varphi_{1k} + \varepsilon^2\varphi_{2k} + \ldots \quad (4.6.25)$$

is useful. Following the procedure described above and using the orthogonality
of the eigenfunctions

$$\int_a^b \varphi_{0k} \left[\lambda_{1k} \mathcal{N}\{\varphi_{0k}\} - \mathcal{S}\{\varphi_{0k}\} \right] dx = 0, \tag{4.6.26}$$

the first perturbation of the eigenvalue is given by

$$\lambda_{1k} = \frac{\int_a^b \varphi_{0k} \mathcal{S}\{\varphi_{0k}\} dx}{\int_a^b \varphi_{0k} \mathcal{N}\{\varphi_{0k}\} dx}, \tag{4.6.27}$$

so that the perturbed eigenvalue is given by

$$\lambda_k \equiv \lambda_{0k} + \varepsilon \lambda_{1k} \tag{4.6.28}$$

and the eigenfunctions of first order are determined by

$$\mathcal{L}\{\varphi_{1k}\} - \lambda_{0k} \mathcal{N}\{\varphi_{1k}\} = \lambda_{1k} \mathcal{N}\{\varphi_{0k}\} - \mathcal{S}\{\varphi_{0k}\},$$
$$\mathcal{R}\{\varphi_{1k}\} = 0. \tag{4.6.29}$$

In section 4.5 we treated a pillar under load P and deformed by its own
weight $\rho q g$. The problem described by (4.5.54) had been solved for two cases:

1. no load P. Equation (4.5.58) had been solved and the critical length
   (4.5.63) had been derived,

2. no weight, but with a load P, see (4.5.64) and problem 6 of section 4.5.

Now we shall handle the full problem (4.5.54) with the boundary conditions
(4.5.55) and (4.5.57). We rewrite (4.5.54) in the form

$$u'''' + \varepsilon \left[(l - x)u'' - xu' \right] + \lambda u'' = 0, \tag{4.6.30}$$

where

$$\varepsilon = \rho g q / EI, \quad \lambda = P / EI \tag{4.6.31}$$

denote the perturbation parameter and the eigenvalue respectively. First we
solve the unperturbed equation

$$u'''' = -\lambda u'' \tag{4.6.32}$$

with the four homogeneous boundary conditions (4.5.55) and (4.5.56). We
obtain

$$u_0(x) = 1 - \cos \sqrt{\frac{4n\pi^2}{l^2}} \, x \tag{4.6.33}$$

and

$$\lambda_{0n} = 4n\pi^2/l^2, \quad n = 1, 2 \ldots. \tag{4.6.34}$$

Finally, (4.6.27) yields

$$\lambda_n = l/2, \qquad \lambda_n \approx 4n\pi^2/l^2 + \varepsilon l/2 \tag{4.6.35}$$

and

$$P = \frac{4\pi^2}{l^2} EI - \frac{\rho g q l}{2} \tag{4.6.36}$$

for EULER'S *critical load*. The buckling length is now given by

$$l = \sqrt[3]{\frac{8\pi^2 EI}{\rho g q}} \approx 4.3 \sqrt[3]{\frac{EI}{\rho g q}}, \tag{4.6.37}$$

compare (4.5.63).

Perturbation theory has also been applied to investigate small alterations of the boundary (FEENBERG *method*) or of the boundary conditions [1.2].

Problems of this kind are handled today by numerical methods. In the examples treated here, we assumed that the limit $\varepsilon \to 0$ does not considerably modify the differential equation. This is not the case for *singular perturbation problems*. Consider

$$-\varepsilon^2 u'' + c(x, u) = 0, \quad u(0) = 0, \quad u(1) = 0, \tag{4.6.38}$$

$$\varepsilon \Delta\Delta\psi(x, y) - \psi_x(x, y) = \sin y. \tag{4.6.39}$$

This equation describes a viscous plane flow [3.9]. Problems of this type become even more complicated, if the boundary curve does not coincide with curves of a coordinate system in which the pertinent partial differential equation is separable. Thus, the boundary conditions may read

$$\psi = 0, \qquad \partial\psi/\partial x = 0 \tag{4.6.40}$$

for a boundary described by $y = ax$, $0 \le x \le x_1$ and also for a boundary $y = bx + c$, $x_0 \le x \le x_1$. Furthermore

$$\psi = 0, \qquad \partial^2\psi/\partial y^2 = 0 \quad \text{for} \quad y = 0. \tag{4.6.41}$$

The unperturbed solution of (4.6.38) is now given by

$$\psi_0(x, y) = -x \sin y + f(y) \sin y. \tag{4.6.42}$$

The basic idea is now the assumption of an *inner* and *outer solution*. The outer solution is assumed to be responsible for locations far outside of the closed triangular inner domain determined by (4.6.40) and (4.6.41). The inner solution has to describe the behavior within the closed domain. To satisfy the assumption that the outer solution decreases with increasing distance normal to the three boundaries, it is of advantage to introduce coordinates ξ, η orthogonal to the boundaries. As an example, we consider the line $y = ax$. Then this line should be vertical to the ξ axis and its normal is defined by

$$\xi = \varepsilon^n(y - ax), \quad \eta = y, \quad \varphi_1 = \varphi_1(\xi, \eta). \tag{4.6.43}$$

ξ must vanish at the boundary $y = ax$. Then the pertinent boundary condition reads

$$\psi_0(ay, y) + \varphi_1(0, \eta) = 0, \quad \frac{\partial \psi_0(ay, y)}{\partial x} + \varepsilon^n \frac{\partial \varphi_1(0, \eta)}{\partial \xi} = 0. \tag{4.6.44}$$

Then (4.6.39) reads

$$\left(1 + a^2\right)^2 \varepsilon^{4n+1} \varphi_{1\xi\xi\xi\xi} - \varepsilon^n \varphi_{1\xi} = 0. \tag{4.6.45}$$

For the contributions of the two leading terms to be of the same order of magnitude in ε, one must have $4n + 1 = n, \quad n = -1/3$ and (4.6.45) becomes

$$\left(1 + a^2\right)^2 \varphi_{1\xi\xi\xi\xi} - \varphi_{1\xi} = 0. \tag{4.6.46}$$

This may be solved by an expansion with respect to powers $\varepsilon^{1/3}$

$$\varphi_1 = G_1(\eta) \exp\left[\gamma\xi e^{2\pi i/3}\right] + G_2(\eta) \exp\left[\gamma\xi e^{4\pi i/3}\right] \tag{4.6.47}$$

with $\gamma = (1 + a^2)^{-3/2}$. If the other two boundaries are treated analogously the solution to (4.6.39) reads [3.9]

$$\psi(x, y) = -\left[x + f(y)\right] \sin y + G_1(y) \exp\left[\gamma e^{2\pi i/3}\xi\right]$$
$$+ G_2(y) \exp\left[\gamma e^{4\pi i/3}\xi\right] + G_3(y) \exp\left[\gamma\tilde{\xi}\right], \tag{4.6.48}$$

$$\tilde{\xi} = \varepsilon^n(x + x_0 - ay). \tag{4.6.49}$$

Now G_1, G_2, G_3 have to be chosen to satisfy all boundary conditions for $\psi(x, y)$.

Other authors prefer to solve singular perturbation problems using *spline methods* [4.12]. The LIE *series* method mentioned in (2.5.44) and (2.7.68) represents another perturbation method. In the NASA report [2.3] all details are given on the solution of the *three body problem* Sun, Jupiter and 8th Moon of Jupiter. It could be demonstrated that perturbation methods based on (2.7.68) converge very rapidly [4.13], [2.11]. The perturbation method applied on the Jupiter Moon has been tested by the program LIESE [2.4] and the results obtained had been compared by KOVALEVSKY, Paris, to the results obtained by the COWELL *method*. Computations with an IBM650 gave deviations in the coordinates and velocities, obtained with a calculating step $\Delta t = 5$ d for 100 days forward and backward, less than $50 \cdot 10^{-10}$ L (L =

astronomical unit $1495.042\,01\cdot10^{10}$ cm). With the LIE method calculations for again 100 days forward and backward with a step $\Delta t = 1$ d, the local deviation has been $15\cdot10^{-10}$ L in Aachen, Germany, on a SIE2002 computer and has later been improved at the University of Wisconsin.

Problems

1. We know that the solutions of the membrane equation are described in an unambiguous way by two homogeneous boundary conditions. Now the question may arise if perturbation theory admits a third boundary condition. Let us investigate if this is possible. We assume a square membrane of lateral length a and situated in such a manner that the point of intersection of the two diagonals coincide with the origin of the cartesian coordinate system. Let us now assume the following three boundary conditions:

$$u\left(\pm\frac{a}{2}, y\right) = 0, \quad u\left(x, \pm\frac{a}{2}\right) = 0, \quad u(0,0) = 0. \tag{4.6.50}$$

The last condition has the meaning that the center of the membrane is fixed. We express this by a modification of the membrane equation

$$\Delta u + k^2 u(1 - \lambda\delta(x)\delta(y)) = 0. \tag{4.6.51}$$

$$u(x,y) = u_0(x,y) + \lambda u_1(x,y) + \lambda^2 u_2(x,y) + \ldots,$$

$$k^2 = E_0 + \lambda E_1 + \lambda^2 E_2 + \ldots,$$

$$u_0 = \frac{2}{a}\cos\frac{\pi x}{a}\cos\frac{\pi y}{a},$$

$$u_{nm}(x,y) = \frac{2}{a}\cos\frac{2n+1}{a}\pi x\cos\frac{2m+1}{a}\pi y. \tag{4.6.52}$$

Solution:

$$E_0 = \frac{2\pi^2}{a^2}, \quad E_{mn} = E_{nm} = \left(\frac{2n+1}{a}\right)^2\pi^2 + \left(\frac{2m+1}{a}\right)^2\pi^2,$$

$$u_1(0,0) = -\sum_{n,m}\frac{2}{a^2}\frac{1}{n^2+m^2+n+m} = -\frac{2}{a^2}\sum_n\frac{1}{n^2+n}\sum_m\frac{1}{1+\dfrac{m^2+m}{n^2+n}}$$

$$< -\frac{2}{a^2}\frac{1}{2}\sum_n\frac{n}{n^2+n} = -\frac{1}{a^2}\sum_n\frac{1}{n+1} \to -\infty. \tag{4.6.53}$$

Discuss the fact that u_1 diverges since

$$\sum_m \frac{1}{1 + \dfrac{m^2 + m}{n^2 + n}} > \frac{n}{2}.$$

2. The SCHROEDINGER *equation* for an electron m_e in a potential U

$$\Delta\psi + \frac{8\pi^2 m_e}{h^2}(E - U)\psi = 0 \qquad (4.6.54)$$

describes the complex scalar field $\psi(x, y, z)$, which gives the probability $\psi\psi^* d\tau$ to localize an electron at the point x, y, z within $d\tau$ with the total energy E. h is PLANCK'S *constant* $6.626 \cdot 10^{-34}$ Js. For an electron in an hydrogen atom, the attractive potential between the electron and the hydrogen nucleus is given by $U = +e^2/r$. Using the FROBENIUS method for the r-dependence and the knowledge of spherical functions one may derive the spherical solution

$$\psi(r, \varphi, \vartheta) = \sum_{m,l} L^{2l+1}_{B-1-l}(\rho) P_l^m(\cos\vartheta) \exp(im\varphi), \qquad (4.6.55)$$

where

$$\frac{8\pi^2 m_e E}{h^2} = -\frac{1}{r_0^2}, \quad \frac{r_0}{2}\frac{8\pi^2 m_e e^2}{h^2} = B, \quad \frac{2r}{r_0} = \rho \qquad (4.6.56)$$

and the L are LAGUERRE *polynomials* (2.3.24). The eigenvalues are given by (BALMER *formula*)

$$E_n = -\frac{2\pi^2 m_e e^4}{n^2 h^2}. \qquad (4.6.57)$$

Due to

$$\sum_{l=0}^{n-1}(2l + 1) = n^2 \qquad (4.6.58)$$

the hydrogen problem is $(n^2 - 1)$fold degenerated. Show that an hydrogen atom placed into an homogeneous constant magnetic field H gains energy $E \to E - ehHm/4\pi m_e$ and that the eigenvalue will be perturbed

$$\Delta E = \pm\frac{ehH}{4\pi m_e}\cdot m = -\mu_B Hm, \quad -l \le m \le l$$

and the degeneration is partially suspended. μ_B is BOHR'S *magneton*, m now the magnetic quantum number.

3. In a crystal lattice the potential acting on an electron may be periodic: $U = U_0 + U_1$, $U_1 = -A\cos 2\pi x/d$, where d is the unit cell dimension

(*lattice parameter*). Show that the setup $\psi(x) = exp2\pi i(by + cz)$ yields $E_1 = E - U_0 - h^2(b^2 + c^2)/2m_e$, $\xi = \pi x/d$, $\eta = 8d^2 m_e E_1/h^2$, $\gamma = 8d^2 m_e A/h^2$ and finally a MATHIEU *equation*

$$\frac{d^2 u}{d\xi^2} + (\eta + \gamma \cos 2\xi)u = 0 \qquad (4.6.59)$$

is obtained. Due to its *stability regions* interesting physical consequences appear.

(a) $\eta < -\gamma$ or $E_1 < -A$: no electron energy values exist that are smaller than the potential energy minimum,

(b) if $-\gamma < \eta < +\gamma$ or $-A < E_1 < +A$: electrons are bound in potential wells and oscillate,

(c) if $\eta > \gamma$ or $E_1 > A$: electrons move freely

(d) if $E_1 < A$, a band spectrum (BRILLOUIN *zones*) exists that broadens with increasing η [4.14].

4. Solve

$$y'' + (1 + x^2)y + 1 = 0, \quad y(\pm 1) = 0 \qquad (4.6.60)$$

using a power series [1.7]

$$y(x) = \sum_{\nu=0}^{\infty} a_\nu x^{2\nu}. \qquad (4.6.61)$$

Equation (4.6.60) describes the bending of a clamped *strut* with a distributed transverse load P. In (4.6.61) only even powers are included due to the symmetry of the differential equation. Insertion of (4.6.61) into (4.6.60) results in

$$\sum_{\nu=1}^{\infty} [(2\nu + 2)(2\nu + 1)a_{\nu+1} + a_\nu + a_{\nu-1}] x^{2\nu} + a_0 + 2a_1 + 1 = 0, \quad (4.6.62)$$

so that

$$1 + a_0 + 2a_1 = 0, \qquad (4.6.63)$$

$$a_{\nu-1} + a_\nu + (2\nu + 2)(2\nu + 1)a_{\nu+1} = 0, \quad \nu = 1, 2, 3, \ldots \qquad (4.6.64)$$

while the boundary condition demands

$$y(1) = \sum_{\nu=0}^{p} a_\nu = 0 \qquad (4.6.65)$$

and yields an equation for a_0. Here p may be $1, 2, \ldots 7$ (or more). Expressing all coefficients by a_0, then (4.6.63) and (4.6.64) yield

p	a_p	$\sum\limits_{\nu=0}^{p} a_\nu$	a_0
1	$-(1+a_0)/2$	$(a_0-1)/2$	1
2	$(1-a_0)/24$	$11(a_0-1)/24$	1
3	$(11+13a_0)/720$	$(343a_0-319)/720$	$319/343$

5. Solve the hyperbolic equation $u_{xx} - u_{tt} = 0$, $u(x,0) = \cos x$, $u_t(x,0) = 0$, $u(\pm\pi/2, t) = \sin t$ by a double power series

$$u(x,t) = \sum_{m,n=0}^{\infty} a_{mn} x^m t^n. \tag{4.6.66}$$

Solution:

$$a_{mn+2} = \frac{(m+2)(m+1)}{(n+2)(n+1)} a_{m+2n}. \tag{4.6.67}$$

The initial condition $u_t(x,0) = 0$ yields $a_{m1} = a_{02n+1} = 0$. Finally, the condition $u(x,0) = \cos x$ determines

$$a_{k0} = a_{0k} = \begin{cases} 0 & \text{odd } k, \\ (-i)^q/(2q)! & \text{even } k = 2q \end{cases}.$$

6. Solve

$$u''(x) - \alpha^2 u(x) = \beta[A(x)u(x) + B(x)v(x)],$$

$$v''(x) - \left(\alpha^2 + \frac{6}{x^2}\right) v(x) = \beta[B(x)u(x) + C(x)v(x)],$$

where α, β are known constants and $A(x), B(x), C(x)$ are known functions. These equations describe the *deuteron* for a regular proton-neutron potential [5.22] and are subjected to the boundary conditions

$$u(0) = 0, \quad v(0) = 0, \quad \lim_{x\to\infty} u(x) = \lim_{x\to\infty} v(x) = 0,$$

$$\int_0^{\infty} (u^2(x) + v^2(x))\, dx = 1. \tag{4.6.68}$$

Solve the problem using the iteration

$$u''_{\nu+1} - \alpha^2 u_{\nu+1} = \beta\left(Au_\nu + Bv_\nu\right),$$

$$v''_{\nu+1} - \left(\alpha^2 + \frac{6}{x^2}\right) v_{\nu+1} = \beta\left(Bu_\nu + Cv_\nu\right).$$

This results in

$$u'' - \alpha^2 u = f(x), \quad v'' - \left(\alpha^2 + \frac{6}{x^2}\right) v = g(x).$$

Integration yields

$$u(x) = c \sinh x + \frac{1}{\alpha} \int_0^x \sinh \alpha(x - \xi) f(\xi) \mathrm{d}\xi, \qquad (4.6.69)$$

where c follows from $u(\infty) = 0$. Furthermore,

$$v = c_1 \exp[\alpha x] p_1(x) + c_2 \exp[-\alpha x] p_2(x)$$

$$+ \frac{1}{2\alpha^2} \int_0^x \Big\{ \exp[\alpha(x - \xi)] p_1(x) p_2(\xi)$$

$$- \exp[-\alpha(x - \xi)] p_1(\xi) p_2(x) \Big\} g(\xi) \mathrm{d}\xi. \qquad (4.6.70)$$

c_1, c_2 are determined by (4.6.68) and for p_1, p_2 one obtains

$$p_1(x) = \alpha^3 - \frac{3\alpha^2}{x} + \frac{3\alpha}{x^2}, \quad p_2(x) = \alpha^3 + \frac{3\alpha^2}{x} + \frac{3\alpha}{x^2}.$$

The integrals contained in (4.6.69) and (4.6.70) are quite laborious. It is advantageous to expand $f(x)$ and $g(x)$ according to a system of orthonormal functions

$$f(x) = \sum_{n=0}^{\infty} C_n \psi_{1n}(x), \quad g(x) = \sum_{n=0}^{\infty} D_n \psi_{1n}(x).$$

$$\int_0^{\infty} \psi_{1n}(x) \psi_{1m}(x) \mathrm{d}x = \delta_{nm}$$
$$\delta_{nm} = \begin{cases} 0 & \text{for} \quad n \neq m, \\ 1 & \text{for} \quad n = m \end{cases}$$
$$\int_0^{\infty} \psi_{3n}(x) \psi_{3m}(x) \mathrm{d}x = \delta_{nm}$$

and

$$c_n = \int_0^{\infty} f(x) \psi_{1n}(x) \mathrm{d}x, \quad d_n = \int_0^{\infty} g(x) \psi_{1n}(x) \mathrm{d}x; \quad n = 0, 1, 2 \ldots.$$

Using LAGUERRE *polynomials* with $\alpha \to 2\alpha$, $k = 0$ and using the ROD-GRIGUEZ formula (2.3.24) one obtains

$$\psi_{1n}(x) = \frac{\sqrt{2\alpha}}{n!} \exp[\alpha x] \frac{\mathrm{d}^n}{\mathrm{d}x^n} \left(\exp[-2\alpha x] x^n \right) = \sqrt{2\alpha} \, L_n^k(x; 2\alpha),$$

$$\psi_{3n}(x) = \frac{2\alpha\sqrt{2\alpha}}{n!\sqrt{(n+1)(n+2)}} \frac{\exp[\alpha x]}{x} \frac{\mathrm{d}^n}{\mathrm{d}x^n} \left(\exp[-2\alpha x] x^{n+2} \right);$$

$$n = 0, 1, 2, \ldots,$$

$$u_0(x) = v_0(x) = \frac{1}{\sqrt{2}} \psi_{10}(x).$$

Using the ψ functions one can solve the differential equations iteratively or using collocation methods.

4.7 Variational calculus

Variational methods allow the development of new procedures to solve various mathematical problems. In section 1.4, equation (1.4.6), we briefly discussed the fundamental problem of the calculus of variation and we used the EULER *equation* (1.4.7). Now it is time to derive these equations. According to (1.4.6) for a given functional $F(x, y, y'(x), y''(x) \dots y^n(x))$, where $y(x)$ is unknown, the integral

$$J = \int_{x_1}^{x_2} F(x, y, y', y'' \dots) dx \qquad (4.7.1)$$

should be made an extremum, so that $y(x)$ may be determined. The existence of an extremum of J is guaranteed by the WEIERSTRASS *theorem*. This states that each continuous function possesses an extremum within a closed interval.

Let us consider a simple example [4.15]. The FERMAT *principle* of geometrical optics states that light propagating within an inhomogeneous (or homogeneous) medium of refractive index n takes that path that is the shortest in time. Formulated as a variational problem the principle reads $J \rightarrow$ extremum

$$J = \int dt = \int \frac{ds}{c(x, y, z)} = c_0 \int n(x, y, z) ds. \qquad (4.7.2)$$

To solve problems of this type we "variate" J:

$$\delta J = \delta \int_{x_1}^{x_2} F(x, y, y' \dots y^{(n)}) dx = 0. \qquad (4.7.3)$$

Assuming that such a function $y(x)$ exists, we write

$$y(x) + \varepsilon \delta y(x), \quad \delta y(x_1) = \delta y(x_2) = 0. \qquad (4.7.4)$$

It is apparent that the limit $\varepsilon \rightarrow 0$ induces the functional. Therefore

$$J = \int_{x_1}^{x_2} F\left(x, \ y + \varepsilon \delta y, \ y' + \varepsilon \delta y', \dots y^{(n)} + \varepsilon \delta y^{(n)}\right) dx \qquad (4.7.5)$$

and

$$\left(\frac{dJ}{d\varepsilon}\right)_{\varepsilon=0} = 0. \qquad (4.7.6)$$

Due to the smallness of ε, we expand the functional F into a TAYLOR series

$$J = \int\limits_{x_1}^{x_2} \left\{ F(x, y, y' \ldots) + \varepsilon \frac{\partial F}{\partial y} \, \delta y + \varepsilon \frac{\partial F}{\partial y'} \, \delta y' \right.$$

$$\left. + \ldots + \varepsilon \frac{\partial F}{\partial y^{(n)}} \, \delta y^{(n)} + \ldots \varepsilon^2 + \ldots \right\} \mathrm{d}x. \qquad (4.7.7)$$

Performing $(\mathrm{d}J/\mathrm{d}\varepsilon)_{\varepsilon=0}$ and integrating we obtain

$$\int\limits_{x_1}^{x_2} \frac{\partial F}{\partial y'} \, \delta y' \mathrm{d}x = \left[\frac{\partial F}{\partial y'} \, \delta y \right]_{x_1}^{x_2} - \int\limits_{x_1}^{x_2} \delta y \frac{\mathrm{d}}{\mathrm{d}x} \frac{\partial F}{\partial y'} \mathrm{d}x. \qquad (4.7.8)$$

Due to $\delta y(x_1) = 0$, $\delta y(x_2) = 0$, the [ ] term vanishes

$$\int\limits_{x_1}^{x_2} \delta y \left\{ \frac{\partial F}{\partial y} - \frac{\mathrm{d}}{\mathrm{d}x} \frac{\partial F}{\partial y'} + \frac{\mathrm{d}^2}{\mathrm{d}x^2} \frac{\partial F}{\partial y''} + \ldots + (-1)^n \frac{\mathrm{d}^n}{\mathrm{d}x^n} \frac{\partial F}{\partial y^{(n)}} \right\} \mathrm{d}x = 0. \qquad (4.7.9)$$

This integral vanishes then and only if the expression in braces vanishes. This gives the EULER *equation* (1.4.7). For one function $y(x)$ it reads

$$\frac{\partial F}{\partial y} - \frac{\mathrm{d}}{\mathrm{d}x} \frac{\partial F}{\partial y'} + \ldots + (-1)^n \frac{\mathrm{d}^n}{\mathrm{d}x^n} \frac{\partial F}{\partial y^{(n)}} = 0. \qquad (4.7.10)$$

These functionals are closely connected to important differential equations of physics and engineering. The functional

$$F(u, u_x, u_y) = \frac{1}{2} \left(u_x^2 + u_y^2 - k^2 u^2 \right) \qquad (4.7.11)$$

yields the HELMHOLTZ *equation* $u_{xx} + u_{yy} + k^2 u = 0$. It can be shown [3.13], [4.16] that every ordinary differential equation of second order $y'' = F(x, y, y')$ has an assigned functional reproducing the differential equation.

The so-called *direct methods* offer a direct solution of the variational problem (4.7.1) without any reference to the pertinent differential equation. Thus the direct methods offer methods to solve boundary value problems even in such cases, that an analytic solution of the differential equation is not possible.

The RITZ *method* is one of the simplest direct methods. It uses the successive setup

$$u_n(x, y) = u_0(x, y) + \sum_{\nu=1}^{n} c_{n\nu} u_\nu(x, y). \qquad (4.7.12)$$

Now one may interpret J as a functional of the new variables $c_{n\nu}$ and using $\partial J/\partial c_{n\nu} = 0$ yields the RITZ equations for the determination of the $c_{n\nu}$. In the RITZ method the functional F must be known. According to GALERKIN this is not necessary. One may use

$$\frac{\partial J}{\partial c_{n\nu}} = \int\limits_a^b\int\limits_c^d \left(\frac{\partial F}{\partial u}\frac{\partial u}{\partial c_{n\nu}} + \frac{\partial F}{\partial u_x}\frac{\partial u_x}{\partial c_{n\nu}} + \frac{\partial F}{\partial u_y}\frac{\partial u_y}{\partial c_{n\nu}} \right) dy dx$$

$$= \int\limits_a^b\int\limits_c^d \left(\frac{\partial F}{\partial u} u_n + \frac{\partial F}{\partial u_x} u_{nx} + \frac{\partial F}{\partial u_y} u_{ny} \right) dy dx. \tag{4.7.13}$$

Partial integration yields

$$\frac{\partial J}{\partial c_{n\nu}} = \frac{\partial F}{\partial u_x} u \Big|_a^b + \frac{\partial F}{\partial u_y} u \Big|_c^d + \int\limits_a^b\int\limits_c^d \left[\frac{\partial F}{\partial u} - \frac{d}{dx}\frac{\partial F}{\partial u_x} - \frac{d}{dy}\frac{\partial F}{\partial u_y} \right] u\, dy dx, \tag{4.7.14}$$

since the $u_n(x,y)$ have to satisfy the rectangular boundary conditions $u_n = 0$ for $x = a, b$, $y = c, d$. The brackets within the integral (4.7.14) vanish and are identical with the partial differential equation $\mathcal{L}\{u\} = 0$. Thus we have the GALERKIN *equations*

$$\int\limits_a^b\int\limits_c^d \mathcal{L}\{u\} u_n(x,y) dy dx = 0. \tag{4.7.15}$$

The coefficients $c_{n\nu}$ have to be determined in such a manner, that the differential equation $\mathcal{L}\{u\}$ is orthogonal to the *trial functions* u_n. To solve $\Delta u = 0$ for an ellipse with the semi-axes a and b with the inhomogeneous boundary condition $u_0 = x^2 + y^2$ one may use the RITZ trial functions

$$u_n(x,y) = x^2 + y^2 + \left(\frac{x^2}{a^2} + \frac{y^2}{b^2} - 1 \right)$$
$$\cdot \left(c_{n0} + c_{n1} x + c_{n2} y + c_{n3} x^2 + \ldots \right). \tag{4.7.16}$$

The first approximation yields

$$u_1 = x^2 + y^2 + c_{10} \left(\frac{x^2}{a^2} + \frac{y^2}{b^2} - 1 \right),$$

$$\frac{\partial u_1}{\partial x} = 2x \left(1 + \frac{c_{10}}{a^2} \right), \quad \frac{\partial u_1}{\partial y} = 2y \left(1 + \frac{c_{10}}{b^2} \right). \tag{4.7.17}$$

Then

$$J = \iint \left[4x^2 \left(1 + \frac{c_{10}}{a^2} \right)^2 + 4y^2 \left(1 + \frac{c_{10}}{b^2} \right)^2 \right] dx \, dy. \qquad (4.7.18)$$

Executing the integrations

$$\iint dx \, dy = \pi ab, \quad \iint x^2 dx \, dy = \frac{a^3 b \pi}{4}, \quad \iint y^2 dx \, dy = \frac{ab^3 \pi}{4} \qquad (4.7.19)$$

and applying (4.7.14) one obtains

$$c_{10} = -\frac{2a^2 b^2}{a^2 + b^2},$$

$$u_1(x, y) = x^2 + y^2 - \frac{2a^2 b^2}{a^2 + b^2} \left(\frac{x^2}{a^2} + \frac{y^2}{b^2} - 1 \right). \qquad (4.7.20)$$

In section 1.2, Table 1.1 we saw that the DIRICHLET problem for an elliptic equation like $\Delta u = 0$ is solvable for *one closed* boundary. No analytic solution seems to exist for *two closed* boundaries. But direct methods of variational calculus are able to solve a boundary problem of a two-fold connected domain [4.17]. The domain is given by a square of length $a = 4$ with a cut-off circle of radius $R = 1$, see Figure 4.14. Assuming the boundary values

$$u(x, y) = x^2 + y^2 \qquad (4.7.21)$$

on the boundaries as well as of the square *and* the circle, the appropriate trial functions will be

$$u_n(x, y) = x^2 + y^2 + (x^2 + y^2 - 1)(x^2 - 4)(y^2 - 4) \sum_{\nu=0}^{n} c_\nu u_\nu(x, y). \qquad (4.7.22)$$

The boundary conditions (4.7.21) are satisfied by these trial functions for the square boundaries $x = \pm 2$, $y = \pm 2$ and the circle boundary $x^2 + y^2 = 1$. Taking into account the symmetries of the problem, one may write

$$\sum_{\nu=0}^{5} c_\nu u_\nu(x, y) = c_0 + c_1(x + y) + c_2 \left(x^2 + y^2 \right)$$

$$+ c_3 xy + c_4 \left(x^3 + y^3 \right) + c_5 \left(x^2 y + xy^2 \right). \qquad (4.7.23)$$

Using the symmetry of the outer boundary and the integrals

$$I(n) = \iint x^n dx \, dy = \begin{cases} \dfrac{2^{n+2}}{n+1} - \dfrac{(n-1)!!}{(n+2)!!} \dfrac{\pi}{2} & \text{for } n \text{ even,} \\[3mm] \dfrac{2^{n+2}}{n+1} - \dfrac{(n-1)!!}{(n+2)!!} & \text{for } n \text{ odd} \end{cases} \qquad (4.7.24)$$

one obtains the numerical values presented in Table 4.1.

Table 4.1. Results from (4.7.23)

approximation	Nr 0	Nr 1	Nr 2	Nr 3
c_0	0.0625	0.1027	0.1812	0.0969
c_1	-	−0.0166	−0.0710	0.0521
c_2	-	-	−0.0005	−0.0673
c_3	-	-	0.0442	−0.0474
c_4	-	-	-	0.0131
c_5	-	-	-	0.0186
J	141.5520	140.7540	139.6440	139.6170

Using the values of the last column the solution may be written
```
f[x_,y_]=x^2+y^2+(x^2+y^2-1)*(x^2-4)*(y^2-4)*
(0.0969+0.0521*(x+y)-0.0673*(x^2+y^2)-0.0474*
x*y+0.0131*(x^3+y^3)+0.0186*(x^2*y+x*y^2))
P1=Plot3D[f[x,y],{x,-2,2},{y,-2,2},
Shading->False,ClipFill->None,
PlotRange-> {0.,8.},PlotPoints->100]
```
(4.7.25)

Function (4.7.25) has been plotted in Figure 4.14 together with the two boundaries. In problem 1 of this section we will discuss the methods to create Figure 4.14. Due to the quite rough approximation an additional small domain has been cut out.

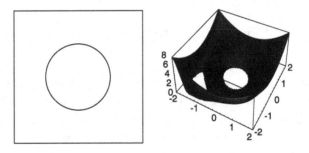

Figure 4.14
Two boundaries for the LAPLACE **equation**

The RITZ method is apparently capable to find a good approximation to the eigenfunctions. The RAYLEIGH *method* is better suited to find eigenvalue approximations [1.2], [3.14], [1.7]. We consider an equation and its functional

$$\frac{d}{dx}\left[p(x)\frac{du}{dx}\right] - q(x)u(x) + \lambda\rho(x)u(x) = 0,$$

$$J = \int[pu'^2 + qu^2 - \lambda\rho u^2]\,dx.$$

(4.7.26)

Then the smallest eigenvalue λ is given by the RAYLEIGH *coefficient*

$$\lambda = \int\limits_a^b (pu'^2 + qu^2)\,\mathrm{d}x \Big/ \int\limits_a^b \rho u^2 \mathrm{d}x. \tag{4.7.27}$$

Let us try an example. According to (2.4.45) a circular membrane of radius R has the lowest eigenvalue $\sqrt{\lambda} = 2.404\,825/R$. Now for the HELMHOLTZ equation in polar coordinates one may choose the trial function

$$u = 1 - r/R. \tag{4.7.28}$$

Inserting $u(r)$ into

$$\int\limits_0^{2\pi}\int\limits_0^R u'^2(r) r\mathrm{d}r\mathrm{d}\varphi, \quad \int\limits_0^{2\pi}\int\limits_0^R u^2(r) r\mathrm{d}r\mathrm{d}\varphi \tag{4.7.29}$$

one gets

$$\sqrt{\lambda} = \sqrt{\frac{\pi}{\pi R^2/6}} = \sqrt{6}/R = 2.449/R. \tag{4.7.30}$$

A third procedure is the TREFFTZ *method*. Instead of domain integrals like (4.7.29), only boundary integrals have to be calculated. This method is similar to the *boundary element method* [4.18]. Another procedure is the *least squares method* looking for a minimum of $(\int (u_0 - u)^2 \mathrm{d}x)^{1/2}$, where $u_0 - u$ is the difference between the solution of the differential equation and the trial function [1.7]. Here the *boundary maximum principle* [1.7] helps. For elliptic partial differential equations it states that the difference $u_0 - u$ is equal or smaller than the difference between the approximate solution and the given boundary values. Let us consider an example for the TREFFTZ method:

$$\Delta u(x,y) = b,$$

$$u(1,y) = u(x,1) = 0. \tag{4.7.31}$$

First the POISSON equation will be homogenized:

$$u(x,y) = b(x^2 + y^2)/4 + v(x,y), \quad \Delta v = 0,$$

$$v = -b(x^2 + y^2)/4 \quad \text{for} \quad x = \pm 1,\, y = \pm 1. \tag{4.7.32}$$

Now we define the trial function for a square

$$v = (x^4 - 6x^2 y^2 + y^4)\alpha. \tag{4.7.33}$$

Use of the second GREEN *theorem*

$$\int (v\nabla u - u\nabla v)\mathrm{d}f = \int (v\Delta u - u\Delta v)\mathrm{d}\tau \tag{4.7.34}$$

and of $\Delta u = 0$, $\Delta v = 0$ leads to the boundary integral over the square

$$\int \left(v \frac{\partial}{\partial x} \delta v - \frac{\partial v}{\partial x} \delta v \right) dx = 0 \qquad (4.7.35)$$

or

$$\int_0^1 \left[\frac{1}{4} b \left(1 + y^2 \right) \left(4 - 12 y^2 \right) \delta \alpha + \alpha \left(4 - 12 y^2 \right) \left(1 - 6 y^2 + y^4 \right) \delta \alpha \right] dy = 0,$$

so that $\alpha = 7b/144$ results.

Variational methods may solve integral equations too. In wave mechanics the connection between a potential V between the particles and the *scattering cross section* is important. From scattering experiments one may calculate the potential [4.19]. Let us consider the integral equation

$$y(r) = y_0(r) + \varepsilon \int_0^\infty K(r, r') y(r') dr'. \qquad (4.7.36)$$

Here ε is a physical constant, the kernel $K(r, r')$ may be symmetric and expressible by a scattering potential V and a GREEN function. Now two integrals will be defined:

$$I_1 = \int_0^\infty y^2(r) dr - \varepsilon \int_0^\infty \int_0^\infty y(r) K(r, r') y(r') dr dr', \qquad (4.7.37)$$

$$I_2 = \int_0^\infty y(r) y_0(r) dr. \qquad (4.7.38)$$

If $y(r)$ solves (4.7.36), then $I_1 = I_2$. This can be shown by multiplying (4.7.36) by $y(r)$ and integration. Now let us variate $y(r) \to y(r) + \delta y(r)$, then

$$\delta I_1 = \int_0^\infty 2 y(r) \delta y(r) dr - \varepsilon \int_0^\infty \int_0^\infty \delta y(r) K(r, r') y(r') dr dr', \qquad (4.7.39)$$

$$\delta I_2 = \int_0^\infty \delta y(r) y_0(r) dr, \qquad (4.7.40)$$

since $\delta y^2 = 2 y \delta y$. (The variation symbol δ is subjected to the same rules as the differential). If $y(r)$ is a solution of the integral equation, then $\delta I_1 = 2 \delta I_2$ or $\delta(I_1 - 2I_2) = 0$ or according to the SCHWINGER *variational principle*:

$$\delta \left(\frac{I_1}{I_2^2} \right) = 0. \qquad (4.7.41)$$

Such variational methods are able to solve nonseparable partial differential equations.

Problems

1. The reader is invited to produce the right-hand side of Figure 4.14 by typing

```
P1=Plot3D[f[x,y],{x,-2,2},
{y,-2,2},Shading->False,ClipFill->None,
PlotRange-> {0.,8.},PlotPoints->100]
```
(4.7.42)

```
ClipFill->None
```

avoids the closing of holes.

The circle in this figure may be created by

```
Clear[CG]; CG = Graphics[Circle[{0., 0.}, 1.],
AspectRatio -> Automatic];
Show[CG]
```

where **AspectRatio -> Automatic** avoids perspective deformation.

The square will be created by

```
Clear[Q];
Q=Line[{{-2.,-2.},{2.,-2},{2.,2.},{-2.,2.},
{-2.,-2.}}]
```

Then the combination of circle and square can be done by the command

```
QG=Graphics[Q];
QQG=Show[QG,CG,AspectRatio->Automatic]
```

Finally the command

```
Show[GraphicsArray[{QQG, P1}]]
```

produces Figure 4.14.

2. Consider the transversal oscillations of a heavy rod with no load P described by (4.5.54) in the form

$$c^2 u_{xxxx} - \omega^2 u + g(l-x)u_{xx} - gu_x = 0.$$
(4.7.43)

Calculate

$$J_1 = \iint u \left[c^2 u_{xxxx} + g(l-x)u_{xx} - gu_x \right] \mathrm{d}x\mathrm{d}y$$

$$J_2 = -\omega^2 \iint u\mathrm{d}x\mathrm{d}y.$$

Combining RAYLEIGH with GALERKIN write

$$\lambda = \omega^2 = \int_0^l u \left[c^2 u_{xxxx} + g(l-x)u_{xx} - gu_x \right] \mathrm{d}x \bigg/ \left(-\int_0^l u^2 \mathrm{d}x \right).$$
(4.7.44)

Assume the trial function,

$$u(x) = a(l^3 x - 2lx^3 + x^4), \tag{4.7.45}$$

which satisfies $u(0) = 0$, $u(l) = 0$, $u''(0) = 0$, $u''(l) = 0$. Calculate

$$\lambda = \frac{9.877}{l^2} \sqrt{\frac{EI}{\rho q}} \left(1 - 0.051 \frac{\rho g q l^3}{EI}\right)$$

or

$$\lambda = \frac{9.877}{l^2} \sqrt{\frac{EI}{\rho q}} \quad \text{for} \quad g = 0.$$

3. Solve the membrane vibrations $c^2(u_{xx} + u_{yy}) + \omega^2 u(x,y) = 0$ for $u(-l,y) = 0$, $u(l,y) = 0$, $u(x,-l) = 0$, $u(x,l) = 0$, using the trial function $a(l^2 - x^2)(l^2 - y^2)$ (RITZ: $\omega \approx 2.236c/l$).

4. Solve $\Delta\Delta u + k^2 u = 0$, $u(-l,y) = 0$, $u(l,y) = 0$, $u(x,-l) = 0$, $u(x,l) = 0$, $\Delta u(-l,y) = 0$, $\Delta u(l,y) = 0$, $\Delta u(x,-l) = 0$, $\Delta u(x,l) = 0$ (NAVIER solutions) using $(l^2 - x^2)(l^2 - y^2)$ (RITZ: $\omega \approx 5.244c/l^2$).

5. Solve

$$\Delta u(x,y) = 0 \quad \text{for} \quad -1 \le x \le +1, -1 \le y \le +1 \tag{4.7.46}$$

with

$$u = 1 + y^2 \text{ for } x = \pm 1, \quad u = x^2 + 1 \text{ for } y = \pm 1. \tag{4.7.47}$$

The solutions $\cos \alpha x \cosh \alpha y$ of the LAPLACE equation cannot be used as trial functions, since $\cosh(2n+1)y/2$ increases with increasing n. Use symmetric *harmonic polynomials*.

$$v_1 = \text{Re}(x + iy)^4 = x^4 - 6x^2y^2 + y^4, \tag{4.7.48}$$

$$v_2 = \text{Re}(x + iy)^8 = x^8 - 28x^6y^2 + 70x^4y^4 - 28x^2y^6 + y^8 \tag{4.7.49}$$

and the trial function

$$v = a_0 + a_1 v_1 + a_2 v_2. \tag{4.7.50}$$

Using the boundary maximum principle one may demand

$$|(x^2 + 1) - (a_0 + a_1(x^4 - 6x^2 + 1))| \quad \to \text{Min} \tag{4.7.51}$$

as first appproximation. If the error may be 0.025 and assuming the two *collocation points* (0,1) and (1,1) one has from (4.7.51)

$$x = 0: \quad 1 - (a_0 + a_1) = 0.025,$$

$$x = 1: \quad 2 - (a_0 + a_1 - 6a_1 + a_1) = 0.025,$$

so that $a_0 = 1.175$, $a_1 = -0.2$. Show that the next approximation demanding an error ≤ 0.005 results in $a_0 = 1.174\,465$, $a_1 = -0.184\,738$, $a_2 = 0.005\,087\,44$. Plot $u = x^2 + y^2$ and the trial function (4.7.50).

6. In problem 5 the differential equation (5.4.1) had been solved exactly and the boundary conditions had been satisfied by the trial function (*outer collocation*). Now try *interior collocation* . Satisfy the boundary conditions exactly and the trial function should approximately satisfy the differential equation. Use

$$u = x^2 + y^2 + (1 - x^2)(1 - y^2) = 1 + x^2 y^2,$$

$$v = u + (1 - x^2)(1 - y^2) \sum_{n,m} a_{nm} x^{2n} y^{2m}.$$

Choose the collocation points $(0,0)$, $(0,1)$, $(1,0)$, $(1,1)$ and $\Delta v = 0$. Result: $a_{00} = a_{10} = a_{01} = 0.142\,857$, $a_{11} = 0$.

4.8   Collocation methods

If a partial differential equation is separable or reducible into ordinary differential equations as discussed in chapter 3 and if the boundaries coincide with coordinate lines of a coordinate system in which the partial differential equation is separable, then the solution of a boundary problem is easy. There occur however many problems that are not well posed or unseparable. In principle, however, the general solution of a partial differential equation must contain the solution of any boundary problem. Not only theoretical considerations but also the success of numerical methods like finite differences, finite elements or boundary elements lead to this conclusion. However, very often the problem is how to find the solution. In some cases, if for instance the boundary has sharp corners or large curvature, the grid of discrete points, the meshes, must be very fine and very large matrices appear in the respective code giving numerical problems and causing long and time-wasting programming efforts. There are problems for instance in plasma physics, where the numerical three-dimensional solution of a plasma confinement exists, but mathematicians deny the possibility of solving the pertinent equations.

Having in mind the theorem that an analytic solution of an elliptic equation exists only for one closed boundary, let us find a classification scheme of such boundary value problems:

1. The partial differential equation is *separable*:

1.1 one smooth boundary: the problem is well posed and the solution is analytic; degeneration of eigenfunctions is admitted and definition of nodal lines as boundary is possible, see (4.4.6),

1.2 two smooth boundaries like in an annular circular membrane: the solution is no longer analytic and contains *singularities*,

1.3 two smooth closed boundaries belonging to two *different* separable coordinate systems ("nonuniform problems"), holes within the domain like in (4.7.21), see Figure 4.14,

1.4 three boundary conditions for an elliptic equation of second order delivering a divergent approximate solution as in problem 1 of section 4.6,

1.5 no smooth JORDAN *curves* as boundaries, but corners etc., see Figures 4.2 and 4.3,

1.6 a special situation is given, if the partial differential equation is separable, but the boundary conditions cannot be satisfied in the actual coordinate system, see section 5.1

2. The partial differential equation is NOT *separable*:

2.1 well posed, no holes, no corners, only one boundary

2.2 two or more boundaries, corners.

In cases 1.2 to 1.5 and for 2., approximate or variational methods might help, but very often only *numerical methods* can solve the problem:

a. *finite differences* [11], [1.7], [3.36], [4.20] to [4.24]

b. *finite element methods* (FEM) [4.25] to [4.27], [4.29], [4.30] (appropriate for the whole domain)

c. *boundary element methods* (BEM) [4.18], [4.27], [4.28] (applicable only if the domain is homogeneous, no density distribution or locally varying coefficients in the differential equation

d. *collocation methods* [1.7], [4.31], [4.32], [4.42] as used in section 1.3 and in problem 5 of section 4.7

e. LIE *series methods* [4.13], [4.20]

f. variational methods, see (4.7.22).

Up to now we mainly considered analytical solutions. Let us now investigate the influence of a *singularity* in the solution. We first discuss a simple homogeneous problem of the LAPLACE *equation* for a rectangle $a \times b$

$$\Delta U(x,y) = 0, \tag{4.8.1}$$

$$U(\pm a, y) = 0, \quad -b \le y \le +b, \tag{4.8.2}$$

$$U(x, \pm b) = f(x) = \cos(\pi x/a), \quad -a \le x \le +a, \tag{4.8.3}$$

where we set later $a = 2$, $b = 1$, see Figure 4.15. The solution to (4.8.1) to (4.8.3) is given by

$$U(x, y) = \sum_{n=0}^{\infty} A_n \cos \frac{(2n+1)\pi x}{2a} \cosh \frac{(2n+1)\pi y}{2a}, \tag{4.8.4}$$

where

$$A_n = \frac{1}{a \cosh \dfrac{(2n+1)\pi b}{2a}} \int_{-a}^{+a} f(\xi) \cos \frac{(2n+1)\pi \xi}{2a} d\xi. \tag{4.8.5}$$

The solution (4.8.4) is shown in Figure 4.15.

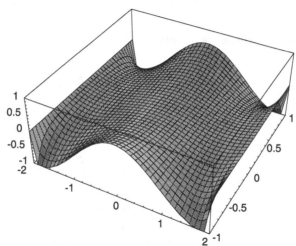

Figure 4.15
Inhomogeneous boundary problem (4.8.4)

The following *Mathematica* command realized the calculations (4.8.5) and generated Figure 4.15.

```
A[n_]=Integrate[Cos[Pi*x/2.]*Cos[(2*n+1)*Pi*x/4.],
{x,-2.,2.}]/(2*Cosh[(2*n+1)*Pi/4.]);
Clear[U];U[x_,y_]=Sum[A[n]*Cos[(2*n+1)*Pi*x/4.]
*Cosh[(2*n+1)*Pi*y/4.],{n,50}];

A[17]
3.836 24×10⁻¹⁴
Plot3D[U[x,y], {x,-2.,2.},{y,-1.,1.},
PlotRange->{-1.,1.0},PlotPoints->40]
```

The transition to a homogeneous boundary problem $f(x) \to 0$ results immediately in the *trivial solution* $U(x,y) = 0$, since all $A_n \to 0$. This confirms the theorem that the LAPLACE equation has only the trivial (analytic) solution for homogeneous boundary conditions. Could a singularity modify the situation? It is known that the logarithm of a square root is a singular particular solution of (4.8.1), see problem 1 in this section. We add such a term to (4.8.4) so that

$$U(x,y) = \sum_{n=0}^{\infty} A_n \cos \frac{(2n+1)\pi x}{2a} \cosh \frac{(2n+1)\pi y}{2a}$$

$$+ \sum_{n=0}^{\infty} B_n \cos \frac{(2n+1)\pi y}{2b} \cosh \frac{(2n+1)\pi x}{2b}$$

$$- \ln \sqrt{\frac{x^2+y^2}{a^2+b^2}}. \tag{4.8.6}$$

Then the homogeneous boundary conditions yield [4.33]

$$A_n = \frac{1}{a \cosh \dfrac{(2n+1)\pi b}{2a}} \int_{-a}^{+a} \ln \sqrt{\frac{x^2+b^2}{a^2+b^2}} \cos \frac{(2n+1)\pi x}{2a} \, dx,$$

$$B_n = \frac{1}{b \cosh \dfrac{(2n+1)\pi a}{2b}} \int_{-b}^{+b} \ln \sqrt{\frac{a^2+y^2}{a^2+b^2}} \cos \frac{(2n+1)\pi y}{2b} \, dy. \tag{4.8.7}$$

The logarithmic term satisfies the homogeneous boundary conditions in the corners. For $n = 0, \ldots 3$ the boundary conditions are satisfied with an accuracy 10^{-4}. The same result may be obtained by a *collocation method*. Eight collocation points (x_i, b), $0 \le x_i \le a$, (a, y_i), $0 \le y_i \le b$, $i = 1 \ldots 4$ produce eight linear equations for the A_n (and B_n)

$$\sum_{n=0}^{3} A_n \cos \frac{(2n+1)\pi x_i}{4} \cosh \frac{(2n+1)\pi}{4} = \ln \sqrt{\frac{x_i^2+1}{5}} \tag{4.8.8}$$

which avoids the numerical integrations (4.1.7). Figure 4.16 shows the solution (4.8.6).

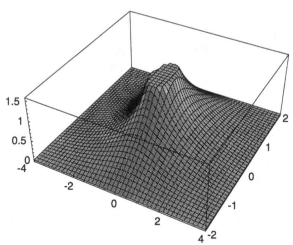

Figure 4.16
Homogeneous boundary problem of the LAPLACE **equation**

Having seen the good results obtained by the collocation method it might be appropriate to discuss this method in detail. Where should the collocation points be chosen? FÉJER answers: for a circle they should be determined by the unit roots in the complex plane and for other boundaries the conformal mapping may help [4.34] to [4.36]. The question of the convergence of collocation methods has been discussed by GRISVARD [4.37] (guarantees the convergence of the solution between collocation points toward the real solution) and EISENSTAT [4.38]. Collocation methods [4.39], [4.40] use mainly exact solutions (like harmonic polynomials etc.) of the pertinent partial differential equations like the LAPLACE or HELMHOLTZ equations.

Collocation methods may also be used to define nodal lines as new boundaries. It is easy to verify that

$$U(x, y) = A \cos kx + B \cos \left(\sqrt{k^2 - \beta^2}\, x \right) \cos \beta y \qquad (4.8.9)$$

is a solution of the two-dimensional HELMHOLTZ *equation* $U_{xx} + U_{yy} + k^2 U = 0$, compare (1.1.55). Now the *geometrical locus* of all points (x_i, y_i) at which $U(x_i, y_i) = U(x_i, y_i(x_i)) = 0$ represents a nodal line. From (4.8.9) one obtains a nodal line

$$y(x) = \frac{1}{\beta} \arccos \left[\frac{-A \cos kx}{B \cos(\sqrt{k^2 - \beta^2} x)} \right]. \qquad (4.8.10)$$

Playing with the three parameters $A/B, k, \beta$ may generate various boundaries. On the other hand we can assume a *circular boundary* $y = \sqrt{R^2 - x^2}$ in *cartesian coordinates* and we could try to derive the eigenvalue k of a circular membrane [4.31]. Let us consider a circular membrane of radius $R (= 1\text{cm})$. The eigenvalue can be calculated exactly in cylindrical coordinates r, φ. For axial symmetry, or $\partial u/\partial \varphi = 0$ one obtains $k = 2.405$. Can we reproduce this value with the new method? We choose the cartesian solution of $u_{xx} + u_{yy} + k^2 u = 0$ which is symmetric in x and y. Writing a FORTRAN program for (4.8.10) with two DO-loops (for variables k and β) we find for $R = 1$ that $\beta = 2.0780$ and $k = 2.4019$ and that (4.8.10) describes a circle. Now we have to control if (4.8.10) describes a continuous function with no singularity and if the accuracy with which it describes the circle $x^2 + y^2 = R^2$ is satisfying, which was the case. When we solved with two "separation constants" and with the collocation points $(0,1)$ $(1,0)$ $(0.7071, 0.7071)$ we obtained the results $\beta_1 = 2.222\,096\,473$, $\beta_2 = 0.919\,496\,181$ and $k = 2.404\,825\,557\,415$. The exact values k given from the zero of the BESSEL function J_0 is $2.404\,825\,557\,069\,6$. We thus also have a method to calculate higher zeroes of BESSEL functions:

Number of collocation points	eigenvalue	exact values
2	2.404 19	2.404 825 557 6
3	2.404 825 557 4	2.404 825 557 6
3	5.520 078 110 3	5.520 078 110 3
4	8.653 727 912 91	8.653 727 912 9
5	11.791 534 439 01	11.791 534 439 1
6	14.930 917 708 49	14.930 917 708 6
7	18.071 063 967 91	18.071 063 967 9
8	21.211 636 629 88	21.211 636 629 9

For a greater number of collocation points, the methods to solve a set of transcendental equations fail or give only the trivial solution. We are therefore using another method that avoids the solution of nonlinear or transcendental equations.

We now consider a generalization of (4.8.9) in the form

$$\sum_{n=1}^{N} A_n \cos\left(\sqrt{k^2 - \beta_n^2}\, x_i\right) \cos\left(\beta_n \sqrt{1 - x_i^2}\right) = 0, \; i = 1\ldots P. \qquad (4.8.11)$$

With the boundary condition on a circle and a solution in cartesian coordinates we have

$$u(r = R = 1) = 0, \quad y = \sqrt{1 - x^2},$$

$$u(x,y) = \sqrt{1 - x^2} = \sum_{n=1}^{N} A_n \cos\left(\sqrt{k^2 - \beta_n^2}\, x_i\right) \cos\left(\beta_n \sqrt{1 - x_i^2}\right) = 0.$$

$$(4.8.12)$$

k: eigenvalue, N: number of partial solutions and the A_n are the amplitudes of the partial modes. The number P of collocation points $x_i, y_i, i = 1\ldots P$

on the circle $x_i^2 + y_i^2 = 1$. For $A_1 = 1$ and the remaining $2N$ unknowns $A_2, \ldots A_n, \beta_1 - \beta_N$ and k by use of a computer code for $N = 2, P = 2N = 4$, we obtain $\beta_1 = 2.22209, \beta_2 = 0.91949, A_2 = 1$ and $k = 2.4048255$ (exact value up to 6 digits). For fixed P the accuracy remains fixed. Choosing arbitrarily the $\beta_n, n = 1, 2 \ldots N$ and $A_1 = 1$, only the $P = N$ unknowns $A_n, n = 2 \ldots N$ and the eigenvalue k remain to be determined. Then (4.8.11) constitutes a system of homogeneous linear equations for the A_n. To find a nontrivial solution, the determinant $D(k)$

$$D = |D_{in}| = |\cos\left(\sqrt{k^2 - \beta_n^2}\, x_i\right) \cos\left(\beta_n \sqrt{1 - x_i^2}\right)| = 0,$$

$$i = 1 \ldots P, n = 1 \ldots P \quad (4.8.13)$$

has to vanish. This determines k. Solving now for the $A_n, n = 2 \ldots N, A_1 = 1$ one obtains the solution. The β_n had been chosen arbitrarily, $\beta_n^2 < k^2$.

A simple FORTRAN *program* does the job to solve $u_{xx} + u_{yy} + k^2 u = 0$ for a clamped circular membrane. The program COLLOC.f reads for $P = 8$ collocation points and 8 arbitrary separation constants BETA. The number P should not be very large (< 20). If the number of collocation points is large, inaccuracies, oscillations may appear and the accuracy goes down.

```
C23456PROGRAM COLLOC
       IMPLICIT NONE
       INTEGER P,I,J,III,IT,L
       PARAMETER (IT=80,P=8)
       INTEGER INDX(P)
       DOUBLE PRECISION  DETN(P,P),DET(P,P),EIGENVALUE,
      *BETA(P),DELTA
       DOUBLE PRECISION DEIGENVALUE,D,DETV(IT),X0(P),
      *Y0(P)
       OPEN (UNIT=2, FILE="COLLOCRes")
C      GUESS OF EIGENVALUE
C      L=1
C      IF(L.GT.1) GOTO 22
       WRITE(6,*)'GIVE GUESS OF EIGENVALUE'
       READ (5,*) EIGENVALUE
  22   DEIGENVALUE=EIGENVALUE/10
       WRITE(2,*)'FIRST EIGENVALUE= ',EIGENVALUE
       DELTA=EIGENVALUE/P
       BETA(1)=EIGENVALUE-0.0001
       DO 44 I=1,P-1
          BETA(I+1)=BETA(I)-DELTA
  44   CONTINUE
       DO 444 I=1,P
          WRITE(2,*)' BETA(',I,')=',BETA(I)
```

```
 444    CONTINUE
        CALL DATA (P,X0,Y0,L)
C       WRITE(2,*) ' EIGENVALUE = , DETERMINANT = '
        DO 999 III=1,IT
           CALL FILLUP (DETN,EIGENVALUE,BETA,X0,Y0,P)
           CALL LUD(DETN,P,INDX,D)
           DO 20 J=1,P
              D=D*DETN(J,J)
 20     CONTINUE
        DETV(III)=D
        IF(III.EQ.1)GOTO 999
C       IS THERE A CHANGE OF SIGN ?
        IF(DETV(III)*DETV(III-1).LT.0.)THEN
        DEIGENVALUE=-DEIGENVALUE/2
        ENDIF
C       WRITE(2,77) EIGENVALUE,DETV(III)
C 77    FORMAT(1X, D19.13,3X,D19.13)
        EIGENVALUE=EIGENVALUE+DEIGENVALUE
 999    CONTINUE
        WRITE(2,*)'LAST EIGENVALUE:  '
        WRITE(6,*)'LAST EIGENVALUE:  '
 55     FORMAT(F16.10)
        CALL FILLUP (DETN,EIGENVALUE,BETA,X0,Y0,P)
C       WRITE(2,55) EIGENVALUE-DEIGENVALUE
        WRITE(6,55) EIGENVALUE-DEIGENVALUE
C       FILL DET WHICH  WILL BE MODIFIED IN LUD AND
C       DETN WILL BE NEEDED
        DO 1 I=1,P
           DO 2 J=1,P
        DET(I,J)=DETN(I,J)
 2      CONTINUE
 1      CONTINUE
        CALL LINEQ (DET,DETN)
C       LOOK FOR NEXT EIGENVALUE
C       L=L+1
C       IF(L.GT.1)THEN
C       EIGENVALUE=EIGENVALUE+0.1
C       IF(L.GT.6)GOTO 100
C       GOTO 22
C       ENDIF
 100    WRITE(6,*)'I HAVE FINISHED'
        WRITE(6,*) CHAR(7)
        STOP
        END
```

```
      SUBROUTINE DATA (P,X0,Y0,L)
C23456CALCULATION OF COLLOCATION POINTS ON BOUNDARY
      IMPLICIT NONE
      INTEGER P,I,L
      DOUBLE PRECISION DTH,TH,AR,X0(P),Y0(P)
      AR=1.
      DTH=3.1415926535/P
      TH=DTH/2
      DO 10 I=1,P
      X0(I)=AR*COS(TH)
      Y0(I)=AR*SIN(TH)
      IF(L.EQ.1) WRITE(2,*)'X,Y',X0(I),Y0(I)
      TH=TH+DTH/2
 10   CONTINUE
      RETURN
      END


      SUBROUTINE FILLUP (DETN,EIGENVALUE,BETA,X0,Y0,P)
C23456FILLING OF DETERMINANT
      IMPLICIT NONE
      INTEGER P,I,N
      DOUBLE PRECISION TT,EIGENVALUE,DETN(P,P),X0(P),
     *Y0(P),BETA(P)
      DO 1 I=1,P
         DO 2 N=1,P
      TT=SQRT(EIGENVALUE**2-BETA(N)**2)
      DETN(I,N)=COS(BETA(N)*Y0(I))*COS(TT*X0(I))
 2    CONTINUE
 1    CONTINUE
      RETURN
      END


      SUBROUTINE LUD(A,P,INDX,D)
C23456CALCULATES DETERMINANT  BY LU DECOMPOSITION
      IMPLICIT NONE
      INTEGER NMAX, P,K,IMAX,J,I, INDX
      PARAMETER (NMAX=100)
      DOUBLE PRECISION A(P,P),D,AAMAX,SUM,DUM, TINY,
     *VV(NMAX)
      PARAMETER(TINY=1.0E-40)
      DIMENSION INDX(P)
      D=1.
      DO 12 I=1,P
         AAMAX=0.
      DO 11 J=1,P
```

```
          IF (ABS(A(I,J)).GT.AAMAX) AAMAX=ABS(A(I,J))
11     CONTINUE
          IF (AAMAX.EQ.0.) PAUSE 'Singular matrix.'
          VV(I)=1./AAMAX
12     CONTINUE
       DO 19 J=1,P
          DO 14 I=1,J-1
          SUM=A(I,J)
             DO 13 K=1,I-1
             SUM=SUM-A(I,K)*A(K,J)
13     CONTINUE
       A(I,J)=SUM
14     CONTINUE
       AAMAX=0.
       DO 16 I=J,P
          SUM=A(I,J)
             DO 15 K=1,J-1
             SUM=SUM-A(I,K)*A(K,J)
15     CONTINUE
       A(I,J)=SUM
       DUM=VV(I)*ABS(SUM)
       IF (DUM.GE.AAMAX) THEN
       IMAX=I
       AAMAX=DUM
       ENDIF
16     CONTINUE
       IF (J.NE.IMAX)THEN
       DO 17 K=1,P
          DUM=A(IMAX,K)
          A(IMAX,K)=A(J,K)
          A(J,K)=DUM
17     CONTINUE
       D=-D
       VV(IMAX)=VV(J)
       ENDIF
       INDX(J)=IMAX
          IF(A(J,J).EQ.0.)A(J,J)=TINY
          IF(J.NE.P)THEN
          DUM=1./A(J,J)
          DO 18 I=J+1,P
          A(I,J)=A(I,J)*DUM
18     CONTINUE
       ENDIF
19     CONTINUE
       RETURN
```

```
      END

      SUBROUTINE LINEQ (DET,DETN)
C23456SOLVES SYSTEM HOMOG. LINEAR EQUATIONS
      IMPLICIT NONE
      INTEGER P,I,J,JF,L,K,NF,IF
      PARAMETER(P=8)
      DOUBLE PRECISION AF(P-1,P-1),BF(P-1),A(P),
     *DETN(P,P),BOUNDARY(P)
      DOUBLE PRECISION C(P-1,P-1),X(P-1),PQ,DET(P,P)
      INTEGER INDX(P-1)
      A(P)=1.000000
C     REDUCTION OF SIZE OF HOMOG. LINEAR SYSTEM
      NF=P-1
      DO 501 IF=1,NF
         BF(IF)=-DET(IF,P)
         DO 501 JF=1,NF
            AF(IF,JF)=DET(IF,JF)
  501 CONTINUE
      DO 12 L=1,NF
         DO 11 K=1,NF
            C(K,L)=AF(K,L)
   11 CONTINUE
   12 CONTINUE
      CALL LUD(C,NF,INDX,PQ)
      DO 13 L=1,NF
         X(L)=BF(L)
   13 CONTINUE
      CALL LUB(C,NF,INDX,X)
      DO 33 I=1,NF
         A(I)=X(I)
   33 CONTINUE
      WRITE(2,*) 'PARTIAL AMPLITUDES ARE: '
      WRITE(6,*) 'PARTIAL AMPLITUDES ARE: '
      WRITE(2,'(1X,6F12.4)') (A(L), L=1,P)
      WRITE(6,'(1X,6F12.4)') (A(L), L=1,P)
      DO 10 I=1,P
         BOUNDARY(I)=0.
      DO 10 J=1,P
         BOUNDARY(I)=BOUNDARY(I)+DETN(I,J)*A(J)
   10 CONTINUE
      DO 4 I=1,P
      WRITE(2,*)'BOUNDARY(',I,')= ', BOUNDARY(I)
      WRITE(6,*)'BOUNDARY(',I,')= ', BOUNDARY(I)
    4 CONTINUE
```

```
        RETURN
        END

        SUBROUTINE LUB(A,NF,INDX,B)
C23456SOLVES REDUCED LINEAR EQUATIONS
        IMPLICIT NONE
        INTEGER NF,I,LL,J,II,INDX(NF)
        DOUBLE PRECISION A(NF,NF),B(NF),SUM
        II=0
        DO 12 I=1,NF
           LL=INDX(I)
           SUM=B(LL)
           B(LL)=B(I)
           IF (II.NE.0)THEN
              DO 11 J=II,I-1
              SUM=SUM-A(I,J)*B(J)
11      CONTINUE
        ELSE IF (SUM.NE.0.) THEN
        II=I
        ENDIF
        B(I)=SUM
12      CONTINUE
        DO 14 I=NF,1,-1
           SUM=B(I)
           IF(I.LT.NF)THEN
           DO 13 J=I+1,NF
              SUM=SUM-A(I,J)*B(J)
13      CONTINUE
        ENDIF
        B(I)=SUM/A(I,I)
14      CONTINUE
        RETURN
        END
```

The executable file may be called circmem and the results are written to the file COLLOCRes that reads

Output on the screen from program circmem

```
GIVE GUESS OF EIGENVALUE
2.4
LAST EIGENVALUE:
2.4048255577
PARTIAL AMPLITUDES ARE:
        .1773      .4925    -.6585   1.8660  -2.7769   3.4474
```

```
        -2.4165      1.0000
BOUNDARY( 1)= -4.440892098500626E-16
BOUNDARY( 2)=   .0
BOUNDARY( 3)=   .0
BOUNDARY( 4)=  1.387778780781446E-17
BOUNDARY( 5)= -8.326672684688674E-17
BOUNDARY( 6)=   .0
BOUNDARY( 7)=  1.110223024625157E-16
BOUNDARY( 8)=  1.271205363195804E-13
```

I HAVE FINISHED

The method presented may be subjected to the following criticism:

1. Is it true that one obtains the correct eigenfrequency in the case of a membrane with a boundary that is not a circle and of which the eigenvalue is unknown? Here one has some control from the FABER *theorem*, which states [4.1] that from all membranes (plates) of arbitrary form but with the same surface or same circumference the circular membrane (plate) has the lowest eigenvalue. Slightly modifying a circular shape into an elliptic or CASSINI curve shape gives a control. The implicit plotting of the homogeneous boundary conditions like (4.8.12) also gives confidence in the correctness of the solution, since (4.8.12) must deliver the original boundary shape, in our case a circle. On the other hand, GRISVARD'S *theorem* guarantees the convergence of the solution between collocation points toward the real solution.

2. Finally, for noncircular boundaries the shapes may be accepted as nodal lines of a larger circular boundary. Then COURANT'S *theorem* guarantees an acceptable solution for the lowest eigenvalue. It states [4.41]: For a self-adjoint, second-order differential equation and a homogeneous boundary condition along a closed domain, the nodal lines of the n-th eigenvalue divide the domain into no more than n subdomains. The lowest eigenvalue does not divide the domain.

3. One may also criticize that the arbitrary choice of the "separation constants" β could lead to wrong or only very approximate results.

If no increasing functions like cosh (appears in the solution of the plate equation, see (4.5.84)) are present, collocation methods yield acceptable results. To discuss the dependence of the results on the arbitrary choice of the separation constants $\beta_n(\beta_n^2 < k^2)$ in (4.8.12) it is necessary to make several calculations with varying sets of β_n. For this purpose we present a *Mathematica* program. It is executed in cartesian coordinates and solves the membrane equation $u_{xx} + u_{yy} + k^2 u = 0$ for the homogeneous boundary conditions of a clamped circular membrane.

The *Mathematica* program Colmeigv. reads as follows. First n collocation points are chosen on the circular boundary of radius 1. In order that **Pi** will be used with 6 digits accuracy one may write **N[Pi,6]**

```
(* Colmeigv.: Eigenvalue problem for
circular membrane.Calculation in cartesian coordinates.

1.step: give n collocation points and separation
constants b *)
n=4; dth=N[Pi/(2*n),6];
Table[x[l]=Cos[l*dth],{l,1,n}];
Table[y[l]=Sin[l*dth],{l,1,n}];
Table[{x[l],y[l]},{l,1,n}];
(* Are the collocation points on the circle? *)
Table[x[l]^2+y[l]^2,{l,1,n}]
```

After control if the chosen n collocation points x_l, y_l are really situated on the boundary $x_l^2 + y_l^2 = 1$, an arbitrary first guess of the eigenvalue, the step in the eigenvalue is given and the separation constants b_l are chosen.

```
firsteigenv=2.000000000000; delta=firsteigenv/n;
b[1]=firsteigenv-0.0000001;
Table[b[l+1]=b[l]-delta,{l,1,n-1}];                      (4.8.14)
Table[b[l],{l,1,n}];
```

In the second step the eigenvalue is calculated from the homogeneous boundary condition

```
(* 2.step: calculate the eigenvalue from
homogeneous boundary conditions *)
Clear[eigenvalue];
f[eigenvalue_]=Det[Table[Cos[b[mi]*y[ipi]]*
Cos[Sqrt[eigenvalue^2-b[mi]^2]*x[ipi]],
{ipi,1,n},{mi,1,n}]];
(* 3.step: make a plot to find another guess
for the first eigenvalue.Try 2 methods for
finding roots of the determinant *)
Plot[f[eigenvalue],{eigenvalue,0.,3.}]
```

The plot (not shown here) presents the curve f(eigenvalue), which crosses $f = 0$ at eigenvalue $= 2.4$.

Then
```
FindRoot[f[eigenvalue]==0,
{eigenvalue,2.0}] //Timing                               (4.8.15)
```
$\{0.36 \text{ Second}, \{\text{eigenvalue} \rightarrow 2.40593\}\}$

refines the value and informs the user that this calculations needed 0.36 sec. In order to calculate the unknown partial amplitudes **Amp** the matrix m is defined and calculated

```
(* 4.step: define the matrix M *) eigenvalue=2.40593;
Clear[m];m=Table[j*k,{j,1,n},{k,1,n}];
Table[m[[ipi,mi]]=Cos[b[mi]*y[ipi]]*
Cos[Sqrt[eigenvalue^2-b[mi]^2]*x[ipi]],{ipi,1,n},
```

```
{mi,1,n}];nf=n-1;
Det[m]
```
3.0361×10^{-9}.

Since the value of the determinant is nearly zero (10^{-9}), the system of linear equations for the amplitudes is homogeneous and one has to rewrite the linear system by reducing the order n of the determinant to $n_f = n - 1$. The last rhs column of the matrix m becomes the rhs of the linear system for n_f unknowns and arbitrarily the last unknown A_n is set to 1.

```
(* 5.step: solve the homogeneous linear equations *)
bbf=Table[-m[[ifit,n]],{ifit,1,nf}];
rdutn=Table[m[[ifit,k1fit]],{ifit,1,nf},{k1fit,1,nf}];     (4.8.16)
B=LinearSolve[rdutn,bbf];
An={1.};
A=Table[B[[1k]],{1k,1,nf}];
(*6.step: check satisfaction of the homogeneous
boundary condition.*)
Amp=Join[A,An];
boundary = m . Amp                                         (4.8.17)
{0., -6.93889×10⁻¹⁷, 0., -0.000114775}.
```

The output obtained by inserting the unknowns into the system of linear equations demonstrates that the first three unknowns satisfy exactly the linear system, but as is very well known, the chosen unknown A_n satisfies the system only roughly. In problem 1 of this section we will discuss the new commands in (4.8.15) to (4.8.17). To check if the numerical solution

$$f(x,y) = \sum_{m=1}^{n} Amp_{mi} \cos(b_{mi}y) \cos\left(\sqrt{\text{eigenv}^2 - b_{mi}^2}\, x\right) \qquad (4.8.18)$$

satisfies the boundary conditions, a numerical and a graphical check are made:

```
fxy[x_,y_]:=Sum[Amp[[mi]]*Cos[b[mi]*y]*
Cos[Sqrt[eigenvalue^2-b[mi]^2]*x],{mi,1,n}]
Do[Print[fxy[x[1],y[1]]],{1,n}]
0.
-9.71445146547012⁻¹⁷
-2.220446049250313⁻¹⁶
-0.000114775
plot1=ContourPlot[fxy[x,y],{x,-1.05,1.05},{y,-1.05,1.05},
ContourShading->False,ContourSmoothing->2,
PlotPoints->60,Contours->{0},DisplayFunction->Identity];
plot2=ListPlot[Table[{x[i],y[i]},{i,1,n}],
Frame->True, AspectRatio->1.,
PlotRange->{{-1.,1.},{-1.,1.}},
PlotStyle->PointSize[1/60],DisplayFunction->Identity];
Show[plot1,plot2,DisplayFunction->$DisplayFunction]
```

These commands generate $f(x, y) = 0$ describing the circular boundary and show the location of the collocation points, see Figure 4.17

Other examples of the use of collocation methods will be given in chapter 5. A warning is necessary: when using collocation methods very often a matrix m (and its determinant) have to be calculated. Care is necessary that the matrix be well-conditioned. Arrangements in the matrix should be made such that the HADAMARD *condition number* K should be $0.1 < K < 1$ [4.43] - [4.45] and [2]. See also problem 1 of this section and (5.2.40).

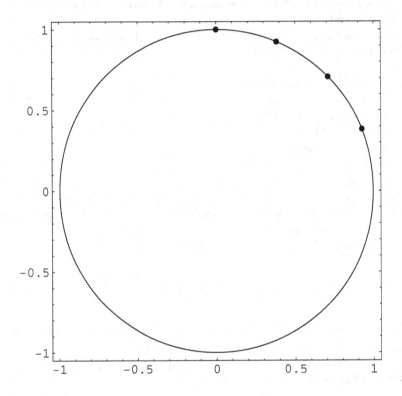

Figure 4.17
Circular membrane in cartesian coordinates

$$K = \frac{\det M}{\alpha_1 \cdot \alpha_2 \ldots \alpha_n}, \qquad (4.8.19)$$

where

$$\alpha_i = \sqrt{m_{i1}^2 + m_{i2}^2 + \ldots + m_{in}^2}. \qquad (4.8.20)$$

More details can be found in problem 1 of this section. Collocation methods should be used with care to avoid disappointments.

Problems

1. The new commands (4.8.15) to (4.8.17) seem to be self explanatory. The first command

   m=Table

   generates the matrix m and the second command

   Table[m[ipi,mi]]

   fills the elements $m_{ipi,mi}$ of the matrix. (4.8.16) may fail or give erronoeous results for not well-conditioned matrices. The necessary *factoring* of a matrix is discussed in [2]. Thus

   LUFactor[m] (4.8.21)

   gives the LU (lower and upper triangularization of a matrix) decomposition and

   LUSolve[lu,b] (4.8.22)

   solves a linear system represented by **lu** and the right-hand side **b**.

   The package **LinearAlgebra`GaussianElimination`** contains the two last commands. Try [2]

   a={{5,3,0},{7,9,2},{-2,-8,-1}} (4.8.23)

   To define the matrix a. You may also use the Table command (4.8.16) or

   MatrixForm[a={{5,3,0},{7,9,2},{-2,-8,-1}}] (4.8.24)

   Then the command **lu=LUFactor[a]** (4.8.25)

   should yield

   $$\left\{\left\{\frac{5}{7}, \frac{12}{19}, -\frac{22}{19}\right\}, \{7, 9, 2,\}\left\{-\frac{2}{7}, -\frac{38}{7}, -\frac{3}{7}\right\}\right\}, \{2, 3, 1\}.$$

   Assume now the existence of an inhomogeneous linear system **lu*x=b**, where **x** are the three unknowns and assume for the rhs vector **b={6,-3,7}**   then **LUSolve[lu,b]** (4.8.26) yields the three unknowns

   $\left\{\frac{75}{44}, -\frac{37}{44}, -\frac{81}{22}\right\}$ To verify the solution write **a.%** (4.8.27)

   which should give $\{6, -3, 7\}$. This method had been used in the form **boundary = m . Amp** (4.8.28)

2. Using *Mathematica* verify these two solutions:
   (4.8.6) solves the LAPLACE equation,

(4.8.9) solves the HELMHOLTZ equation. Use

```
VV[x,y]=Cos[Pi*x/a]*Cosh[Pi*y/a]+Cosh[Pi*x/b]*
Cos[Pi*y/b]+Log[Sqrt[(x^2+y^2)/(a^2+b^2)]];
Chop[Simplify[D[VV[x,y],{x,2}]+D[VV[x,y],{y,2}]]]
0
```

3. Play (and plot) with various values of $A/B, k, \beta$ in (4.8.10).

4. Using the *Mathematica* plot commands, create Figures 4.15 and 4.16.

5. Use the FORTRAN *program* COLLOC and/or the *Mathematica program* Colmeigv with various sets of the separation constants β_n (BETA or b). Establish a table describing how the eigenvalue k depends on the number P or n of collocation points, on the values of $\beta_n (\beta_n^2 < k^2)$ and on the first guess of the eigenvalue EIGENVALUE (resp. firsteigenv).

5

Boundary problems with two closed boundaries

5.1 Inseparable problems

Many problems in physics and engineering that exhibit boundaries that cannot be described by coordinate lines of a coordinate system in which the actual partial differential equation is separable. On the other hand, there are inseparable partial differential equations and the boundary conditions could be described by any coordinate system. In such a situation one can try to find formal solutions to the partial differential equation in cartesian coordinates. Furthermore, some problems may be nonlinear in nature and might become separable by linearization. We give an example.

Two-dimensional plasma equilibria are described by the SHAFRANOV *equation* [2.5], [3.1]. If the magnetic flux surfaces are given *a priori*, the equilibrium equations and the transport equations can be decoupled. But for *a priori* given cross sections of the flux surfaces the constraints to be imposed on the SHAFRANOV equation make the problem awkward because the pressure function $p(\psi)$ and the current $j(\psi)$ must be specified as a function of ψ, whose spatial dependence $\psi(r, z)$ is not yet known. In the case of the linear SHAFRANOV equation particular solutions can be superposed and the equilibrium problem may be reformulated as an eigenvalue problem for an arbitrary cross section of the magnetic surfaces.

The SHAFRANOV equation reads in cylindrical coordinates $r, z; \partial/\partial\varphi = 0$

$$\frac{\partial^2\psi}{\partial z^2} + \frac{\partial^2\psi}{\partial r^2} - \frac{1}{r}\frac{\partial\psi}{\partial r} = -4\pi^2\mu_0 r^2 \frac{\partial p}{\partial\psi} - \frac{\mu_0^2}{2}\frac{\partial j^2}{\partial\psi}. \tag{5.1.1}$$

Here $\psi(r, z)$ is the magnetic flux, $p(\psi)$ is the pressure, and j the current distribution. Making the linearizing setup

$$p = \frac{a}{2}\psi^2, \quad j^2 = \frac{b}{2}\psi^2, \tag{5.1.2}$$

we obtain with

$$\psi(r, z) = \sum_{l=1}^{L}(A_l \cos k_l z + B_l \sin k_l z)R_l(r)$$

from (5.1.1) the equation

$$R_l'' - \frac{1}{r}R_l' + \left(\frac{\mu_0^2 b}{2} - k_l^2\right) R_l + \alpha r^2 R_l = 0, \qquad (5.1.3)$$

where b is the same eigenvalue for several R_l, k_l is the separation constant, which must not be an integer, $\alpha = 4\pi^2 \mu_0 a$ and the $R_l(r)$ are the degenerate eigenfunctions belonging to the same eigenvalue b. The partial differential equation is separable, but the boundary conditions cannot be satisfied on curves of the cylindrical coordinate system r, z.

If we designate the two solutions of (5.1.3) by $R_l^{(1)}$ and $R_l^{(2)}$ for given k_l, then the flux surfaces $\psi = const$ are given by

$$\psi(r, z) = \sum_{l=1}^{L} \left(C_l R_l^{(1)}(r) + D_l R_L^{(2)}(r)\right) (A_l \cos k_l z + B_l \sin k_l z). \qquad (5.1.4)$$

If now the cross section $z(r)$ of the outer magnetic surface $\psi = 0$ (plasma surface) is given *a priori* by a set of N pairs $r_i, z_i (i = 1, \ldots, N)$, then (5.1.4) constitutes a system of N linear homogeneous equations for the $4L = N$ unknown coefficients $A_l C_l, A_l D_l, C_l B_l, B_l D_l$. For a set of given k_l the vanishing of the determinant of the coefficients yields the *eigenvalue* b. Initially b is not known. An estimate has to be made in the beginning. The determinant does not vanish and becomes a transcendental function of b. By trial and error and by repeated integration of (5.1.3), an exact value b can be found [5.1]. Apparently *Mathematica* is not able to solve (5.1.3). The relevant partial differential equation has been separable, but the boundary conditions on the magnetic surface $\psi = 0$ determined by the solution of (5.1.1) could not be satisfied on surfaces of the cylindrical coordinates in which (5.1.1) has been separable.

A similar problem arises when one has to calculate the eigenfrequencies of a membrane of arbitrary form. The HELMHOLTZ *equation* itself is separable in cartesian coordinates, but an arbitrary membrane boundary cannot be described by the straight lines of a cartesian coordinate system. Again collocation methods may help. The FABER *theorem* [4.1] is also useful: any membrane of arbitrary form but same area should have a first eigenvalue near but greater than $2.404\,825$ of a circular membrane, see (2.4.45). This offers a criterion concerning the eigenvalue. We now investigate a membrane with a boundary described by a CASSINI *curve*. This algebraic curve of fourth order is described by

$$F(x, y) = (x^2 + y^2)^2 - 2c^2(x^2 - y^2) - (a^4 - c^4) = 0, \quad c > 0, a > 0, a > c, \quad (5.1.5)$$

or in polar coordinates

$$r(\varphi) = \sqrt{c^2 \cos(2\varphi) \pm \sqrt{a^4 - c^4 \sin^2(2\varphi)}}. \qquad (5.1.6)$$

For $c = 0$, (5.1.5) becomes a circle of radius a. We first need the location of the four vertex points $(xmax, 0)$; $(xmin, 0)$; $(0, ymax)$; $(0, ymin)$, where

$$xmax = +\sqrt{a^2 + c^2}, \quad ymax = +\sqrt{a^2 - c^2}. \qquad (5.1.7)$$

Giving various values to a and c, we may find the area of the membrane by numerical integration

```
NIntegrate[F[x,y],{x,-xmax,xmax},{y,-ymax,ymax}]     (5.1.8)
```

Now let *Mathematica* do the work. We wrote a full program that is an extension of the program Colmeigv discussed in section 4.8. First we verify the solution of the membrane equation.

```
(* Program Cassmem.nb: Clamped Cassini membrane in
Cartesian coordinates. Eigenvalue problem
of the homogeneous equation.Verify solution*)
Clear[u,x,y,A,b,n,k]

u[x,y]=A[n]*Cos[Sqrt[k^2-b[n]^2]*x]*Cos[b[n]*y];
Simplify[D[u[x,y],{x,2}]+D[u[x,y],{y,2}]+k^2*u[x,y]]
```

and giving $= 0$
proves the solution in cartesian coordinates. Then we define the boundary given by (5.1.5) and (5.1.6) respectively and plot it using the arbitrary values $a = 1.0$, $c = 0.85$

```
(* Step 1: Define the Cassini boundary, the
4 vertex points and n collocation points, x,y *)
Clear[F,b,xmin,xmax,ymax,ymin,Fy,Fx,x,y,n,a,c];
n=8;a=1.0;c=0.85;
F[x_,y_]:= (x^2+y^2)^2-2*c^2*(x^2-y^2)-a^4+c^4;
Fy[x_]=InputForm[Solve[F[x,y]==0,y]];
ymax=Sqrt[Sqrt[a^4] - c^2];
Fx[y]=InputForm[Solve[F[x,y]==0,x]];
Fx[y_]:= Sqrt[c^2-y^2+Sqrt[a^4-4*c^2*y^2]]
xmax=Fx[0];xmin=-xmax;ymin=-ymax;dx=xmax/(n-1);
x[1]=xmin;Table[x[l]=xmin+(1-1)*dx,{1,1,n}];
dth=N[Pi/(2*n)];
r[phi_]:=Sqrt[c^2*Cos[2*phi]+Sqrt[a^4-c^4*
(Sin[2*phi])^2]]
Table[x[l]=r[l*dth]*Cos[l*dth],{1,1,n}];
Table[y[l]=r[l*dth]*Sin[l*dth],{1,1,n}];
TXY=Table[{x[l],y[l]},{1,1,n}];
<<Graphics`ImplicitPlot`
Clear[pl1,pl2,pl3];Off[General::spell]
pl1=ImplicitPlot[F[x,y]==0,{x, xmin, xmax},
AspectRatio->1,DisplayFunction->Identity];
pl2=ListPlot[TXY, DisplayFunction->Identity];
pl3=Show[pl1,pl2,DisplayFunction->
$DisplayFunction,Prolog->AbsolutePointSize[6]];
Off[General::spell]
```

Due to the signs of the two square roots contained in $y(x)$ according to (5.1.5) we also use (5.1.6) to calculate the vertex points. The x-coordinates of the $n(= 8)$ collocation points may be calculated from (5.1.5) or (5.1.6), but the y-coordinates must be calculated from (5.1.6), otherwise when using (5.1.6) multivalued or even complex y are resulting. The combined plot **pl1** (describing the boundary (5.1.5) in x, y coordinates) and plot **pl2** (showing the n collocation points) is shown by **pl3** in Figure 5.1.

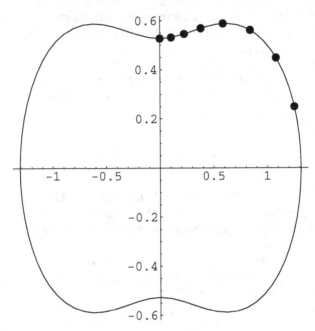

Figure 5.1
Cassini boundary with 8 collocation points

Now we calculate the separation constants b and fill the matrix MM satisfying the boundary condition

```
(* Step 2: Calculate the separation constants b. *)
Clear[fst,delta,b,tb];fst=2.3;delta=fst/n;
b[1]=fst-0.0001;
Table[b[k+1]=b[k]-delta,{k,1,n}];tb=Table[b[k],{k,1,n}];

(* Step 3:Fill matrix MM for the boundary condition *)
Clear[M,MM,k,W];M=Table[ip*li,{li,1,n},{ip,1,n}];
MM=Table[M[[li,ip]]=Cos[Sqrt[k^2-b[ip]^2]*x[li]]*
Cos[b[ip]*y[li]],{li,1,n},{ip,1,n}]; W[k_]=Det[MM];
//Timing
```

In this part of the program we defined **fst=2.3** as the lowest possible eigen-value (remember the FABER theorem) and as the limit of the largest separation constant **b[1]**. Then we start to calculate the eigenvalue as the root of the determinant of the matrix **MM**:

```
(* Step 4: Find the eigenvalue k. Do not forget to define
  the result as k *)
```

```
Clear[k];FindRoot[W[k]==0,{k,{fst,3.}}] //Timing
```

$\{41.74 \text{ Second}, \{k \to 2.99572\}\}$
The result $k = 2.99572$ must now be communicated to *Mathematica*
k=2.99572;
Having this eigenvalue, which makes the linear system homogeneous for the unknown amplitudes A, we create an inhomogeneous rhs term **bbf** for the $n - 1 = 7$ linear equations **rdutn=bbf**, fix the last (n-th) amplitude $A_n = 1$ and solve

```
(* Step 5: Calculate the partial amplitudes A[n] *)
nf=n-1;bbf=Table[-MM[[ifit,n]],{ifit,1,nf}];
rdutn=Table[MM[[ifit,klfit]],
{ifit,1,nf},{klfit,1,nf}];B=LinearSolve[rdutn,bbf];
An={1};A=Table[B[[lk]],{lk,1,nf}];
```

Then we first check on the correctness of the solution of the linear equations by inserting and using the symbol **Amp** for all n unknown amplitudes:

```
(* Step 6: Check satisfaction of the boundary
condition *)
Amp=Join[A,An];boundary = MM . Amp
```

The result is satisfying, since 10^{-16} is numerically equivalent to zero for the computer and also the last term due to the chosen $A_n = 1$ is satisfying:
$\{0., -3.33067 \times 10^{-16}, -3.33067 \times 10^{-16}, 1.38778 \times 10^{-16}, 2.77556 \times 10^{-16}, 2.22045 \times 10^{-16}, 1.11022 \times 10^{-16}, -2.6098 \times 10^{-12}\}$
Now we have to double-check the satisfaction of the CASSINI boundary condition in cartesian coordinates:

```
fxy[x_,y_]=Sum[Amp[[l]]*Cos[b[l]*y]*
Cos[Sqrt[k^2-b[l]^2]*x],{l,1,n}];
Do[Print[fxy[x[l],y[l]]],{l,1,n}]
```

The first (numerical) check is positive:
  0.
-5.87421×10^{-16}
-5.07291×10^{-16}
  1.45825×10^{-16}

2.12504×10^{-17}

3.21358×10^{-16}

$- 3.96167 \times 10^{-16}$

$- 2.60987 \times 10^{-12}$

Apparently our solution satisfies with an accuracy $<10^{-12}$ the CASSINI boundary. But we make a second (graphic) check. We type

```
Clear[p14];p14=ContourPlot[fxy[x,y],{x,xmin,xmax},
{y,ymin,ymax},
ContourShading->False,ContourSmoothing-> 2,
PlotPoints->60]
Show[p12,p14,AspectRatio->1,DisplayFunction->
$DisplayFunction,Prolog->AbsolutePointSize[6]]
Show[p13,p14,AspectRatio->1,DisplayFunction->
$DisplayFunction,Prolog->AbsolutePointSize[6]]
```

and obtain Figure 5.2.

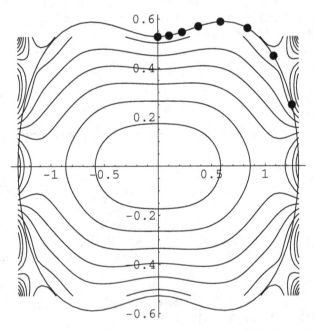

Figure 5.2
Cassini membrane

Boundary element methods [4.18] may solve this problem, but only finite element or collocation methods are suitable if the domain is inhomogeneous (partial differential equation with variable coefficients). An example is given by a membrane with varying surface mass density or by *neutron diffusion* in a *nuclear reactor* [4.6]. In problem 6 of section 4.4 an analytic solution for a circular membrane with radially varying density $\rho(r) = \rho_0[(R^2 - r^2) + 1]$ has been treated. After separation of the membrane equation two ordinary differential equations (4.4.77) and (4.4.78) result. One may then use a collocation method to satisfy the conditions and the density distribution. The following program does the job.

```
(* VARMEM: Circular membrane of radius 1 with symmetric
surface density. For constant density (alph=0) the
eigenvalue must be om=2.404825. Step 1: Define the
boundary and Plot a density distribution *) R=1.;
Clear[n,pl1,pl2,dth,x,y];n=6;dth=Pi/(2*n);Table[x[l]=
R*Cos[l*dth],{l,1,n}];Table[y[l]=R*Sin[l*dth],{l,1,n}];
pl1=ListPlot[Table[{x[l],y[l]},{l,1,n}],Frame->True,
AspectRatio->1.,PlotStyle->PointSize[1/40],
DisplayFunction->Identity];pl2=Plot[y=Sqrt[R^2-x^2],
{x,0,1.}, AspectRatio->1.,DisplayFunction->Identity]
Show[pl1,pl2,DisplayFunction->$DisplayFunction]
```

These commands generate Figure 5.3.

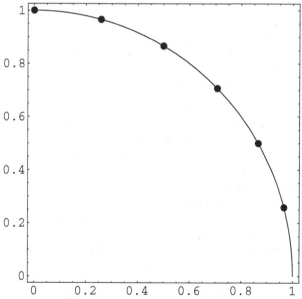

Figure 5.3
Circle with collocation points

Next, the density distribution will be defined:

```
Clear[alph,rho0,pro,rho,R,PL,x,y];
rho0=1.;alph=0.5;R=1.;
```

```
rho[x_,y_]=rho0*(alph*(R^2-x^2-y^2)+1.);
pro[x_,y_]=rho[x,y]/; x^2+y^2<=R^2;
Off[Plot3D::plnc];Off[Plot3D::gval];Clear [x,y]
PL=Plot3D[rho[x,y],{x,-R,R},{y,-R,R},PlotPoints->30]
```

Figure 5.4 depicts this distribution:

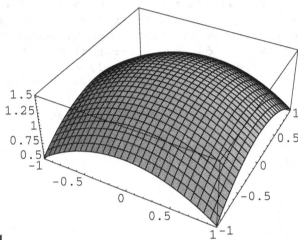

Figure 5.4
Density distribution

Since we do not use the analytic solution (AIRY *functions*), but solve (4.4.77) and (4.4.78) numerically (so that the program may be used for quite arbitrary density distributions), we continue:

```
(* Step 2: Arrange four-fold loop for eigenvalue om (q),
number of iteration (p),for l and k for the matrix.
Start iteration with n=4 and iterat=6. Then refine the
search interval for om by modifying omlast as
soon as det changes sign *)
Clear[om,omlast,x,y,Eyoung,rho0,alph,R,dth];
Eyoung=1.;rho0=1.;alph=0.1;R=1.;dth=Pi/(2*n);
Table[x[l]=R*Cos[l*dth],{l,1,n}];Table[y[l]=
R*Sin[l*dth],{l,1,n}];
Clear[sol1,ty11,X1,FX1,Y1,FY1,M,omeg,detf,Detfct];
M=Table[l*k,{l,1,n},{k,1,n}];
For[q=1, q < 2,
om1=2.3;omlast=4.;iterat=6;deltaom=(omlast-om1)/iterat;
om=om1;b[1]=om1-0.00001;delta=om1/n;
Table[b[k+1]=b[k]-delta,{k,1,n-1}];
(*Print["beta's = ",Table[b[k],{k,1,n}]];*)
For[p=1, p < iterat+1,
For[l=1, l < n+1,
```

```
For[k=1, k < n+1,
(*Print[{"omega= ",om,"k= ",k,"l= ",l,"beta= ",b[k]}];*)
sol1=NDSolve[{X1''[x]+X1[x]*(-b[k]^2*Eyoung+om^2*rho0*
(1.+alph*(R^2-x^2)))/Eyoung==0,X1'[0.]==0.0,X1[0.]==1.},
X1,{x,0.,2.*Pi}];FX1[x_,b[k]]=X1[x] /. First[sol1];
ty11=NDSolve[{Y1''[y]+b[k]^2*Y1[y]-Y1[y]*om^2*rho0*alph*
y^2/Eyoung==0,Y1'[0]==0.0, Y1[0]==1.},Y1,{y,0.,2.*Pi}];
FY1[y_,b[k]]=Y1[y] /. First[ty11];
Table[M[[l,k]]=FX1[x[l],b[k]]*FY1[y[l],b[k]]];
;k++];
;l++];
omeg[p]=om;detf[p]=Det[M];
Print[{"iteration= ",p,"omega= ",omeg[p],"Det= ",
detf[p]}];om=om+deltaom;
p++];
q++];
```

This constitutes a four-fold loop that outputs the following results:
$\{iteration =, 1, omega =, 2.3, \quad Det =, \quad 1.91469 \times 10^{-13}\}$
$\{iteration =, 2, omega =, 2.58333, Det =, -2.18576 \times 10^{-12}\}$
$\{iteration =, 3, omega =, 2.86666, Det =, -2.91506 \times 10^{-12}\}$
$\{iteration =, 4, omega =, 3.14999, Det =, -2.66164 \times 10^{-12}\}$
$\{iteration =, 5, omega =, 3.43333, Det =, -1.58279 \times 10^{-12}\}$
$\{iteration =, 6, omega =, 3.71666, Det =, -1.17786 \times 10^{-12}\}$

To check the change of sign of the determinant a plot of $\omega(p)$ is made, where p is the iteration parameter $1 \leq p \leq$ iterat, $p <$ iterat $+ 1$. This is effectuated by the command

```
Clear[Detfct];Detfct=Table[{omeg[p],detf[p]},{p,1,3}];
ListPlot[Detfct,PlotJoined->True]
```

resulting in Figure 5.5:

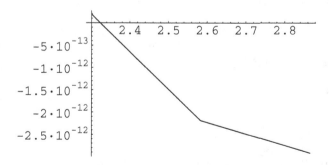

Figure 5.5
Zero of the determinant as function of p

The guess of the eigenvalue $2.3 \le$ om ≤ 26 is now improved:

```
(* Step 3: Check and refine result *)
Clear[HI,om];HI=Interpolation[Detfct];
Plot[HI[om],{om,2.3,2.6}];
FindRoot[HI[om]==0,{om,{2.3,2.6}}]
```

and yields Figure 5.6.

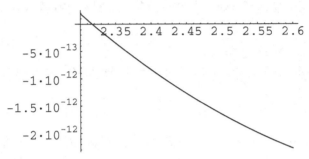

Figure 5.6
Improved eigenvalue om

The eigenvalue found **om** $= 2.3233$ is conform to the FABER *theorem* and is entered into the program: **om=2.3233** 2.3233.

If one wants to see the next steps clearly, it is recommended to give the command **MatrixForm[M]** and to omit the concluding semicolon (;) in the next 6 commands

```
nf=n-1;bbf=Table[-M[[ifit,n]],{ifit,1,nf}];
rdutn=Table[M[[ifit,klfit]],
{ifit,1,nf},{klfit,1,nf}];
B=LinearSolve[rdutn,bbf];
An={1.};
A=Table[B[[lk]],{lk,1,nf}];
```

By the next two commands the solution for the amplitudes **Amp** is tested by a calculation of the boundary values, which includes a test of the solution of the linear equations for **Amp**

```
(* Step 4: Check satisfaction of boundary condition *)
Amp=Join[A,An]
```

$\{-0.0456859, \quad 0.383603, \quad -1.34147, \quad 2.5064, \quad -2.50196, \quad 1.\}$
boundary = M . Amp
$\{1.11022 \times 10^{-16}, 0., 1.11022 \times 10^{-16}, 5.55112 \times 10^{-17},$
$\quad - 3.33067 \times 10^{-16}, 0.0000171437\}$ Finally, the boundary values are tested by

```
(* Step 5: Check the satisfaction of the boundary
condition *)
Clear[fxy];fxy[x_,y_]=
Sum[Amp[[k]]*FX1[x,b[k]]*FY1[y,b[k]],{k,1,n}];
(*/;x^2+y^2<=R^2 *);Do[Print[fxy[x[l],y[l]]],{l,1,n}]
```

2.22045×10^{-16}	0.	-2.22045×10^{-16}
-1.66533×10^{-16}	-2.22045×10^{-16}	0.0000171437

Nearly all toroidal problems are not separable. Apparently only the LAPLACE *equation* is separable in toroidal coordinates. The general solution reads

$$U(\eta, \vartheta, \psi) = \left[A\mathcal{P}^m_{p-1/2}(\cosh \eta) + B\mathcal{Q}^m_{p-1/2}(\cosh \eta) \right]$$
$$\cdot \left[C \sin p\vartheta + D \cos p\vartheta \right] \cdot \left[E \sin m\psi + F \cos m\psi \right], \quad (5.1.9)$$

where the generalized spherical functions $\mathcal{P}, \mathcal{Q}$ satisfy the LEGENDRE equation

$$\frac{d^2\mathcal{Q}(\eta)}{d\eta^2} + \coth \eta \frac{d\mathcal{Q}(\eta)}{d\eta} + \left(\frac{1}{4} - p^2 - \frac{m^2}{\sinh^2 \eta} \right) \mathcal{Q}(\eta) = 0. \quad (5.1.10)$$

Mathematica uses the toroidal coordinate system u, v, phi, a, which is obtained by rotating bipolar coordinates u, v, z, a about an axis perpendicular to the axis connecting the two foci of bipolar coordinates. The angle φ parameterizes the rotation. Bipolar coordinates are built around two foci separated by $2a$. Holding u fixed produces a family of circles that pass through both foci. Holding v fixed produces a family of degenerate ellipses about one of the foci. The coordinate z describes the distance along the common foci (default value $a = 1$, [2]). The bipolar coordinates in the $z = 0$ plane are shown in Figure 4.11.

Using a trick, even the *vector* HELMHOLTZ *equation* (3.1.80) can be solved [5.2]. Expressing F_φ and F_r by their cartesian coordinates $F_\varphi = -F_x \sin \varphi + F_y \cos \varphi$, $F_r = F_x \cos \varphi + F_y \sin \varphi$ and deriving both components with respect to φ yields

$$\frac{\partial F_\varphi}{\partial \varphi} = -\frac{\partial F_x}{\partial \varphi} \sin \varphi + \frac{\partial F_y}{\partial \varphi} \cos \varphi - F_r, \quad (5.1.11)$$

$$\frac{\partial F_r}{\partial \varphi} = \frac{\partial F_x}{\partial \varphi} \cos \varphi + \frac{\partial F_y}{\partial \varphi} \sin \varphi + F_\varphi. \quad (5.1.12)$$

Applying the LAPLACE operator on F_φ and using $\Delta F_x = -k^2 F_x$, $\Delta F_y = -k^2 F_y$, $\Delta \sin \varphi = -\frac{1}{r^2} \sin \varphi$, $\Delta \cos \varphi = -\frac{1}{r^2} \cos \varphi$ one obtains

$$\Delta F_\varphi = -k^2 F_\varphi - \frac{1}{r^2} F_\varphi - \frac{2}{r^2} \left(\frac{\partial F_x}{\partial \varphi} \cos \varphi + \frac{\partial F_y}{\partial \varphi} \sin \varphi \right)$$

$$= -k^2 F_\varphi - \frac{1}{r^2} F_\varphi - \frac{2}{r^2} \frac{\partial F_r}{\partial \varphi} + \frac{2}{r^2} F_\varphi \quad (5.1.13)$$

and analogously,

$$\Delta F_r = -k^2 F_r + \frac{1}{r^2} F_r + \frac{2}{r^2} \frac{\partial F_\varphi}{\partial \varphi}. \tag{5.1.14}$$

Due to the necessary uniqueness $\vec{F}(r, z, \varphi + 2\pi) = \vec{F}(r, z, \varphi)$ one has to assume a dependence $\sim \exp(im\varphi)$ so that (5.1.13) becomes

$$\frac{r^2}{2} \left(\Delta_t F_\varphi + k^2 F_\varphi - \frac{1}{r^2} F_\varphi - \frac{m^2}{r^2} F_\varphi \right) = -imF_r, \tag{5.1.15}$$

where

$$\Delta_t = \Delta + m^2/r^2 \tag{5.1.16}$$

is the transversal Laplacian operator. Its transversal coordinates q_1, q_2 describing the cross section of the torus laying in the cylindrical coordinate system q_1 (may be r), q_2 (may be z), φ may describe an arbitrary torus cross section. Insertion of F_r from (5.1.15) into (5.1.14) yields

$$\left(\Delta_t - \frac{m^2 + 1}{r^2} + k^2 \right) \left(\Delta_t - \frac{m^2 - 1}{r^2} + k^2 \right) F_\varphi - \frac{4m^2}{r^4} F_\varphi$$

$$= \left(\Delta_t - \frac{1}{r^2}(m+1)^2 + k^2 \right) \left(\Delta_t - \frac{1}{r^2}(m-1)^2 + k^2 \right) F_\varphi = 0. \tag{5.1.17}$$

This structure recalls (4.5.40). We thus have two solutions:

$$F_\varphi(q_1, q_2) = C_1 F_\varphi^+ + C_2 F_\varphi^-, \tag{5.1.18}$$

where

$$\left(\Delta_t + k^2 - \frac{(m \pm 1)^2}{r^2} \right) F_\varphi^\pm = 0. \tag{5.1.19}$$

Assuming now $q_1 = r$, $q_2 = z$ and $\sim \exp(i\gamma z)$ one obtains

$$\frac{d^2 F_\varphi^\pm}{dr^2} + \frac{1}{r} \frac{dF_\varphi^\pm}{dr} + (k^2 - \gamma^2) F_\varphi^\pm - \frac{(m \pm 1)^2}{r^2} F_\varphi^\pm = 0 \tag{5.1.20}$$

and

$$F_\varphi(r, z) = \left[C_1 Z_{m+1} \left(\sqrt{k^2 - \gamma^2}\, r \right) + C_2 Z_{m-1} \left(\sqrt{k^2 - \gamma^2}\, r \right) \right] \exp(i\gamma z). \tag{5.1.21}$$

Here Z are cylinder functions, either BESSEL + NEUMANN or HANKEL. For a dependence $\sim \exp(im\varphi)$ one obtains FOCK *equations*:

$$F_r(r, z, \varphi, t) = \frac{A_{mk} \exp(i\omega t + im\varphi)}{r^2 k^2 - m^2} \left(im \frac{\partial r F_\varphi(r, z)}{\partial r} - rk \frac{\partial r F_\varphi(r, z)}{\partial z} \right), \tag{5.1.22}$$

$$F_z(r, z, \varphi, t) = \frac{A_{mk} \exp(i\omega t + im\varphi)}{r^2 k^2 - m^2} \left(im \frac{\partial r F_\varphi(r, z)}{\partial z} + rk \frac{\partial r F_\varphi(r, z)}{\partial r} \right). \tag{5.1.23}$$

These solutions allow the investigation of electromagnetic waves in toroids. If the electric conductivity of the torus wall is high, then the boundary condition $B_n = 0$ is valid for the magnetic field normal component. Then for an exciting frequency $\omega = kc$ a solution reads

$$B_\varphi(r, z, \varphi) = \sum_{m,l} \left[a_{ml} J_m \left(\sqrt{k^2 - \gamma_l^2}\, r \right) + b_{ml} N_m \left(\sqrt{k^2 - \gamma_l^2}\, r \right) \right]$$

$$\cdot \cos \gamma_l z \cos m\varphi \tag{5.1.24}$$

and analogously for $B_r(r, z, \varphi)$, $B_z(r, z, \varphi)$. Then the unknown partial amplitudes a_{ml}, b_{mk} have to be chosen to satisfy $\operatorname{div} \vec{B} = 0$. Integrating n times the differential equations for the field lines

$$\frac{dr}{B_r} = \frac{dz}{B_z} = \frac{r d\varphi}{B_\varphi} \quad \text{or} \quad \frac{dz}{d\varphi} = \frac{r B_z}{B_\varphi}, \quad \frac{dr}{d\varphi} = \frac{r B_r}{B_\varphi} \tag{5.1.25}$$

around the torus $0 \le \varphi \le n \cdot \pi$, one may find the cross section of a torus. The totality of all points crossing the $z - r$ plane draws a picture of the cross section in the $z - r$ plane (POINCARÉ *map*). If after n ($n < \infty$) revolutions of the field line around the z axis an earlier point is hit, then the toroidal surface drawn is called *resonant toroidal surface*.

Many toroidal problems have been solved either by collocation methods or by numerical methods [5.1] to [5.4]. In plasma physics arbitrary shapes of the meridional cross section of the torus are of interest. One possibility to solve such problems is to construct the boundary curve after having received a general solution; the second possibility is offered by the collocation method. We first discuss the possibility of constructing a given shape afterward. One may express the electromagnetic fields $\vec{E}$ and $\vec{B}$ by a complex quantity

$$\vec{F} = \vec{E} - i\vec{B}/\sqrt{\epsilon_0 \mu_0}. \tag{5.1.26}$$

Then MAXWELL'S equations give for cylindrical geometry $\partial/\partial\varphi = 0$, where φ is the angle around the z axis the solution

$$B_\varphi(r, z) = \gamma^2 [b_0 J_1(\gamma r) + c_0 N_1(\gamma r)] + \gamma \sqrt{\gamma^2 - k^2}$$
$$\cdot \left[b_1 J_1 \left(\sqrt{\gamma^2 - k^2}\, r \right) + c_1 N_1 \left(\sqrt{\gamma^2 - k^2}\, r \right) \right] \cos kz,$$

$$B_r(r, z) = k \sqrt{\gamma^2 - k^2}$$
$$\cdot \left[b_1 J_1 \left(\sqrt{\gamma^2 - k^2}\, r \right) + c_1 N_1 \left(\sqrt{\gamma^2 - k^2}\, r \right) \right] \sin kz,$$

$$B_z(r, z) = \gamma^2 [b_0 J_0(\gamma r) + c_0 N_0(\gamma r)] + (\gamma^2 - k^2)$$
$$\cdot \left[b_1 J_0 \left(\sqrt{\gamma^2 - k^2}\, r \right) + c_1 N_0 \left(\sqrt{\gamma^2 - k^2}\, r \right) \right] \cos kz, \tag{5.1.27}$$

where J_p and N_p are BESSEL and NEUMANN functions, respectively.

Now the differential equations for the field lines in the r, z plane are

$$\frac{dr}{B_r} = \frac{dz}{B_z} \tag{5.1.28}$$

Inserting for B_r and B_z we can integrate. There is, however, another possibility. We can express B_r and B_z by B_φ. This gives

$$\frac{\partial(B_\varphi r)}{\partial r}dr + \frac{\partial(B_\varphi r)}{\partial z}dz = d(B_\varphi r) = 0. \qquad (5.1.29)$$

Thus, the lines $B_\varphi r = \text{const} = D$ are *identical* in form with the B_r, B_z field lines in the r, z plane, i.e., identical with the cross section of the toroidal resonator. To obtain a toroidal resonator of major radius R and nearly circular cross section of minor radius a of the form $f(r,z) = (r - R^2) + z^2 - a^2 = 0$ we have to determine the constants b_0, c_0, b_1, c_1, D in B_φ from

$$z = \frac{1}{k}\arccos$$

$$\left[\frac{D - b_0 r\gamma^2 J_1(\gamma r) - c_0\gamma^2 r N_1(\gamma r)}{\gamma\sqrt{\gamma^2 - k^2}\left(b_1 r J_1\left(\sqrt{\gamma^2 - k^2}\,r\right) + c_1 r N_1\left(\sqrt{\gamma^2 - k^2}\,r\right)\right)}\right]. \qquad (5.1.30)$$

By choosing for a given k a set of collocation points $z_i = z(r_i), i = 1,\ldots,5$ we may generate various arbitrary cross sections.

Using similar methods the calculation of exact analytical force-free three-dimensional toroidal equilibria of arbitrary cross section is possible.

To demonstrate that the results are extremely insensitive to the arbitrary choice of the "separation constants" α_n, we solved the equations for an axisymmetric force-free equilibrium

$$\frac{\partial B_r}{\partial z} - \frac{\partial B_z}{\partial r} = \gamma B_\varphi, \qquad B_r = -\frac{1}{\gamma}\frac{\partial B_\varphi}{\partial z}, \qquad B_z = \frac{1}{\gamma}\frac{1}{r}\frac{\partial}{\partial r}r B_\varphi \qquad (5.1.31)$$

in the cylindrical coordinate systems

$$B_\varphi = \sum_{n=1}^{N} \gamma\cos(\alpha_n z)\left[A_n J_1\left(\sqrt{\gamma^2 - \alpha_n^2}\,r\right) + B_n N_1\left(\sqrt{\gamma^2 - \alpha_n^2}\,r\right)\right]. \qquad (5.1.32)$$

We assume a toroidal container of major radius R and some effective minor radius a and defined by a meridional cross section curve $z = z^*(r)$ in the r, z plane. This cross section curve may be a circle given by $z^* = \sqrt{a^2 - (r - R)^2}$, an ellipse or a CASSINI curve $[(r - R)^2 + z^2]^2 + 2b^2[(r - R)^2 - z^2] - a^4 + b^4 = 0$. The boundary condition along the toroidal surface cross section curve $z^*(r)$ is $r B_\varphi = \text{const}$.

One obtains for a circle:

N	α_1	α_2	α_3	α_4	α_5	α_6	$(R/a = 2)$	$(R/a = 5)$	$(R/a = 20)$
1	0.001	0.4	0.6	1.7	1.9	2.0	2.446 927 246	2.411 138 020	2.405 216 280
2	0.001	0.5	1.0	1.8	2.0	2.1	2.446 926 960	2.411 137 800	2.405 215 400
3	0.01	0.1	1.0	1.5	2.0	2.1	2.446 927 093	2.411 137 800	2.405 216 082
4	0.01	0.3	0.7	1.5	1.9	2.0	2.446 927 422	2.411 138 193	2.405 216 500
5	0.1	0.5	0.6	1.3	1.8	2.0	2.446 927 620	2.411 138 372	2.405 216 500
6	1.0	1.2	1.4	1.6	1.8	2.0	2.446 927 097	2.411 137 800	2.405 216 082
7	1.0	1.3	1.5	1.7	1.9	2.1	2.446 926 960	2.411 137 800	2.405 215 400

A similar toroidal boundary problem with arbitrary cross section has been solved for the radial part of the vector HELMHOLTZ equation for the toroidal electric field $E_\varphi \equiv u$

$$u_{rr} + \frac{1}{r}u_r + u_{zz} - \frac{1}{r^2}u + k^2 u = 0. \tag{5.1.33}$$

To satisfy the homogeneous boundary condition $u(r, z_R) = 0$ on the surface of an axisymmetric toroid with the meridional cross section curve $z = z_R(r)$ we use $\gamma_1 = 0$, $\gamma_2 = \gamma$ and the four particular solutions namely

$$u(r, z) = A_1 J_1(kr) + A_2 N_1(kr)$$
$$+ \left[B_1 J_1 \left(\sqrt{k^2 - \gamma^2} r \right) + B_2 N_1 \left(\sqrt{k^2 - \gamma^2} r \right) \right] \cos \gamma z. \tag{5.1.34}$$

The boundary condition $u(r, z_R) = 0$ on the cross section curve $z = z_R(r)$ yields

$$z_i = \arccos \left[\frac{-A_1 J_1(kr_i) - A_2 N_1(kr_i)}{B_1 J_1 \left(\sqrt{k^2 - \gamma^2} r_i \right) + B_2 N_1 \left(\sqrt{k^2 - \gamma^2} r_i \right)} \right]. \tag{5.1.35}$$

A workstation delivers the result $\gamma = 7.4$, $k = 8.176\,299\,6497$, $A_1 = 0.173\,255$, $A_2 = -0.779\,360$, $B_1 = 0.445\,031$, $B_2 = 1$, if an oval defined numerically by

i	1	2	3	4
r_i	0.6	0.83	1.10	1.3
z_i	0	0.28	0.20	0

is chosen as meridional cross section curve $z_i = z_R(r_i)$. It seems that this analytical method works quicker than any numerical finite differences or finite element method.

The collocation method has been applied to many toroidal *plasma* problems. Thus the electromagnetic field propagation and eigenfrequency in anisotropic homogeneous and isotropic inhomogeneous toroidal plasmas of arbitrary cross section as well as for anisotropic inhomogeneous axisymmetric plasmas has been investigated. In such a plasma the elements of the *dielectric tensor* are functions of space [2.5].

For a plasma magnetized by a toroidal magnetic field $\vec{B}_0 = B_0 \vec{e}_\varphi$ one has

$$\epsilon_{rr} = 1 + \sum_s \frac{\omega_{Ps}^2}{\Omega_s^2 - \omega^2} \equiv \epsilon_r,$$

$$\epsilon_{\varphi\varphi} = 1 - \sum_s \frac{\omega_{Ps}^2}{\omega^2} \equiv \epsilon_\varphi,$$

$$\epsilon_{rz} = \sum_s \frac{i\Omega_s}{\Omega_s^2 - \omega^2} \frac{\omega_{Ps}^2}{\omega} \equiv \epsilon_z. \tag{5.1.36}$$

In an inhomogeneous magnetized plasma Ω_s and n_0 (and ω_{Ps}) depend on space. We have for a toroidal plasma (s is the species index)

$$\Omega_s = \Omega_{s0} R/r, \qquad \Omega_{s0} = e_s B_0/m_s, \qquad (5.1.37)$$

where R is the major radius of the torus. In toroidal experiments the plasma density n_0 is mainly a parabolic function of the distance ρ from the magnetic axis. Expressing n_0 in cylindrical coordinates r, z we have

$$n_0(r, z) = n_1 - n_0 z^2 - n_0 (r - R)^2, \qquad (5.1.38)$$

since $\rho^2 = z^2 + (r - R)^2$. Here n_1 is the maximum density on the magnetic axis ($r = R, z = 0$) and $n_0 = n_1/a^2$ is determined by the minor toroidal radius a. At $z = 0$ and $r = R \pm a$ we have the wall ($\rho = a$) of the circular toroidal vessel and the density vanishes. In the axisymmetric case the system of electromagnetic equations splits into three equations for the TE mode with B_φ, E_r, E_z and for the TM mode with E_φ, B_r, B_z.

For *axisymmetric* modes we obtain for the TM mode

$$B_r = -\frac{i}{\omega} \frac{\partial E_\varphi}{\partial z}, \qquad B_z = \frac{i}{\omega} \frac{1}{r} \frac{\partial}{\partial r} (r E_\varphi), \qquad (5.1.39)$$

$$\frac{\partial E_\varphi}{\partial z^2} + \frac{\partial^2 E_\varphi}{\partial r^2} + \frac{1}{r} \frac{\partial E_\varphi}{\partial r} - \frac{1}{r^2} E_\varphi + \omega^2 \epsilon_0 \mu_0 \epsilon_\varphi E_\varphi = 0. \qquad (5.1.40)$$

Here $E_\varphi(r, z)$ is a toroidal electric mode, depending on $\epsilon_\varphi(r, z; \omega)$.

The solutions have to satisfy the electromagnetic boundary conditions on the wall of the toroidal vessel. On a perfectly conducting wall the tangential electric component E_t and the normal magnetic component B_n must vanish. In our rotational coordinates the form of an arbitrary cross section of the toroidal vessel is given by a function $z = z(r)$. Projecting into the meridional plane a vector $\vec{A}$ (which signifies $\vec{E}$ or $\vec{B}$) on this curve $z(r)$, we have

$$A_t = A_r \cos a + A_z \sin a,$$
$$A_n = -A_r \sin a + A_z \cos a, \qquad (5.1.41)$$

where $\tan a = dz/dr$. The *electric boundary condition* $E_t = 0$ therefore yields

$$E_r + E_z \frac{dz}{dr} = 0. \qquad (5.1.42)$$

The *magnetic boundary condition* $B_n = 0$ yields (5.1.28), which represents the equation for the magnetic field lines (due to $B_n = 0$, the wall *is* a special magnetic field line), and which is a condition for E_φ. Inserting we obtain after integration

$$r E_\varphi = \text{const.} \qquad (5.1.43)$$

This describes the projection of the magnetic field lines onto the meridional plane and is an expression for the cross section $z(r)$ of the wall. (Here E_φ

is tangential and vanishes on the conductor.) Inserting $\epsilon_\varphi, \omega_{Ps}$ and n_0 into (5.1.40) we obtain

$$\frac{\partial^2 E_\varphi}{\partial z^2} + \frac{\partial^2 E_\varphi}{\partial r^2}$$

$$+ \frac{1}{r}\frac{\partial E_\varphi}{\partial r} - \frac{1}{r^2}E_\varphi + \gamma^2 E_\varphi + (a + br + cr^2 + cz^2)E_\varphi = 0, \quad (5.1.44)$$

where

$$\gamma^2 = \epsilon_0\mu_0\omega^2,$$

$$a = \frac{e^2\mu_0}{m_I m_E}\left(m_I n_1 - m_I n_0 R^2 + m_E n_1 - m_E n_0 R^2\right),$$

$$b = \frac{2n_0 e^2 \mu_0 R}{m_I m_E}(m_I + m_E), \qquad R = 0.95 \text{ m}, \qquad a = 0.35 \text{ m},$$

$$c = \frac{e^2\mu_0 n_0}{m_I m_E}(-m_I - m_E), \qquad n_0 = 10^{16} \text{ m}^{-3}, \quad B_0 = 1 \text{ kG}.$$

Here m_I and m_E are the proton (ion) and electron mass, respectively. To solve, we make the ansatz $E_\varphi(r, z) = R(r)U(z)$ and obtain

$$R'' + (1/r)R' - (1/r^2)R + (\gamma^2 - k^2 + a + br + cr^2)R = 0, \qquad (5.1.45)$$

which is a BÔCHER *equation*, and

$$U'' + cz^2 U + k^2 U = 0, \qquad (5.1.46)$$

of the same kind as the WEBER *equation*. Here k is the separation constant, which must *not* be an *integer*. To satisfy the boundary conditions we integrate (5.1.45) and (5.1.46) numerically. To do this we need a first (arbitrary) approximation of the eigenvalue γ. This can be taken from the empty resonator for $a = b = c = 0$. Since both equations are homogeneous, the eigenvalue does not depend on the initial conditions. We choose therefore two different arbitrary initial conditions $R_1(r = R - a), R_1'(r = R - a), U_1(0), U_1'(0)$, and $R_2(r = R - a), R_2'(r = R - a), U_2(0), U_2'(0)$. We thus obtain a solution in the form

$$E_\varphi(r, z) = (AR_1 + BR_2)(CU_1 + DU_2). \qquad (5.1.47)$$

Choosing different values k_i of the separation constant $(i = 1, \ldots, N)$ we can obtain a full set of solutions $E_\varphi^{(i)}(r, z)$ so that $\sum_{i=1}^{N} E_\varphi^{(i)}(r, z)$ is a solution containing $4N$ arbitrary constants $A^{(i)}, B^{(i)}, C^{(i)}, D^{(i)}$. Defining the arbitrarily given cross section $z(r)$ by assuming the coordinates r_p, z_p of $p = 1, \ldots, P$ points lying on $z(r)$, we obtain $P = 4N$ linear homogeneous equations:

$$\sum_{i=1}^{N} \left[A^{(i)} R_1^{(i)}(r_p) + B^{(i)} R_2^{(i)}(r_p) \right] \cdot \left[C^{(i)} U_1^{(i)}(z_p) + D^{(i)} U_2^{(i)}(z_p) \right] = 0. \quad (5.1.48)$$

The $R_l^{(i)}, U_l^{(i)} (l = 1, 2)$ have been obtained by numerical integration, assuming an approximate value of the still unknown eigenvalue γ. The system (5.1.48) has then and only then a nontrivial solution for the $4N$ unknown coefficients $A^{(i)}, B^{(i)}, C^{(i)}, D^{(i)}$ if the determinant of the coefficients of the system vanishes. Since γ had been assumed arbitrarily, the determinant $D(\gamma)$ will not vanish. Calculating $D(\gamma)$ for varying γ (by integrating (5.1.45) and (5.1.46) several times) and using the *regula falsi* method, a better value of γ can be obtained. Renewed integration with the better value and repetition of the procedure finally yields the exact value of γ. As soon as the correct value of γ is found, the system (5.1.48) can be solved for the $4N$ unknowns $A^{(i)}, B^{(i)}, C^{(i)}$, whereas, for example, one $D^{(i)}$ will be chosen to be unity out of $i = 1, \ldots, N - 1$.

We assume $a = 319.5$, $b = -672.7$, $c = 354.1$, or $B_0 = 0.1$ T, $n_1 = 1.225 \cdot 10^{15}$ m^{-3}, $\omega_{PE}^2 = 3.88 \cdot 10^{18}$, $\omega_{PI}^2 = 2.12 \cdot 10^{15}$, $\Omega_E^2 = 3.08 \cdot 10^{20}$, $\Omega_I^2 = 9.17 \cdot 10^{13}$.

Various aspect ratios of circular toroidal vessels and other noncircular cross sections can be described. For $i = 1, 2, 3, 4$ we have $r = 0.60, 1.30, 0.83, 1.10$, and $z = 0, 0, 0.28, 0.20$, respectively, and $k = 7.4$ we obtain, for instance, an ovaloid cross section. Now $\gamma = 7.655\,598\,7648$ or $\omega/2\pi = 3.653\,82 \times 10^8$ cycles and $A = -0.557\,43$, $B = 0.800\,47$, $C = -0.650\,90$, $D = 1$. For $k = 6$ we obtain $\gamma = 7.276\,150\,90$ and another oval. Many other forms have been produced by assuming various r_i, z_i.

The collocation method can also be used for the low-frequency MHD *waves*. Starting from the equations of continuity, of motion, Ohm's law and Maxwell's equations, we used the usual linearization. Elimination of $B_r, B_z, B_\varphi, v_\varphi$ and ρ, a time dependence $\exp(i\omega t)$ yields two ordinary differential equations of first order with variable coefficients

$$i\frac{dv_r}{dr} \left(c_A^2 + n c_s^2 \right) + k v_z \left[\left(c_A^2 + n c_s^2 \right) - \frac{\omega^2}{k^2} + \frac{m^2}{r^2 k^2 c_A^2} \right]$$

$$+ \frac{i v_r}{r} \left[\left(n c_s^2 \left(r \frac{1}{\rho_0} \frac{d\rho_0}{dr} + 1 \right) - c_A^2 \right) \right] = 0, \quad (5.1.49)$$

$$\frac{dv_z}{dr} \left(r^2 \omega^2 - m^2 c_A^2 \right) + 2i \frac{dv_r}{dr} k r c_A^2 + v_z \left(r^2 \omega^2 \frac{1}{\rho_0} \frac{d\rho_0}{dr} + 2 r k^2 c_A^2 + 4 m^2 \frac{c_A^2}{r} \right) +$$

$$i k v_r \left(r^2 \omega^2 - m^2 c_A^2 - 2 c_A^2 \right) = 0. \quad (5.1.50)$$

Here n is an abbreviation for $n = \omega^2 r^2/(\omega^2 r^2 - m^2 c_s^2)$. m is the toroidal mode number in the φ direction, and the ALFVÉN *speed* is given by $c_A^2 = B_0^2(r)/\mu_0 \rho_0(r)$ so that n and c_A depend on r. Due to the gyrotropy of a magnetized plasma the phase factor $\exp(i\pi/2) = i$ appears so that we make a "snap-shot" at time $t = 0$. We then have two possibilities to make the differential equations real: modes of type 1 (replace $v_r = i\bar{v}_r$) and modes of type 2 (replace $v_z = i\bar{v}_z$). $\vec{B}_0, \rho_0, \vec{v}_0 = 0$ are the equilibrium quantities, c_s is the adiabatic sonic speed. Taking circular cylinder coordinates we assumed $\rho_0 = \rho_0(r)$, $\vec{B}_0(r) = B_0(r)\vec{e}_\varphi$, $B_0(r) = RB_0(R)/r$. R is the major and a the minor torus radius. Furthermore a dependence $\exp(im\varphi - ikz)$ has been assumed for all wave quantities.

If the meridional cross section in the r, z plane of the containing toroidal surface is described by a curve $z^* = z(r)$ and if $\tan\alpha = dz^*/dr$, then $v_n = -v_r \sin\alpha + v_z \cos\alpha = 0$. Inserting the real solutions v_z and $\bar{v}_r$ in the form for mode type 1 we get

$$v_z(r, \varphi, z) = \sum_{k,m,s}^{K,M} A_{km}^s v_{zs}^{(k,m)}(r) \cos kz \cdot \cos m\varphi, \qquad (5.1.51)$$

$$\bar{v}_r(r, \varphi, z) = \sum_{k,m,s}^{K,M} A_{km}^s \bar{v}_{zs}^{(k,m)}(r) \sin kz \cdot \cos m\varphi, \qquad (5.1.52)$$

where $s = 1, 2$ indicates two different but arbitrary initial conditions used in the numerical integration of (5.1.49), (5.1.50). One obtains for P collocation points r_i, z_i, $i = 1 \ldots P$ a system of P equations for the unknown partial amplitudes A_{km}^s. Since the sum over $s = 1, 2$ gives two values and if summation over m gives M values and over k one might have K values, the number N of unknown partial amplitudes A_{km}^s is given by $P = 2(M-1) \cdot K$. The number P of collocation points determines the accuracy of the solution. For fixed P it has been shown earlier that the accuracy of the results is practically independent of the arbitrary choice of the separation constants k. According to our experience with axisymmetric MHD modes we make the choice $k = 1/a, 2/a, 3/a \ldots K/a$, where a is the "effective" minor torus radius for arbitrary cross section. When the k-values are given, then the condition $v_n = 0$ with $\bar{v}_r, v_z$ inserted constitutes a system of P linear homogeneous equations for the $2(M-1) \cdot K$ unknowns A_{km}^s. To solve this system the determinant D of the coefficients (known at r_i, z_i) must vanish. This condition determines the global eigenfrequency ω. In order to be able to integrate (5.1.49), (5.1.50) the density distribution of the plasma and the frequency ω have to be known. We thus integrate the differential equations (which deliver the elements of the determinant) with an initial guess for ω and calculate $D(\omega)$. Looking for a root of the function $D(\omega)$ we obtain an improved value for ω and a new integration yields improved coefficients and an improved ω, etc.

It seems that the influence of the plasma density inhomogeneity on the global eigenfrequency is exhibited best by the distribution

$$\bar{\rho}(x) = A \exp(-\beta x^2), \qquad x = (r - R)/a.$$

As soon as the solutions (5.1.51), (5.1.52) are known it is possible to calculate the electric field $E_r(r, z, \varphi)$ and $E_z(r, z, \varphi)$. When they are known, the differential equations for the electric field lines $dz/E_z(r, z) = dr/E_r(r, z)$ have to be integrated numerically. The results obtained are summarized. Here the following parameters have been chosen: major radius $R = 3$ m, $B_0 = 5$ Tesla, $T = 10$ keV, $c_s^2 = \gamma^2 10^3 eT/m, a = 0.7$ m.

Curve b	m	β	Frequency type 1	Frequency type 2
Circle $= a$	0	1.4	3.777E7 Hz	3.350E7 Hz
Circle $= a$	0	2.8	3.987E7 Hz	3.125E7 Hz
Circle $= a$	1	1.4	3.814E7 Hz	3.376E7 Hz
Circle $= a$	1	2.0	3.883E7 Hz	3.264E7 Hz
Ellipse 1.2	0	2.4	2.556E7 Hz	2.849E7 Hz
Ellipse 1.8	0	2.4	1.794E7 Hz	-

We see that for mode type 1 the frequency increases but for mode type 2 decreases with increasing inhomogeneity. With increasing mode number m the frequency decreases for type 1 but increases for type 2. Probably for rotating patterns the frequencies converge. Also the dependence of the global eigenfrequency on the aspect ratio has been investigated. As it is expected, for large aspect ratio $A \geq 20$ one obtains the solution for a cylindrical plasma.

Problems

1. Modify the program Cassmem to describe a circular membrane of radius $R = 1$ and calculate the first four eigenvalues. Compare the results with the values of the roots of the BESSEL functions given earlier.

2. Demonstrate that the program VARMEM yields om $= 2.404\,825$ for constant surface density of the membrane.

3. Modify VARMEM for an elliptic membrane of semi-axes $a = 5, b = 3$ and constant density. Is it possible to reproduce the eigenvalues of the MATHIEU functions?

4. In a nuclear reactor the *geometric buckling* B assumed to be a linear function of x, y and z. Solve $\Delta n(x, y, z) + B^2(x, y, z)u(x, y, z) = 0$ for $n(\text{boundary}) = 0$ for a cube and a sphere.

5. Use the following collocation program Homcircplate to calculate the lowest eigenvalue of a circular plate of radius $R = 1$ in cartesian coordinates. Read in section 4.5 equations (4.5.40) - (4.5.44) and remember the result $3.196\,220\,612$. Is the collocation program able to reproduce this value? Execute the plot commands. Try to vary the separation constants within the interval $0 < b_n < fst\,(= 3.00)$.

```
(* Program Homcircplate: Clamped circular plate in
Cartesian coordinates, no load. Eigenvalue problem
of the homogeneous equation with two homogeneous
boundary conditions. *)
Clear[u,m,x,y,A,B,b,n]

u[x,y]=A[n]*Cos[Sqrt[k^2-b[n]^2]*x]*Cos[b[n]*y]+
B[n]*Cosh[b[n]*y]*Cosh[Sqrt[k^2-b[n]^2]*x];
Simplify[D[u[x,y],{x,4}]+2*D[u[x,y],{x,2},{y,2}]+
D[u[x,y],{y,4}]-k^4*u[x,y]]

(* Later on we need: *)
ux=InputForm[D[u[x,y],x]]

uy=InputForm[D[u[x,y],y]]

(* START HERE: Step 1: Define the radius R of the
circular plate and nn collocation points, x,y *)
Clear[x,y,k,R,nn];
nn=14; dth=N[Pi/(2*nn),6];R=1.;
Table[x[l]=R*Cos[l*dth],{l,1,nn,2}];
Table[y[l]=Sqrt[R-x[l]^2],{l,1,nn,2}];
Table[x[l]=R*Cos[l*dth-0.001],{l,2,nn,2}];
Table[y[l]=Sqrt[R-x[l]^2],{l,2,nn,2}];
TA=Table[{x[l],y[l]},{l,1,nn}];
pl1=ListPlot[TA, AspectRatio->1,
Prolog->AbsolutePointSize[7]]

(* Step 2: Calculate the b. Do not modify fst *)
fst=3.00000;Clear[b]
delta=fst/nn; b[1]=N[fst-0.00001];
Table[N[b[n+1]=b[n]-delta],{n,1,nn}];
tb=Table[b[n],{n,1,nn}];
```

```
(* Step 3: Fill matrix for the boundary
conditions *)
Clear[MM,k];(*Clear[x,y,b,A,B]*)
MM=Table[1*n,{1,1,nn},{n,1,nn}];
Table[MM[[1,n]]=Cos[Sqrt[k^2-b[n]^2]*x[1]]*
Cos[b[n]*y[1]],{1,1,nn,2},{n,1,nn/2}];

Table[MM[[1,n]]=Cosh[Sqrt[k^2-b[n]^2]*x[1]]*
Cosh[b[n]*y[1]],
{1,1,nn,2},{n,nn/2+1,nn}];

Table[MM[[1,n]]=-x[1]*Sqrt[k^2 - b[n]^2]*
Cos[y[1]*b[n]]*Sin[x[1]*Sqrt[k^2 - b[n]^2]]-
y[1]*b[n]*Cos[x[1]*Sqrt[k^2 - b[n]^2]]*
Sin[y[1]*b[n]],{1,2,nn,2},{n,1,nn/2}];

Table[MM[[1,n]]=x[1]*Sqrt[k^2 - b[n]^2]*
Cosh[y[1]*b[n]]*Sinh[x[1]*Sqrt[k^2 - b[n]^2]]+
y[1]*b[n]*Cosh[x[1]*Sqrt[k^2 - b[n]^2]]*
Sinh[y[1]*b[n]],
{1,2,nn,2},{n,nn/2+1,nn}];

Table[Det[MM],{k,3.1962183,3.196219,0.0000001}]

Clear[k,W];W[k_]:=Det[MM]
Plot[W[k],{k,fst,4.}] //Timing

FindRoot[W[k]==0,{k,2.6}] //Timing

k=3.1962183;
nf=nn-1;bbf=Table[-MM[[ifit,nn]],{ifit,1,nf}];
rdutn=Table[MM[[ifit,klfit]],
{ifit,1,nf},{klfit,1,nf}];
B=LinearSolve[rdutn,bbf];
An={1};
A=Table[B[[lk]],{lk,1,nf}];
(* Check satisfaction of boundary cond.*)
Amp=Join[A,An];
boundary = MM . Amp

fxy[x_,y_]:=Sum[N[Amp[[n]]*Cos[b[n]*y]*
Cos[Sqrt[k^2-b[n]^2]*x]],{n,1,nn/2}]+
Sum[N[Amp[[n]]*Cosh[b[n]*y]*
Cosh[Sqrt[k^2-b[n]^2]*x]],
{n,nn/2+1,nn}]
```

```
Do[Print[fxy[x[l],y[l]]],
{1,1,nn/2,2}]

pl2=ContourPlot[fxy[x,y],{x,-1.,1.},{y,-1.,1.},
ContourShading->False,ContourSmoothing-> 2,
PlotPoints->60, DisplayFunction-> Identity]
Show[pll,pl2, DisplayFunction->$DisplayFunction]
gxy[x_,y_]:=fxy[x,y]/; x^2+y^2<=R^2
Off[Plot3D::plnc];Off[Plot3D::gval];
Plot3D[-gxy[x,y],{x,-1.,1.},{y,-1.,1.},
PlotPoints->50]
```

Does the FABER theorem hold for plates too? (Yes).
Compare the k=3.1962183 above with (4.5.45).

5.2 Holes in the domain. Two boundaries belonging to different coordinate systems

The theorem that the solution of an elliptic partial differential equation is uniquely determined by ONE closed boundary is valid for analytic solutions only. A partial differential equation of second order has however two arbitrary functions in its general solution. The LAPLACE *equation* has two solutions (4.1.63) and the HELMHOLTZ equation has two solutions (4.4.10), J_p and N_p, but we excluded the latter due to its singularity at $r = 0$. Now let us investigate the combination of the two particular solutions.

We first consider the LAPLACE *equation* (1.1.31) in polar coordinates r, φ. Assuming the inhomogeneous boundary conditions on a circular ring of radii R_1 and R_2, the conditions read

$$(\partial U/\partial r)_{r=R_1} = f_1(\varphi), U(r = R_2, \varphi) = f_2(\varphi). \qquad (5.2.1)$$

The solution is then given by

$$U(r,\varphi) = a_{02} + a_{01} R_1 \ln \frac{r}{R_2} \qquad (5.2.2)$$

$$+ \sum_{k=1}^{\infty} \frac{\left(a_{k1} R_2^{-k} + k R_1^{-k-1} a_{k2}\right) r^k + \left(k R_1^{-k-1} a_{k2} - R_2^k a_{k1}\right) r^{-k}}{k \left(R_1^{k-1} R_2^{-k} + R_2^k R_1^{-k-1}\right)} \cos k\varphi$$

$$+ \sum_{k=1}^{\infty} \frac{\left(b_{k1} R_2^{-k} + k R_1^{-k-1} b_{k2}\right) r^k + \left(k R_1^{k-1} b_{k2} - R_2^k b_{k1}\right) r^{-k}}{k \left(R_1^{k-1} R_2^{-k} + R_2^k R_1^{-k-1}\right)} \sin k\varphi,$$

where

$$a_{01} = \frac{1}{2\pi} \int\limits_{-\pi}^{+\pi} f_1(\varphi) \mathrm{d}\varphi, \qquad a_{k1} = \frac{1}{\pi} \int\limits_{-\pi}^{+\pi} f_1(\varphi) \cos k\varphi \mathrm{d}\varphi,$$

$$a_{k2} = \frac{1}{\pi} \int\limits_{-\pi}^{+\pi} f_2(\varphi) \cos k\varphi \mathrm{d}\varphi, \qquad a_{02} = \frac{1}{2\pi} \int\limits_{-\pi}^{+\pi} f_2(\varphi) \mathrm{d}\varphi,$$

$$b_{k1} = \frac{1}{\pi} \int\limits_{-\pi}^{+\pi} f_1(\varphi) \sin k\varphi \mathrm{d}\varphi, \qquad b_{k2} = \frac{1}{\pi} \int\limits_{-\pi}^{+\pi} f_2(\varphi) \sin k\varphi \mathrm{d}\varphi.$$

One may remember the Laplacian singularity $\ln r$, which we know from equation (4.8.6).

A sector of a circular ring with the radii $R_1, R_2 > R_1$ and the central angle α and subjected to the inhomogeneous boundary conditions

$$U(R_1,\varphi) = 0, \ \ U(R_2,\varphi) = U_0, \ \ U(r,0) = 0, \ \ U(r,\alpha) = 0 \qquad (5.2.3)$$

has for the LAPLACE *equation* the solution

$$U(r,\varphi) = \frac{4U_0}{\pi} \sum_{n=0}^{\infty} \frac{(r/R_1)^{\gamma_n} - (R_2/r)^{\gamma_n}}{(R_2/R_1)^{\gamma_n} - (R_1/R_2)^{\gamma_n}} \cdot \frac{\sin(\gamma_n \varphi)}{2n+1}, \ \ \gamma_n = \frac{2n+1}{\alpha} \pi. \tag{5.2.4}$$

The solutions of the HELMHOLTZ *equation* (1.1.35) are of greater interest. We consider the homogeneous boundary problem of a *circular ring membrane* with the radii $r = R_1, R_2 > R_1$. There are now two boundary conditions

$$u(R_1,\varphi) = 0, \quad u(R_2,\varphi) = 0. \tag{5.2.5}$$

In plane polar coordinates r, φ *Mathematica* has given the solution (4.4.10) which we shall use now. The NEUMANN *function* Y is the second solution of the BESSEL equation (2.4.43). It may be defined by

$$Y_\nu(x) = \frac{J_\nu(x) \cos \pi\nu - J_{-\nu}(x)}{\sin \pi\nu}, \tag{5.2.6}$$

where ν is not an integer. For $\nu = p$ (integer) the HOSPITAL *rule* yields

$$Y_p(x) = \frac{2}{\pi} \left(0.557 + \ln \frac{x}{2} \right) \cdot J_p. \tag{5.2.7}$$

For $\partial u/\partial \varphi = 0, p = 0$. The result (5.2.7) may also be derived from (2.2.43). The two boundary conditions (5.2.5) demand

$$A J_p \left(\frac{\omega}{c} R_1 \right) + B Y_p \left(\frac{\omega R_1}{c} \right) = 0,$$

$$A J_p \left(\frac{\omega}{c} R_2 \right) + B Y_p \left(\frac{\omega R_2}{c} \right) = 0. \tag{5.2.8}$$

This homogeneous linear system for the partial amplitueds A, B can only then be solved, if the determinant vanishes

$$J_p\left(\frac{\omega R_1}{c}\right) Y_p\left(\frac{\omega R_2}{c}\right) - Y_p\left(\frac{\omega}{c}R_1\right) J_p\left(\frac{\omega}{c}R_2\right) = 0. \qquad (5.2.9)$$

For given R_1, R_2 and c *Mathematica* gives the solution for $p = 0$ by

```
a=1.;b=2.;
FindRoot[BesselJ[0,a*x]*BesselY[0,b*x]-
BesselY[0,a*x]*BesselJ[0,b*x]==0,{x,4.0}]
```

This gives $x = 3.123\,03$.

For a sector of a circular ring membrane the solution analogous to (5.2.4) is found by replacing the powers of r by BESSEL functions. Thus an annulus $a \le r \le b$ with a sector $a \le \varphi \le \pi\beta$ has solutions of the form

$$J_{m/\beta}\left(\frac{j_{mn}r}{a}\right) \sin\left(\frac{m\varphi}{\beta}\right). \qquad (5.2.10)$$

BESSEL functions of fractional order appear and the order is determined by the angle β [5.14], [5.16].

Now we understand the role of singularities: they cut out a hole within the domain circumscribed by a closed boundary. The method worked quite well in plane polar coordinate systems r, φ which has a "natural" singularity at $r = 0$. Although we had some success with the solution function cosh of the LAPLACE equation in Figure 4.16 and equation (4.8.6) we are sceptic, since the function cosh tends to assume large values. Therefore we still use polar coordinates. Let *Mathematica* do the work:

```
<<Calculus`VectorAnalysis`
SetCoordinates[Cylindrical]
```

which gives Cylindrical[Rr,Ttheta,Zz]

Now we make a setup and solve the LAPLACE equation

```
u[Rr_,Ttheta_,0]:=R[Rr]*Cos[m*Ttheta]
LL=Laplacian[u[Rr,Ttheta,0]]
Collect[LL,Cos[m*Ttheta]]
```

O.K., the result looks nice. We obtain an ordinary differential equation for the r-dependence

$$\text{Cos}[m\ \text{Ttheta}]\left(-\frac{m^2\ R[Rr]}{Rr} + R'[Rr] + Rr\ R''[Rr]\right)/Rr$$

and ask *Mathematica* to solve it.

```
DSolve[R''[r]+R'[r]/r-m^2*R[r]/r^2==0, R[r],r]
```
which yields a not very nice expression $R[r] \to C[1]\ \text{Cosh}[m\ \text{Log}[r]] + i\ C[2]\ \text{Sinh}[m\ \text{Log}[r]]$

So we would like to test our own solution

```
Clear[U,a,b];U[r,φ]=
a0+c0*Log[r]+Sum[(c[n]*r^n+b[n]*r^(-n))*Cos[n*φ],{n,nn}]
```

$$a0 + c0 \, \mathrm{Log}[r] + \sum_{n=1}^{nn} (c[n] \; r^n + b[n] \; r^{-n}) \mathrm{Cos}[n \; \varphi]. \qquad (5.2.11)$$

To verify that this is really a solution of the LAPLACE equation in polar co-ordinates r, φ we type **Simplify[D[U[r,φ],{r,2}]+D[U[r,φ],r]/r+ D[U[r,φ],{φ,2}]/r^2]** which is apparently too difficult yielding

$$\left(\sum_{n=1}^{nn} \partial_r (\partial_r ((c[n] \; r^n + b[n] \; r^{-n}) \mathrm{Cos}[n \; \varphi])) \right) /r+$$

$$\sum_{n=1}^{nn} \partial_r (\partial_r ((c[n] \; r^n + b[n] \; r^{-n}) \mathrm{Cos}[n \; \varphi]))+$$

$$\left(\sum_{n=1}^{nn} \partial_r (\partial_r + ((c[n] \; r^n + b[n] \; r^{-n}) \mathrm{Cos}[n \; \varphi])) \right) \; r^2$$

O.K., then we simplify and set

Clear[U,a,b];
U[r,φ]=ao+c0*Log[r]+(c[n]*r^n+b[n]*r^(-n))*Cos[n*φ]

and the next command verifies our solution (5.2.11).

Simplify[D[U[r,φ],{r,2}]+D[U[r,φ],r]/r+ D[U[r,φ],{φ,2}]/r^2] yielding 0

Now we are able to solve the LAPLACE equation with two boundaries belonging to two different coordinate systems. As the *outer boundary* we choose a circle of radius $R = 2$ with the homogeneous boundary condition $U(r = R, \varphi) = 0, 0 \le \varphi \le 2\pi$. For the *inner boundary* we choose a square of lateral length 2 with the inhomogeneous boundary condition $U(\text{square}) = U_0$ (const), $-1 \le x \le +1, -1 \le y \le 1$. The homogeneous boundary condition on the circle demands

$$a_0 + c_0 \ln R = 0 \quad \text{and} \quad c_n R^n + b_n R^{-n} = 0, \qquad (5.2.12)$$

so that the solution (5.2.11) becomes

$$U(r, \varphi) = -c_0 \ln \frac{r}{R} + \sum_{n=1} c_n \left(r^n - R^{2n} r^{-n} \right) \cos n\varphi. \qquad (5.2.13)$$

This solves the problem of the *electromagnetic potential* between an infinite square prism and a circular cylinder.

Since the c_0, c_n are arbitrary, they can be used to satisfy the inner in-homogeneous boundary condition on the square. We do this using a col-location method. Avoiding the corner points, we choose the 6 collocation points $x_i \approx 1, i = 1, 2, \ldots 6, y_i = 0.0001, 0, 0.2, 0.4, 0.6, 0.8, 0.98$) and $r_i = \sqrt{x_i^2 + y_i^2}, \varphi_i = \arctan(y_i/x_i)$. Then the inhomogeneous inner boundary con-dition reads

$$U(r_i, \varphi_i) = -c_0 \ln \frac{r_i}{R} + \sum_{n=1}^{5} c_n \left(r_i^n - R^{2n} r_i^{-n} \right) \cos n\varphi_i = U_0, \; i = 1, 2, \ldots 6.$$

$$(5.2.14)$$

Since we have 6 collocation points, we have 6 linear equations that determine the 6 unknowns $c_0, c_1, c_2, c_3, c_4, c_5$. Now let *Mathematica* do the work:

```
(*Program DUM: inner inhomogeneous boundary U0 on
square, outer homogeneous boundary on circle, Laplace
operator*)
(*Step 1: define 6 collocation points on square of
length 2.*) Clear[x,y,r,fi];Table[x[i]=0.9999,{i,1,6}];
y[1]=0.00001;y[2]=0.2;y[3]=0.4;y[4]=0.6;y[5]=0.8;
y[6]=0.99;Clear[L1,GL1,L2,GL2,Ci,CiG,GG];
L1=ListPlot[Table[{x[i],y[i]},{i,1,6}],Prolog->
AbsolutePointSize[6],
DisplayFunction->Identity];GL1=Graphics[L1];
L2=ListPlot[{{-1.0,-1.0},{1.0,-1.0},{1.0,1.0},
{-1.0,1.0},{-1.0,-1.0}},Prolog->AbsolutePointSize[6],
PlotJoined->True,Axes->False,AspectRatio->1,
DisplayFunction->Identity];GL2=Graphics[L2];
Ci=Circle[{0.,0.},2];CiG=Graphics[Ci];
Show[GL1,GL2,CiG,Frame->True,DisplayFunction->
$DisplayFunction,AspectRatio->1]
```

This program generates Figure 5.7.

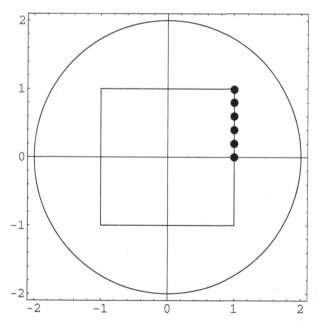

Figure 5.7
Capacitor consisting of square and circle

Since we want to execute the calculations in polar coordinates, we transform the coordinates of the 6 collocation points.

```
(*Step 2: transformation to polar coordinates
and using the solution in these coordinates*)
Clear[U,c0,c];Table[r[i]=Sqrt[x[i]^2+y[i]^2],{i,1,6}];

Table[φ[i]=ArcTan[x[i],y[i]],{i,1,6}]/;
ArcTan[x[i],y[i]]≥ 0; U[r[i],φ[i]]= -c0*Log[r[i]/R]+
Sum[c[n]*(r[i]^n-R^(2*n)*r[i]^(-n))*Cos[n*φ[i]],{n,5}]
```

Now we receive the expression for the solution in the collocation points

$$- c0 \, \text{Log}[1.\, r[i]] + c[1] \, \text{Cos}[\varphi[i]] \left(-\frac{1.}{r[i]} + r[i] \right)$$

$$+ c[2] \, \text{Cos}[2\,\varphi[i]] \left(-\frac{1.}{r[i]^2} + r[i]^2 \right) + c[3] \, \text{Cos}[3\,\varphi[i]] \left(-\frac{1.}{r[i]^3} + r[i]^3 \right)$$

$$+ c[4] \, \text{Cos}[4\,\varphi[i]] \left(-\frac{1.}{r[i]^4} + r[i]^4 \right) + c[5] \, \text{Cos}[5\,\varphi[i]] \left(-\frac{1.}{r[i]^5} + r[i]^5 \right).$$

Now *Mathematica* calculates the matrix and solves the linear system for the unknowns $c0, c[1] \ldots$.

```
(*Step 3: define the matrix and solve the linear system
for the partial amplitudes c*)
```

$$\text{M=Table}[k*1,\{k,1,6\},\{1,1,6\}];\ \text{M =Table}[\{-\text{Log}[\frac{r[i]}{R}],$$

$$\text{Cos}[\varphi[i]] \left(-\frac{R^2}{r[i]} + r[i] \right), \text{Cos}[2\,\varphi[i]] \left(-\frac{R^4}{r[i]^2} + r[i]^2 \right),$$

$$\text{Cos}[3\,\varphi[i]] \left(-\frac{R^6}{r[i]^3} + r[i]^3 \right), \text{Cos}[4\,\varphi[i]] \left(-\frac{R^8}{r[i]^4} + r[i]^4 \right),$$

$$\text{Cos}[5\,\varphi[i]] \left(-\frac{R^{10}}{r[i]^5} + r[i]^5 \right)\},\{i,1,6\}];$$

```
R=1.;U0=10.;
MatrixForm[M];
b={U0,U0,U0,U0,U0,U0};
B=LinearSolve[M,b];
```

We now help *Mathematica*. By copy and paste of the results we inform the program on the results obtained for $c0, c[1]$, etc.

```
(*Step 4: define the amplitudes and check the
satisfaction of both boundary values*)
c0=-3.3801962002952893`*^11;c[1]=-2.854707102125234*^11;
c[2]=8.487964476323438*^10;c[3]=-2.2662483860722004*^10;
c[4]=4.1363773733735595*^9;  c[5]=-3.7129944751173663*^8;
```

```
V[r_,φ_]=-c0*Log[r/R]+
Sum[c[n]*(r^n-R^(2*n)*r^(-n))*Cos[n*φ],{n,5}];
Table[V[r[i],φ[i]],{i,1,6}]
Table[V[R,φ[i]],{i,1,6}]
```
The result of checking the satisfaction of both boundary conditions is delightful:

$$\{10., 10., 9.99999, 10., 10., 10.\}$$
$$\{0., 0., 0., 0., 0., 0.\}$$

Thus we have demonstrated that collocation methods are able to solve an elliptic partial differential equation even for *two closed boundaries* even in the case that one boundary value problem is homogeneous and the other one inhomogeneous!

Now we exchange the boundaries: the outer inhomogeneous boundary is described by a rectangle 8×4 and the inner homogeneous boundary is given by a circle of radius 1. Then the boundary conditions read

$$U(r = 1, \varphi) = 0, \ U(x, y = \pm 2) = \cos \pi x/8, \ U(x = \pm 4, y) = 0. \quad (5.2.15)$$

We use very detailled commands and choose 7 collocation points and identify them using a symbol.

point	(4.,0)	(4.,1.)	(4.,1.5)	(4.,2.)	(1.,2.)	(2.,2.)	(3.,2.)
symbol	q	l	t	u	v	w	p
r_i^2	16	17	18.25	20	5	8	13.

and $r_i = \sqrt{x_i^2 + y_i^2}$, $\varphi_i = \arctan(y_i/x_i)$, $l_n = (\sqrt[n]{17} - 1/\sqrt[n]{17}) \cos(n \arctan(1/4))$. For $\ln r_i$ the abbreviation $a_i = \ln r_i$ will be used. *Mathematica* does the work. At first one has to define a matrix

```
m={{a1,u1,u2,u3,u4,u5,u6},
   {a2,v1,v2,v3,v4,v5,v6},
   {a3,w1,w2,w3,w4,w5,w6},
   {a4,p1,p2,p3,p4,p5,p6},
   {a5,q1,q2,q3,q4,q5,q6},
   {a6,l1,l2,l3,l4,l5,l6},
   {a7,t1,t2,t3,t4,t5,t6}}                          (5.2.16)
```
Then the boundary conditions create the rhs term of the linear system.
```
b=N[{0, Cos[Pi/8],Cos[Pi/4],Cos[3*Pi/8],0,0, 0}]   (5.2.17)
```
Then we solve the system by the command
```
LinearSolve [m,b]
```
yielding c_n. Informing *Mathematica* about these values and the definitions
```
r[x_,y_]:=Sqrt[x^2+y^2]
f[x_,y_]:=N[ArcTan[x,y]] /;y>=0 ;
f[x_,y_]:=2*Pi+N[ArcTan[x,y]] /; y<0              (5.2.18)
```
and plot commands like

```
Plot3D[C1*Log[r[x,y]]+
C2*(r[x,y]^1-r[x,y]^-1)*Cos[1*f[x,y]]+
```

```
C3*(r[x,y]^2-r[x,y]^-2)*Cos[2*f[x,y]]+
C4*(r[x,y]^3-r[x,y]^-3)*Cos[3*f[x,y]]+
C5*(r[x,y]^4-r[x,y]^-4)*Cos[4*f[x,y]]+
C6*(r[x,y]^5-r[x,y]^-5)*Cos[5*f[x,y]]+
C7*(r[x,y]^6-r[x,y]^-6)*Cos[6*f[x,y]]],
{x,-3.99999,3.99999},{y,-1.99999,1.99999},
PlotPoints->60, PlotRange->{0,1.5},ClipFill->None,
Shading->False,AspectRatio->1.]
```
create Figures 5.8 and 5.9.

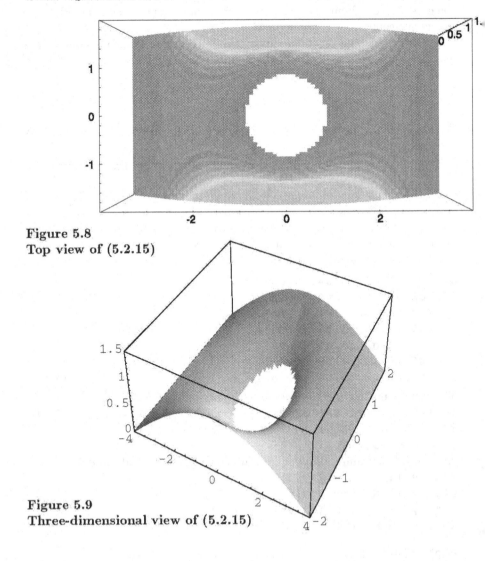

Figure 5.8
Top view of (5.2.15)

Figure 5.9
Three-dimensional view of (5.2.15)

Finally, the question arises whether problems with boundaries belonging to different coordinate systems can be solved by other methods than by collocation methods. A method of solving these *nonuniform boundary problems* has been known since 1936 [5.5], but it is cumbersome. TINHOFER has applied this method on our problem of a rectangle $a \times b$ with a circular hole of radius c. For the LAPLACE equation the following boundary conditions have been assumed:

$$U(x = \pm a, y) = U(x, y = \pm b) = 0,$$

$$U(r = c, \varphi) = \sum_n D_{2n} \cos 2n\varphi. \tag{5.2.19}$$

Now the cartesian solution (4.8.6) will be expressed by the polar solutions (4.1.63)

$$U(r, \varphi) = \frac{a_0}{2} + \sum_{\nu=1} c_\nu r^\nu \left(a_\nu \cos \nu\varphi + b_\nu \sin \nu\varphi\right). \tag{5.2.20}$$

To do this, the formulae $x = r \cos \varphi$, $y = r \sin \varphi$

$$\cos \alpha x = \frac{1}{2}\left(e^{i\alpha x} + e^{-i\alpha x}\right) = \frac{1}{2}\left(e^{i\alpha r \cos \varphi} + e^{-i\alpha r \cos \varphi}\right),$$

$$\cosh \alpha y = \frac{1}{2}\left(e^{\alpha y} + e^{-\alpha y}\right) = \frac{1}{2}\left(e^{\alpha r \sin \varphi} + e^{-\alpha r \sin \varphi}\right),$$

$$\cos \frac{(2n+1)\pi x}{2a} \cosh \frac{(2n+1)\pi y}{2a}$$

$$= \operatorname{Re}\left\{\frac{1}{2}\left(\exp\left[-i\frac{2n+1}{2}\pi\frac{r}{a}e^{i\varphi}\right] + \exp\left[-i\frac{2n+1}{2}\pi\frac{r}{a}e^{-i\varphi}\right]\right)\right\}, \tag{5.2.21}$$

are helpful.

We see that the function cos as well as cosh are both expressed by their common mother function exp; recall the grandmother (*hypergeometric equation*) (2.2.59). Furthermore, one has to use

$$e^x = \sum_{s=0} \frac{x^s}{s!}, \quad e^{i\alpha r e^{-i\varphi}} = \sum_{s=0} i^s \alpha^s r^s e^{-is\varphi} \frac{1}{s!}, \tag{5.2.22}$$

and $s \to 2s$, $i^{2s} = (-1)^s$ so that from (5.2.21) results

$$\frac{1}{2}\operatorname{Re}\left\{\sum_s (-1)^s \frac{i^s}{s!}\left(\frac{2n+1}{2}\pi\frac{r}{a}\right)^s (\cos s\varphi + i \sin s\varphi)\right.$$

$$\left. + \sum_s (-1)^s \frac{i^s}{s!}\left(\frac{2n+1}{2}\pi\frac{r}{a}\right)^s (\cos s\varphi - i \sin s\varphi)\right\}$$

$$= \sum_s (-1)^s \frac{1}{(2s)!}\left(\frac{2n+1}{2}\pi\frac{r}{a}\right)^{2s} \cos 2s\varphi \tag{5.2.23}$$

and

$$\cos \frac{(2n+1)\pi y}{2b} \cosh \frac{(2n+1)\pi x}{2b}$$

$$= \sum_s \left(\frac{2n+1}{2} \pi \frac{r}{b} \right)^{2s} \frac{(-1)^s}{(2s)!} \cos 2s\varphi. \qquad (5.2.24)$$

Now it is possible to rewrite the solution (4.8.6) in the form

$$U(x,y) = U(r,\varphi) = -\ln \sqrt{\frac{x^2+y^2}{a^2+b^2}}$$

$$+ \sum_n A_n \sum_s \left(\frac{(2n+1)\pi r}{2a} \right)^{2s} \frac{(-1)^s}{(2s)!} \cos 2s\varphi$$

$$+ \sum_n B_n \sum_s \left(\frac{(2n+1)\pi r}{2b} \right)^{2s} \frac{(-1)^s}{(2s)!} \cos 2s\varphi, \qquad (5.2.25)$$

where A_n and B_n are known from (4.8.7). Introducing the abbreviations

$$C_{2s} = \sum_n A_n \left(\frac{2n+1}{2} \pi \frac{1}{a} \right)^{2s} \frac{(-1)^s}{(2s)!}$$

$$+ \sum_n B_n \left(\frac{(2n+1)}{2} \pi \frac{1}{b} \right)^{2s} \frac{(-1)^s}{(2s)!} \qquad (5.2.26)$$

the solution (5.2.25) can be rewritten (using $\delta_{s0} = 1$ for $s = 0$, but 0 otherwise) as

$$U(x,y) = U(r,\varphi) = \sum_s \left\{ -\ln \frac{r}{\sqrt{a^2+b^2}} \delta_{s0} + C_{2s} r^{2s} \cos 2s\varphi \right\}. \qquad (5.2.27)$$

To satisfy the boundary condition on the circle, one uses the fact that the derivative of a solution of a differential equation is again a solution of the differential equation. We build new solutions:

$$U_{2\mu}(x,y) = \frac{\partial^{2\mu}}{\partial x^{2\mu}} U(x,y) = -\frac{\partial^{2\mu} \ln r}{\partial x^{2\mu}} \Big|_{a,b}^{x,y} + \sum_s C_{2s}^{(2\mu)} r^{2s} \cos 2s\varphi, , \qquad (5.2.28)$$

where

$$C_{2s}^{(2\mu)} = \sum_n \left\{ A_n^{(2\mu)} \left(\frac{2n+1}{2} \frac{\pi}{a} \right)^{2s} \frac{(-1)^s}{(2s)!} \right.$$

$$\left. + \frac{B_n^{(2\mu)}}{(2s)!} \left(\frac{2n+1}{2} \frac{\pi}{b} \right)^{2s} \frac{(-1)^s}{(2s)!} \right\} \qquad (5.2.29)$$

and $A^{(2\mu)}, B^{(2\mu)}$ have to be redefined to satisfy the boundary conditions:

$$A_n^{(2\mu)} = \frac{1}{a\cosh\dfrac{(2n+1)\pi b}{2a}} \int_{-a}^{+a} \frac{\partial^{2\mu}}{\partial x^{2\mu}} \ln r \Bigg|_{a,b}^{x,y} \cos\frac{2n+1}{2}\pi\frac{x}{a}\,dx,$$

$$B_n^{(2\mu)} = \frac{1}{b\cosh\dfrac{(2n+1)\pi a}{2b}} \int_{-b}^{+b} \frac{\partial^{2\mu}}{\partial x^{2\mu}} \ln r \Bigg|_{a,b}^{x,y} \cos\frac{2n+1}{2}\pi\frac{y}{b}\,dy. \quad (5.2.30)$$

Using

$$\frac{\partial^{2\mu}\ln r}{\partial x^{2\mu}} = -(2\mu-1)!\,r^{-2\mu}\cos 2\mu\varphi, \qquad (5.2.31)$$

the new partial solutions (5.2.28) assume the form

$$U_{2\mu}(x,y) = U_{2\mu}(r,\varphi) = \frac{(2\mu-1)!}{\sqrt{a^2+b^2}^{2\mu}}\cos\frac{\mu\pi}{2} - \frac{(2\mu-1)!}{r^{2\mu}}\cos 2\mu\varphi$$

$$+ \sum_{s=0}^{\infty} C_{2s}^{(2\mu)} r^{2\mu}\cos 2\mu\varphi. \qquad (5.2.32)$$

Using

$$\Phi_0(r,\varphi) = \sum_{s=0}\left[\ln\frac{r}{\sqrt{a^2+b^2}}\delta_{s0} - C_{2s}r^{2s}\right]\cos 2s\varphi,$$

$$\Phi_{2\mu}(r,\varphi) = \left\{\left[\frac{(2\mu-1)!}{\sqrt{a^2+b^2}^{2\mu}}\cos\frac{\mu\pi}{2} - C_0^{(2\mu)}\right]\right. \qquad (5.2.33)$$

$$\left. - \sum_{s=1}^{\infty}\left[\frac{(2\mu-1)!}{r^{2\mu}}\delta_{2\mu 2s} + C_{2s}^{(2\mu)}r^{2s}\right]\right\}\cos 2s\varphi, \quad \mu\neq 0$$

one can write

$$U(r,\varphi) = \sum_{\mu=0}^{\infty} S_{2\mu}\Phi_{2\mu}(r,\varphi). \qquad (5.2.34)$$

Here the expansion coefficients $S_{2\mu}$ are still unknown; they depend on the boundary condition (5.2.19), which reads now $\sum_s D_{2s}\cos 2s\varphi$. The coefficients $S_{2\mu}$ have to be calculated from

$$\sum_{\mu=0}^{\infty} \gamma_{2s}^{2\mu} S_{2\mu} = D_{2s},$$

where

$$\gamma_0^0 = \ln \frac{c}{\sqrt{a^2 + b^2}} - C_0,$$

$$\gamma_0^{2\mu} = \frac{(2\mu - 1)!}{\sqrt{a^2 + b^2}^{2\mu}} \cos \frac{\mu\pi}{2} - C_0^{(2\mu)}, \qquad \mu \geq 1,$$

$$\gamma_{2s} = -C_{2s}c^{2s},$$

$$\gamma_{2s}^{2\mu} = - \left[(2\mu - 1)! c^{-2\mu} \delta_{2\mu 2s} + C_{2s}^{(2\mu)} c^{2s} \right], \qquad \mu \geq 1. \qquad (5.2.35)$$

Some of these nonuniform problems may be very important. In the *nuclear power station* KAHL it has been crucial that the neutron-absorbing control rods could control the chain reaction. In this nuclear reactor the control rods were arranged on the surface of a cone. The boundary value problem of the neutron diffusion equation could not be solved numerically with the necessary accuracy, since two boundaries (on cylinder and cone) had to be satisfied and the vertex of the cone presented a singularity [5.6].

Membrane domains could exhibit holes, too. We first investigate if collocation methods can be avoided. A clamped square membrane $-\frac{a}{2} \leq x \leq +\frac{a}{6}$, $-\frac{a}{2} \leq y \leq \frac{a}{2}$ can again be described by a modified HELMHOLTZ equation of type (4.6.51)

$$\Delta u + E(1 - \lambda\delta(R - r))u = 0, \quad u\left(\pm\frac{a}{2}, y\right) = u\left(x, \pm\frac{a}{2}\right) = 0, \qquad (5.2.36)$$

where λ is a given parameter. This cuts out a circular hole of radius R. Perturbation theory

$$u = u_0 + \lambda u_1 + \lambda^2 u_2 + \dots, \quad E = E_0 + \lambda E_1 + \lambda^2 E_2 + \dots \qquad (5.2.37)$$

with

$$u_0 = \frac{2}{a} \cos \frac{\pi x}{a} \cdot \cos \frac{\pi y}{a}, \quad u_1 = \sum c_{nm} u_{nm},$$

$$u_{nm} = \frac{2}{a} \cos \frac{2n + 1}{a} \pi x \cos \frac{2m + 1}{a} \pi y,$$

$$E_1 = E_0 \frac{4}{a^2} \int \int \delta(R - r) \cos^2 \frac{\pi x}{a} \cos^2 \frac{\pi y}{a} dx dy =$$

$$E_0 \frac{2\pi R}{a} \left[1 + 2J_0 \left(\frac{2\pi R}{a} \right) + J_0 \left(\frac{2\pi\sqrt{2}R}{a} \right) \right] \qquad (5.2.38)$$

might be the answer (E. RIETSCH). For $R = a/4$ one obtains

$$E = \frac{2\pi^2}{a^2} \left(1 + 1.02 \frac{\lambda\pi}{a} \right). \qquad (5.2.39)$$

One might also think to expand the point forces represented by $\delta(R - r)$ with respect to

$$\cos \frac{2\nu + 1}{a} \pi x \cos \frac{2\mu + 1}{a} \pi y.$$

Such annular membranes have been discussed in the literature even with variable density [5.7], [5.8]. Many problems concerning two boundaries need to know how to handle corners in a boundary. Problems will thus be offered in the next section. Anyway, try to solve this problem: The program DUM solves an inner homogeneous boundary on a square and an outer homogeneous bondary on a circle. Modify the program to calculate an inner homogeneous boundary on the square and an outer inhomogeneous boundary on the circle. (Solution: the modifications have to be made in steps 1 and 3. In step 4, the resulting two output lines should then be exchanged numerically).

Problems

1. Solve the inhomogeneous problem of a rectangular membrane 4×2 with an inner homogeneous circular boundary of radius 1. The inhomogeneous values on the rectangle should be 1.

   ```
   (* Membrane-hole. Solution of Helmholtz equation for
   2 boundaries belonging to two different coordinate
   systems. Outer boundary: rectangle 4 x 2, boundary
   value 1, inner boundary circle of radius 1,
   homogeneous condition. No eigenvalue problem.
   Solution in polar coordinates *)

   (* Step 1: define collocation points*) n=18;
   (*on inner circle*)
   x[1]=0.5; y[1]=0.; x[2]=0.45;
   y[2]=Sqrt[0.5^2-x[2]^2];
   x[3]=0.35;y[3]=Sqrt[0.5^2-x[3]^2]; x[4]=0.15;
   y[4]=Sqrt[0.5^2-x[4]^2]; x[5]=0;
   y[5]=0.5;x[6]=-x[4]; y[6]=y[4];
   x[7]=-x[3]; y[7]=y[3]; x[8]=-x[2]; y[8]=y[2];
   x[9]=-x[1]; y[9]=0.;
   (* on rectangle *)
   x[10]=2.;    y[10]=0.;   x[11]=2.;    y[11]=0.5;
   x[12]=2.;    y[12]=1.;   x[13]=1.;    y[13]=1.;
   x[14]=0.;    y[14]=1.;   x[15]=-1.;   y[15]=1.;
   x[16]=-2.;   y[16]=1.;   x[17]=-2.;   y[17]=0.5;
   x[18]=-2.0; y[18]=0.;

   Table[f[l]=N[ArcTan[x[l],y[l]],12],{l,1,n}];
   Table[r[l]=N[Sqrt[x[l]^2+y[l]^2],12],{l,1,n}];
   ```

```
Clear[p1];
p1=ListPlot[Table[{x[l],y[l]},{l,1,18}],
PlotStyle->PointSize[1/40],AspectRatio->0.5,
PlotRange->{{-2.,2.},{0.,2.}}];
```

Here N guarantees that one gets numerical values from the function arctan. The command **ListPlot** generates Figure 5.10 showing the location of all collocation points. **PointSize** determines the diameter of the "points" by 1/40 of the dimension of the plot, **AspectRatio->0.5** fixes the ratio 1:2 of the y to the x dimension and **PlotRange** describes the limits of the plot.

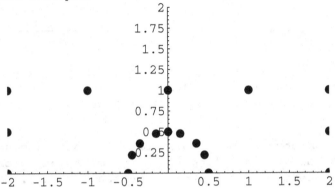

Figure 5.10
List plot of collocation points

Now one has to define the matrix **m** and the two boundary values:

```
(* Step 2: define matrix m to be filled later*)
m=Table[i*j],{i,1,n},{j,1,n}];
Table[m[[l,k]]=BesselJ[(k/2-1/2),r[l]]*
Cos[(k/2-1/2)*f[l]],{l,1,18},{k,1,n,2}];
Table[m[[l,k]]=BesselY[(k/2-1),r[l]]*
Cos[(k/2-1)*f[l]],{l,1,18},{k,2,n,2}];

(* Step 3: define boundary values and solve
linear equ. *)
b={0,0,0,0,0,0,0,0,0,1,1,1,1,1,1,1,1,1};
LinearSolve[m,b];
kolist=% ;Table[A[k]=kolist[[k]],{k,1,n}];
```

It might be that you get a message concerning a badly conditioned matrix. In this case, remember problem 1 of section 4.8:

```
<< LinearAlgebra`GaussianElimination`
LUFactor[m]; LUSolve[lu, m]
```

Another method is to rationalize all numbers: **RM=Rationalize[M,0]**

```
RP=Rationalize[P,0]
BC=LinearSolve[RM,RP]
A=Table[BC[[lk]],[lk,1,n}]                                    (5.2.40)
```

The command **Rationalize[M,p]** transforms all elements of the matrix **M** into rational numbers (fractions) using a tolerance of 10^{-p} in the approximation.

```
(* Step 4: verify the boundary conditions *)
r[x_,y_]:=N[Sqrt[x^2+y^2]];
f[x_,y_]:=N[ArcTan[x,y]];
g1[x_,y_]:=Sum[A[k]*BesselJ[(k/2-1/2),r[x,y]]*
Cos[(k/2-1/2)*f[x,y]],{k,1,n-1,2}];
g2[x_,y_]:=Sum[A[k]*BesselY[(k/2-1),r[x,y]]*
Cos[(k/2-1)*f[x,y]],{k,2,n,2}];
g[x_,y_]:=g1[x,y]+g2[x,y] (*/;x^2+y^2>=1*)
Table[g[x[i],y[i]],{i,1,n}]
```

If all is correct, the result should be {0.,0.,0.,0.,0.,0.,0.,0.,0., 1.,1.,1.,1.,1.,1.,1.,1.,1.}. Observe the condition
/;x^2+y^2>=1 after the definition of the function **g[x_,y_]** which should define the function $g(x,y)$ for values only outside and on the circular inner boundary.

The following commands should generate Figures 5.11 - 5.14:

```
Clear[p2];p2=ContourPlot[g[x,y],{x,-2.0,2.0},
{y,-1.0,1.0},
ContourShading->False,ContourSmoothing->2,
PlotPoints->30,Contours->{0},AspectRatio->0.5]
Show[Graphics[p1], Graphics[p2]]
```

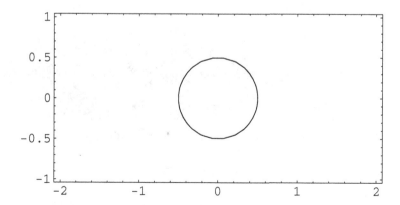

Figure 5.11
Contourplot of rectangle with circular hole

```
Clear[p3];p3=Plot3D[g[x,y],
{x,-2.,2.},{y,-1.,1.},
PlotPoints->60,
PlotRange->{0,1.2},
ViewPoint->{0.,0.,2.},
ClipFill->None,
AspectRatio->0.5,
Shading->True];
Show[p3,ViewPoint->{1.3,-2.4,2.}];
Show[p3,ViewPoint->{0., -2.0, 0.}]
```

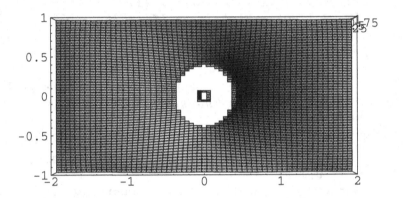

Figure 5.12
Rectangular membrane with circular hole

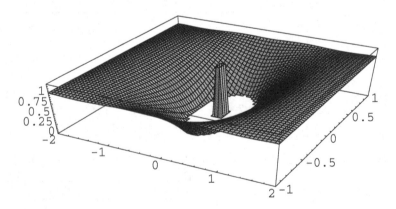

Figure 5.13
Rectangular membrane from default viewpoint

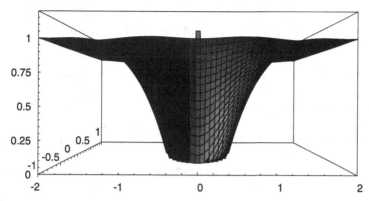

Figure 5.14
Membrane sideways

2. Try to write a program to calculate the eigenvalue of a circular membrane with a rectangular hole. Use polar or cartesian coordinates. Solution: very difficult, approximation very rough. Collocation methods often fail to solve eigenvalue problems with two different boundaries.

5.3   Corners in the boundary

Corners in the boundary of a domain present new mathematical difficulties. In recent years quite a great number of papers have been published on this subject. Due to the large amount of material we only may quote a small selection. A collocation method using boundary points has been discussed for a triangular plate as early as 1960 [5.9] and the same author treated vibrations of a conical bar [5.10]. Triangular and rhombic plates were also discussed [5.11]. VEKUA drew the attention to the fact that smoother boundaries yield better approximations [5.12]. FOX stated that it is necessary to take the origin of the polar coordinate system as the vertex of a corner with the angle π/α and that a function

$$u(r,\varphi) = \sum_{n=1}^{N} C_n J_{n\alpha}(\sqrt{\lambda}\, r)\sin n\alpha\varphi), \qquad (5.3.1)$$

where $n\alpha = \alpha_n$, $n = 1\ldots N$ satisfies the boundary condition on both sides of the angle and has the correct singularity at the corner [5.13]. For an L-shaped membrane one has an angle $3\pi/2$ and $\alpha = 2/3$, so that BESSEL *functions* of *fractional order* appear. Such reentrant corners also have been investigated by DONNELY [5.14]. Survey reviews have been published by KONDRATEV and

KUTTLER [5.15], [5.16]. The various authors use variational methods, collocation methods and others but state that often engineering applications solved
by the *finite element method* are of limited accuracy [5.16]. A collocation
method applied on a nonconvex domain with one boundary generally yields
poor accuracy too [5.17]. It is therefore interesting to know the computable
bounds for eigenvalues of elliptic operators [5.18]. Another review on problems on corner domains has been given by DAUGE [5.19]. Finally, Ukrainian
authors draw the attention to the obstacles hindering the use of variational
methods (as well as of finite difference, finite elements and boundary elements)
for such boundary problems and present the theory of *R functions* [5.20]. In
section 4.1, Figure 4.2 showed a rectangle with an incision. The angle of
that incision is apparently given by $3\pi/2$ (*reentrant corner* [5.14]). We first
consider the LAPLACE *equation* in cartesian coordinates with homogeneous
boundary conditions:

$$u(0, y) = 0, \qquad -1 \le y \le 0, \tag{5.3.2}$$

$$u(x, 0) = 0, \qquad 0 \le x \le 1. \tag{5.3.3}$$

We assume diagonal symmetry around the diagonal straight line $y = x$. For
the remaining boundaries, we assume inhomogeneous conditions:

$$u(1, y) = \sin \frac{\pi}{2} y, \quad 0 \le y \le 1, \tag{5.3.4}$$

$$u(x, 1) = 1, \qquad -1 \le x \le +1. \tag{5.3.5}$$

These conditions guarantee continuity in the points (1,0) and (1,1). Avoiding
cartesian solutions with the disagreable cosh we use polar coordinates. To
satisfy the boundary conditions along the lines $0 \le x \le 1$, $y = 0$ (thus $\varphi = 0$)
and along $x = 0$, $-1 \le y \le 0$, (thus $\varphi = 3\pi/2$, $0 \le r \le 1$) we set up

$$u(x, y) = u(r, \varphi) = \sum_{n=1}^{N} a_n r^{\frac{2}{3}(2n-1)} \sin \frac{2}{3}(2n - 1)\varphi. \tag{5.3.6}$$

This is an exact solution of the LAPLACE equation in polar coordinates, compare similar solutions like (5.2.2).

We now use the partial amplitudes a_n to satisfy the boundary conditions.
For this purpose several methods can be used. We prefer the collocation
method choosing 8 collocation points (1.,0.2), (1.,0.4), (1.,0.7), (1.,0.1), (0.5,1.),
(0,1.), (-0.5,1.), (-1.,1.). Using the abbreviations

$$r_i = \sqrt{x_i^2 + y_i^2},$$

$$\varphi_i = \arctan(y_i/x_i) = f(x_i, y_i),$$

$$s_{in} = \sin 2(2n - 1)\varphi_i/3,$$

$$u(x_i, y_i) = \sum_n a_n r_i^{2(2n-1)/3} s_{in} \tag{5.3.7}$$

and inserting into the boundary condition (5.3.4), (5.3.5) one can solve for the partial amplitudes a_n and plot $u(x, y)$. This can be done by a *Mathematica* program. You should Input **x[i],y[i],r,f,m**

```
b={N[Sin[Pi*y[1]/2]],N[Sin[Pi*y[2]/2]],
N[Sin[Pi*y[3]/2]],N[Sin[Pi*y[4]/2]],1,1,1,1}
```

{0.309017, 0.587785, 0.891007, 0.156434, 1, 1, 1, 1}

LinearSolve[m,b]

{1.15218, 0.224138, 0.0277575, 0.0331684,
0.0216117, -0.000531457, -0.00341928, 0.00109555}

```
Plot3D[
  1.15218*(r[x,y]^(2*(2*1-1)/3))*Sin[2*(2*1-1)*f[x,y]/3]
+0.224138*(r[x,y]^(2*(2*2-1)/3))*Sin[2*(2*2-1)*f[x,y]/3]
+0.0277575*(r[x,y]^(2*(2*3-1)/3))*Sin[2*(2*3-1)*f[x,y]/3]
+0.0331684*(r[x,y]^(2*(2*4-1)/3))*Sin[2*(2*4-1)*f[x,y]/3]
+0.0216117*(r[x,y]^(2*(2*5-1)/3))*Sin[2*(2*5-1)*f[x,y]/3]
-0.000531457*(r[x,y]^(2*(2*6-1)/3))*Sin[2*(2*6-1)*f[x,y]/3]
-0.00341928*(r[x,y]^(2*(2*7-1)/3))*Sin[2*(2*7-1)*f[x,y]/3]
+0.00109555*(r[x,y]^(2*(2*8-1)/3))*Sin[2*(2*8-1)*f[x,y]/3],
{x,-1.,1.},{y,-1.,1}, PlotRange->{0.,1.},
Shading->False,
ClipFill->None,PlotPoints->60]
```

This generates Figure 5.15.

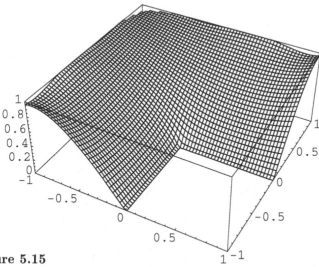

Figure 5.15
Reentrant corner

To solve a similar boundary value problem for the HELMHOLTZ *equation*, we use the corresponding solutions [5.18]

$$u(r, \varphi) = \sum_{n=1}^{N} A_n J_{(2/3)(2n-1)}(kr) \sin \frac{2}{3}(2n - 1)\varphi. \qquad (5.3.8)$$

In an analogous program one would have

```
Table[f[l]=ArcTan[x[l],y[l]],{1,1,n}];
Table[r[l]=Sqrt[x[l]^2+y[l]^2],{1,1,n}];
m=Table[m[[k,l]]=BesselJ[2*(2*k-1)/3,r[l]]*
Sin[2*(2*k-1)*f[l]/3],{1,1,n},{k,1,n}];
```

```
b={Sin[Pi*y[1]/2],Sin[Pi*y[2]/2],
Sin[Pi*y[3]/2],Sin[Pi*y[4]/2],Sin[Pi*y[5]/2],1,1,1,1}
```

Far more complicated are problems with two rectangular boundaries. For the LAPLACE *equation* STILL (private communication) has solved such a problem. For $\Delta u = 0$, he assumes a homogeneous boundary problem for the inner square and an inhomogeneous problem $u = 1$ on the outer square. Due to the symmetry, only half of the first quadrant will be considered. A trial function is taken in the form

$$u(r,\varphi) = a_1 + a_2 \ln r + \sum_{k=1}^{n-1} \left(a_{2k+1}r^{4k} + a_{2k+2}r^{-4k}\right) \cos(4k\varphi). \qquad (5.3.9)$$

A collocation method with $N = 2n$ points showed heavy oscillations between the collocation points. The method obviously suffered from the corners with an angle $\pi/4$, where (5.3.9) is not differentiable. In fact, it allows a singular expansion around this point (r_0, φ_0) by

$$u(r_0, \varphi_0) = r_0^{2/3} \sin(2/3)\varphi_0. \qquad (5.3.10)$$

Therefore, the trial function (5.3.9) had been modified by adding the term

$$\sum_{k=1}^{n} \left(a_{4k}r^{4+2} + a_{4k+2}r^{-4k-2}\right) \sin(4k+2)\varphi$$

$$+a_{4n+3}r_0^{2/3} \sin(2/3)\varphi_0. \qquad (5.3.11)$$

This trial function gave slightly better results. Furthermore, STILL also considered conformal mapping.

An even more general problem has also been considered by STILL [5.21]. The boundary of a triangle may consist of two straight lines $\Gamma_1(\varphi = 0), \Gamma_2(\varphi = \omega)$ and a curve described by $\Gamma_3(g(r, \varphi))$. The LAPLACE equation solution should satisfy the boundary values $g_\nu(r, \varphi)$ on Γ_ν for $\nu = 1, 2, 3$. One can assume

$$g_1(r) = \sum_{n=0}^{N} a_n^{(1)}r^n \text{ for } \varphi = 0, \quad g_2(r) = \sum_{n=0}^{N} a_n^{(2)}r^n \text{ for } \varphi = \omega \qquad (5.3.12)$$

or simpler

$$g_1(r) = a_0^{(1)}(r - R_1) \quad \text{for} \quad \varphi = 0, \quad g_2(r) = a_0^{(2)}R_1 \quad \text{for} \quad \varphi = \omega,$$

$$g_3(r, \varphi) = -a_0^{(1)}R_1 \sin\frac{\pi\varphi}{2\omega} + Br^2 \sin\frac{\pi\varphi}{\omega} \quad \text{for} \quad 0 \le \varphi \le \omega. \qquad (5.3.13)$$

These functions are continuous at the three corners $(r = 0, \varphi = 0)$, $(r = R_1, \varphi = 0)$, $(r = R_2, \varphi = \omega)$. Then the singularity free setup may be suggesting itself:

$$u(r, \varphi) = u_0(r, \varphi) + u_1(r, \varphi) = a_0 + \sum_{n=1}^{N} (a_n \cos n\varphi + b_n \sin n\varphi) \, r^n$$

$$+ \sum_{n=1}^{N} \beta_n r^{\alpha n} \sin n\alpha\varphi, \tag{5.3.14}$$

where $\alpha = \pi/\omega$. If the boundary conditions would contain a discontinuity, the addition of singular terms like $c\varphi$, $cr(\varphi \cos \varphi + \log r \sin \varphi)$, which solve the LAPLACE equation, could help. The β_n in (5.3.14) have to be determined by the boundary condition on Γ_3:

$$g_3(r, \varphi) - u_0(r, \varphi) = \sum_{n=1}^{N} \beta_n r^{\alpha n} \sin n\alpha\varphi, \tag{5.3.15}$$

whilst for Γ_1, Γ_2 one has

$$\Delta u_1 = 0, \quad u_1(r, \varphi) = 0. \tag{5.3.16}$$

If instead of the LAPLACE the HELMHOLTZ equation has to be solved, one has to make the replacement

$$r^n \cos n\varphi \quad \rightarrow \quad J_n(kr) \cos n\varphi. \tag{5.3.17}$$

To execute numerical calculations, we choose $R_1 = 1.$, $R_2 = 1.5$, $a_0^{(1)} = 1.$, $a_0^{(2)} = -1.$, $B = 1.$. Satisfying the boundary conditions results in

$$\varphi = 0 : \ g_1(r) = r - 1 = u(r, 0) = a_0 + \sum_{n=1}^{N} a_n r^n, \tag{5.3.18}$$

$$\varphi = \omega : \ g_2(r) = -1 = u(r, \omega)$$

$$= a_0 + \sum_{n=1}^{N} (a_n \cos n\omega + b_n \sin n\omega) \, r^n, \tag{5.3.19}$$

$$g_3(r, \varphi) = -\sin \frac{\pi\varphi}{2\omega} + Br^2 \sin \frac{\pi\varphi}{\omega} \tag{5.3.20}$$

$$= a_0 + \sum_{n=1}^{N} (a_n \cos n\varphi + b_n \sin n\varphi) \, r^n + \sum_{n=1}^{N} \beta_n r^{n\pi/\omega} \sin n\frac{\pi}{\omega}\varphi.$$

The equations (5.3.18), (5.3.19) determine the coefficients a_0, a_n and b_n in Γ_1 and Γ_2. Equation (5.3.20) may yield the β_n. This may be done using a collocation method. First, (5.3.18) gives $a_0 = -1., a_1 = 1., a_n = 0, n > 1$. From (5.3.19) one obtains for $\omega \neq \pi$

$$b_1 = \frac{-\cos \omega}{\sin \omega}, \quad b_n = 0, \quad n > 1. \tag{5.3.21}$$

Finally (5.3.20) yields

$$- \sin \frac{\pi \varphi_i}{2\omega} + Br_i^2 \sin \frac{\pi \varphi_i}{\omega} = -1 + \cos \varphi_i$$

$$+ \sum_{n=1}^{N} \beta_n r_i^{n\pi/\omega} \sin \frac{n\pi \varphi_i}{\omega}, \quad 0 < \varphi_i < \omega, \ i = 1, 2, \ldots N. \quad (5.3.22)$$

Now choosing $\omega = \pi/6$ and four collocation points on Γ_3 determining $\beta_1 \ldots \beta_4$ one obtains Table 5.1.

Table 5.1 The expansion coefficients

i	x_i	y_i	b_i	β_I
1	1.	0.	-19.0867	-0.104899
2	1.209059	0.193579	10.0928	0.041507
3	1.5557	0.336465	0.	0.000449585
4	1.57821	0.512293	0.	0.0000667401

By similar methods (and fingers crossed) corner problems of this type can be solved. These might include rectangular domains with rectangular or triangular or hexagonal holes or domains without a hole but exhibiting polygonal boundaries.

Problems

1. Find the lowest eigenvalue of an hexagonal membrane with no holes and homogeneous boundary conditions. Assume an hexagon with the same area as a circular membrane of radius 1, so that the FABER *theorem* may help to control your results. You may choose 9 collocation points and a radius of the circumcircle equal to $\sqrt{2\pi/3\sqrt{3}}$ guaranteeing the same area π of the circular membrane. Then you should obtain the eigenvalue 2.43156 (>2.404825 of the circular membrane). Try to find the eigenvalue by plotting Det(m) as function of eigenvalue by narrowing the search interval.

2. Discuss a rectangular membrane with a rectangular hole. Boundary conditions might be value = 1. on the outer (inner) rectangle and homogeneous (clamped membrane) on the inner (outer) rectangle. Use polar coordinates. Make a Plot3D of the solution.

6

Nonlinear boundary problems

6.1 Some definitions and examples

A boundary problem is called nonlinear, if *either* the differential equation *or* the boundary value problem is nonlinear. The concept of a nonlinear differential equation is clear, see sections 2.6 and 3.4. For the solutions of these equations the *superposition principle* does not hold. From the standpoint of the physicist or engineer, nonlinear partial differential equations appear, if the underlying phenomenon itself is nonlinear or if parameters or coefficients in the differential equation depend on the solution function. In electrodynamics very often the material parameter μ depends on the solution of the equations of electromagnetism.

A boundary condition is called nonlinear, if the condition itself depends on the solution function of the actual differential equation.

Let us view some typical *nonlinear differential equations*:

1. the diffusion equation with variable diffusion coefficient (3.2.19) or (3.4.1)

2. the CARRIER equation for a string with varying tension (3.4.55)

3. the EULER equation of motion (3.2.52).

Other examples are the partial differential equations describing the *cooling* of a *nuclear reactor*, where the neutron density $n(x,t)$ and the temperature $T(x,t)$ are coupled through $n = vT_x + T_t$, $n_t = n_{xx} + \rho(T)n$, where v is the flow velocity of the coolant and ρ the coolant density. Another example is given by the equation of motion of *glaciers* (FINSTERWALDER *equation*) [6.20].

$$[(n+1)ku^n - a]\frac{\partial u}{\partial x} + a = \frac{\partial u}{\partial t}. \tag{6.1.1}$$

Here $u(x,t)$ is the vertical thickness of the glacier, which glides downhill on an inclined plane of angle α, where $k = \tan\alpha$ and $n \sim 0.25 - 0.5$.

Nonlinear boundary conditions are found when considering the *superfluidity* of He^3. The pertinent partial differential equation $u_t = u_{xx}$ and the initial condition $u(x,0) = 0$ are linear, but the boundary condition is nonlinear:

$$u_x(0,t) = k\left[u(0,t) - c\sin t\right]^3. \tag{6.1.2}$$

The flow velocity v of *gravity waves* on the surface of a lake are described by $v(x, y, z, t) = -\nabla\varphi$, where $\Delta\varphi = 0$. The partial differential equation is linear, but the boundary condition that there exists atmospheric pressure p_0 on the water surface, is not linear. The BERNOULLI *equation* reads

$$\frac{\partial \vec{v}}{\partial t} + \nabla\left(\frac{\vec{v}^2}{2} + \frac{p}{\rho} + gz\right) = \nabla\left(-\frac{\partial\varphi}{\partial t} + \frac{(\nabla\varphi)^2}{2} + \frac{p}{\rho} + gz\right) = 0. \quad (6.1.3)$$

Here g is the gravitational acceleration, p pressure, ρ density of water and z is the coordinate orthogonal on the water surface. Formal integration with respect to the three space coordinates x, y, z results in

$$-\frac{\partial\varphi}{\partial t} + \frac{(\nabla\varphi)^2}{2} + \frac{p}{\rho} + gz = const. \quad (6.1.4)$$

On the ground $z = -h$ of the lake, the normal component v_n vanishes, so that $(\partial\varphi/\partial z)_{z=-h} = 0$. On the surface of the lake $z = 0$ one has

$$-\frac{\partial\varphi}{\partial t} + \frac{(\nabla\varphi)^2}{2} + \frac{p_0}{\rho} = const. \quad (6.1.5)$$

Integration of $\Delta\varphi$ together with the two boundary conditions results in elliptic functions (*"snoidal"* and *"cnoidal"* waves). A linearized approximate solution of the problem will be discussed in problem 1. Moving boundaries are mainly nonlinear, although very often it is difficult to recognize the nonlinearity. Consider an infinite half-space orthogonal to the x axis, which is filled with water. On the boundary plane $x = 0$ the temperature is cooled down to $T_0 < 0°$ C. Thus the water will freeze there. As time passes on, a *freezing front* will penetrate into the water and freeze it. If we designate by $T_1(x, t)$ the temperature within the ice and by $T_2(x, t)$ within the water and let T_s be the temperature within the front, then continuity demands

$$T_s = T_1(x = X(t), t) = T_2(x = X(t), t) \quad \text{or} \quad dT_1 = dT_2. \quad (6.1.6)$$

Here $X(t)$ designates the actual location of the freezing front. Now we consider the two heat conduction equations. Subscript 1 refers to ice and subscript 2 to water, respectively:

$$\frac{\partial T_1(x, t)}{\partial t} = \frac{\lambda_1}{\rho_1 c_1}\frac{\partial^2 T_1(x, t)}{\partial x^2}, \quad \frac{\partial T_2(x, t)}{\partial t} = \frac{\lambda_2}{\rho_2 c_2}\frac{\partial^2 T_2(x, t)}{\partial x^2}, \quad (6.1.7)$$

where $\rho\,[\mathrm{kg/dm^3}]$ and $c\,[\mathrm{kJ/kg\,K}]$ designate density and specific heat, K is the thermodynamic temperature, $\lambda\,[\mathrm{W/mK}]$ is the *thermal conductivity* (heat conductivity). Boundary conditions exist mainly within the freezing front. Energy conservation demands that the relative rate of heat flow through the front is equal to the melting heat $L\,[\mathrm{kJ/kg}]$ transported away [6.1]:

$$\lambda_1\frac{\partial T_1(x = X(t), t)}{\partial t} - \lambda_2\frac{\partial T_2(x = X(t), t)}{\partial t} = L\rho\frac{dX}{dt}, \quad (6.1.8)$$

where $\rho = (\rho_1 + \rho_2)/2$ (STEFAN *boundary condition*). Furthermore, the conditions at $x = 0$ and $x = \infty$ have to be taken into account:

$$T_1(0, t) = T_0, \quad T_2(\infty, t) = T^*, \quad (6.1.9)$$

where T^* is the initial water temperature:

$$T_2(x,0) = T^* \quad \text{for} \quad x \geq 0. \tag{6.1.10}$$

Having a short look at the foregoing equations, it is not immediately clear that they represent a nonlinear problem. But using

$$dT_i = \frac{\partial T_i}{\partial t} dt + \frac{\partial T_i}{\partial x} dx \quad \text{for} \quad i = 1, 2 \tag{6.1.11}$$

and

$$\frac{dx}{dt} = \frac{dX}{dt} = \left(\frac{\partial T_1}{\partial t} - \frac{\partial T_2}{\partial t} \right) \Big/ \left(\frac{\partial T_2}{\partial X} - \frac{\partial T_1}{\partial X} \right) \tag{6.1.12}$$

and insertion into (6.1.8) yield the *nonlinear boundary condition*

$$\left(\frac{\partial T_2}{\partial X} - \frac{\partial T_1}{\partial X} \right) \left(\lambda_1 \frac{\partial T_1}{\partial X} - \lambda_2 \frac{\partial T_2}{\partial X} \right) = L\rho \left(\frac{\partial T_1}{\partial t} - \frac{\partial T_2}{\partial t} \right). \tag{6.1.13}$$

The nonlinearity is hidden in the fact that the moving front boundary $X(t)$ depends on the temperature $T(x,t)$. We will present a solution of the STEFAN *problem* in the next section.

Another example of a nonlinear boundary problem appears in connection with the disposal of *radioactive waste*. One must know if the heat produced by the radioactivity of the waste may be removed by the surrounding rocks [3.9]. For the time-dependent *thermal diffusivity* $a(t) = \lambda(t)/\rho(t)c(t)$, the heat conduction equation reads

$$\frac{\partial u(x,t)}{\partial t} = a(t) \frac{\partial^2 u(x,t)}{\partial x^2}, \quad 0 < x < \infty, \quad 0 < t < \tau. \tag{6.1.14}$$

At first sight, this looks linear. But the initial condition $u(x,0) = 0$ and the boundary condition

$$u(0,t) = f(t), \quad -a(t) \frac{\partial u(0,t)}{\partial x} = g(t), \tag{6.1.15}$$

where $f(t)$ and $g(t)$ are known, contain nonlinearities. It has been shown that this problem is actually nonlinear [3.9].

Such nonlinear boundary problems occur for ordinary differential equations too. COLLATZ [1.7] gives some examples:

$$y'' = 3y^2/2, \quad y(0) = 4, \quad y(1) = 1, \tag{6.1.16}$$

$$y'' = f(y), \quad y(0) = y_0, \quad y(l) = y_1, \tag{6.1.17}$$

$$y'' + 6y + y^2 = -\frac{3}{2} \cos x, \quad y(0) - y(2\pi) = y'(0) - y'(2\pi) = 0. \tag{6.1.18}$$

Solutions will be discussed in the problems.

Problems

1. Solve the problem of water waves on a lake by linearization. This yields

$$\Delta\varphi = 0, \quad \frac{\partial^2\varphi}{\partial t^2} = -g\frac{\partial\varphi}{\partial z} \quad \text{for} \quad z = 0. \tag{6.1.19}$$

Use the setup

$$\varphi(x, y, z, t) = A \exp[i(k_x x + k_y y - \omega t)]U(z) \tag{6.1.20}$$

and the boundary conditions

$$\frac{\partial U}{\partial z} = 0 \quad \text{for} \quad z = -h, \tag{6.1.21}$$

$$g\frac{\partial\varphi}{\partial t} = \omega^2\varphi \quad \text{for} \quad z = 0. \tag{6.1.22}$$

Solution: $U(z) = \cosh k(z + h)$, $k^2 = k_x^2 + k_y^2$,

$$-\omega^2 U = -g\frac{\mathrm{d}U}{\mathrm{d}z} = -gk\sinh kh = -\omega^2\cosh kh \quad \text{for} \quad z = 0, \tag{6.1.23}$$

so that the *wave dispersion* $\omega(k)$ is given by

$$c_{Ph} = \frac{\omega}{k} = \sqrt{\frac{g}{k}\tanh hk} = \omega(k). \tag{6.1.24}$$

2. Solve (6.1.16). Solution: $y = 4/(1 + x)^2$. (Hint: use the method how to solve the equation of motion $m\ddot{x} = K(x)$.) Try **DSolve**

3. Solve (6.1.17) by the same method. (Equation of motion).

4. Try to solve (6.1.18).

6.2 Moving and free boundaries

It seems to be clear that moving boundaries may appear only for partial differential equations containing the independent variable time. These partial differential equations are mostly of parabolic type like (6.1.7). Free boundaries, however, occur mainly with elliptic partial differential equations and very often describe equilibrium situations.

We first would like to solve the STEFAN *problem*. The heat conduction equations (6.1.7) are exactly of the form (3.3.51) in section 3.3. There we found that the *similarity transformation*

$$\eta = x/2\sqrt{\tau}, \quad \tau = \lambda t/\rho c, \quad a = \lambda/\rho c, \tag{6.2.1}$$

transforms the heat conduction equation into

$$\frac{\mathrm{d}^2 T(\eta)}{\mathrm{d}\eta^2} = -2\eta \frac{\mathrm{d}T(\eta)}{\mathrm{d}\eta}. \tag{6.2.2}$$

Using $u(\eta) = T'(\eta)$ we obtain a separable equation like (1.1.4)

$$\frac{\mathrm{d}u}{\mathrm{d}\eta} = -2\eta u, \tag{6.2.3}$$

which yields after integration

$$\int \frac{\mathrm{d}u}{u} = -\int 2\eta \mathrm{d}\eta = \ln u = -\eta^2 + \ln C,$$

$$u(\eta) = const\, \exp(-\eta^2) \tag{6.2.4}$$

and

$$T(\eta) = const \int_0^\eta \exp(-\eta^2)\mathrm{d}\eta$$

$$= const + A\mathrm{erf}(\eta) = T_0 + \mathrm{erf}\left(\frac{x}{2\sqrt{at}}\right). \tag{6.2.5}$$

Thus, we have for ice

$$T_1(x,t) = T_0 + A\mathrm{erf}\left(\frac{x}{2\sqrt{a_1 t}}\right) \tag{6.2.6}$$

and for the water

$$T_2(x,t) = T^* + B\mathrm{erfc}\left(\frac{x}{2\sqrt{a_2 t}}\right). \tag{6.2.7}$$

Here the function erf is the *error function* and $\mathrm{erfc}(\eta) = 1 - \mathrm{erf}(\eta)$ is the *complement error function*. As is well known [2.9], one has

$$\mathrm{erf}(0) = 0, \quad \mathrm{erfc}(0) = 1, \quad \mathrm{erf}(\infty) = 1, \quad \mathrm{erfc}(\infty) = 0. \tag{6.2.8}$$

The solutions (6.2.6) and (6.2.7) satisfy the boundary conditions (6.1.9). For a solution of (6.2.2.), *Mathematica* yields

$$T(\eta) = C_2 + \frac{\sqrt{\pi}}{2} C_1 \mathrm{erf}(\eta). \tag{6.2.9}$$

Returning to (6.2.7) we find that the initial condition (6.1.10) is also satisfied due to (6.2.8). Now the two unknown integration constants A and B will be used to satisfy the boundary condition (6.1.8). Let us make a setup for the path of the freezing front:

$$X(t) = \alpha t^{1/2}, \quad \frac{\mathrm{d}X}{\mathrm{d}t} = \frac{\alpha}{2\sqrt{t}} = \frac{\alpha^2}{2X}. \tag{6.2.10}$$

Then (6.1.6) and (6.2.10) yield two equations:

$$T_0 + A \, \mathrm{erf} \left(\frac{\alpha}{2\sqrt{a_1}} \right) = T_s = T^* + B \left[1 - \mathrm{erf} \left(\frac{\alpha}{2\sqrt{a_2}} \right) \right]. \qquad (6.2.11)$$

If α is assumed to become known later, we can find an equation for A and B:

$$A = \frac{T_s - T_0}{\mathrm{erf} \left(\alpha/2\sqrt{a_1} \right)}, \quad B = \frac{T_s - T^*}{\mathrm{erfc} \left(\alpha/2\sqrt{a_2} \right)}. \qquad (6.2.12)$$

Now using the setup (6.2.10) we can satisfy (6.1.8). We calculate

$$\frac{\partial T_1(x,t)}{\partial x} \bigg|_{x=X(t)} \quad \text{and} \quad \frac{\partial T_2(x,t)}{\partial x} \bigg|_{x=X(t)}$$

and

$$\frac{\mathrm{d \, erf}(\eta)}{\mathrm{d}\eta} = \frac{2}{\sqrt{\pi}} \mathrm{e}^{-\eta^2}. \qquad (6.2.13)$$

Insertion of these expressions into (6.1.8) results in

$$\frac{\lambda_1 (T_s - T_0) \exp \left(-\alpha^2/4a_1 \right)}{\sqrt{a_1} \, \mathrm{erf} \left(\alpha/2\sqrt{a_1} \right)} + \frac{\lambda_2 (T_s - T_0) \exp \left(-\alpha^2/4a_2 \right)}{\sqrt{a_2} \, \mathrm{erf} \left(\alpha/2\sqrt{a_2} \right)}$$

$$= -\frac{\sqrt{\pi}}{2} L\rho\alpha. \qquad (6.2.14)$$

For given parameters $\lambda_1, \lambda_2, a_1, a_2, L, \rho$ for ice and water and assumptions for $T_s = 0°C, T^* = 5°C$ equation (6.2.14) is a transcendental equation determining α giving something like $\alpha \approx 0.242$, see problem 1. Then the solutions read

$$T_1(x,y) = T_0 + \frac{T_s - T_0}{\mathrm{erf} \left(\alpha/2\sqrt{a_1} \right)} \mathrm{erf} \left(\frac{x}{2\sqrt{a_1 t}} \right), \quad 0 \le x \le X(t), \qquad (6.2.15)$$

$$T_2(x,t) = T^* + \frac{T_s - T^*}{\mathrm{erfc} \left(\alpha/2\sqrt{a_2} \right)} \mathrm{erfc} \left(\frac{x}{2\sqrt{a_2 t}} \right), \quad x \ge X(t). \qquad (6.2.16)$$

Moving boundaries may also occur for elliptic partial differential equations [3.15]. The incompressible, steady-state 2D viscous flow between two plates in a distance $2a$, the so-called HELE-SHAW *problem*, is an example. *Finite elements* have been used to solve it [4.29]. The plates are arranged in such a way that gravity acts parallel to the y axis and main flow is in the x direction. Then the two velocity components u, v are driven by pressure p and gravity g

$$u(x,y) = -\frac{a^2}{12\mu} \frac{\partial p(x,y)}{\partial x}, \quad v(x,y) = \frac{-a^2}{12\mu} \left(\frac{\partial p(x,y)}{\partial y} + \rho g \right), \qquad (6.2.17)$$

and the velocity potential reads

$$\varphi(x,y) = \frac{a^2}{12\mu} (p(x,y) + \rho g y). \qquad (6.2.18)$$

μ is the viscosity. Now the idea is that a fluid 1 is being pushed out completely by fluid 2. What, then, is the boundary condition in a steady state on the interface between the two fluids? Apparently, the mean velocity u_1 and u_2 respectively will be given by

$$u_1 = -\frac{a^2}{12\mu_1}\operatorname{grad}(p + \rho_1 gy) = \operatorname{grad}\varphi_1,$$

$$u_2 = -\frac{a^2}{12\mu_2}\operatorname{grad}(p + \rho_2 gy) = \operatorname{grad}\varphi_2. \qquad (6.2.19)$$

One can assume that the components of u_1, v_1 normal to the interface between the two fluids are continuous. Special applications of this problem are of industrial interest: injection of a fluid into a cell, *electrochemical machining*, *seeping of water* through a dam, *lubrication* of a bearing, *cavitation*, etc., may be formulated as a HELE-SHAW problem [3.15].

The *taste of cheese* depends on the diffusion of chemical substances produced by typical bacteria or mold fungi. Thus, the cheese producers are interested in diffusion problems within solids. The speed of the moving front determines the storage time and consequently price and quality of the product. For the steady state of a diffusion process one has

$$D\frac{\partial^2 c}{\partial x^2} - m = 0. \qquad (6.2.20)$$

$c(x)$ is the local concentration of the chemical substance and m is a production (or absorption) rate (moles or grams/sec). Let $x = 0$ describe the cheese surface and designate by $x_0(t)$ the actual depth of penetration of the diffusing substance. The steady state will be described by

$$c(x,t) = 0, \quad \frac{\partial c}{\partial x} = 0 \text{ for } x > x_0 \qquad (6.2.21)$$

and, during the maturity process, one has on the cheese surface

$$c(0,t) = c_0 = const. \qquad (6.2.22)$$

Apparently,

$$c_1 = \frac{m}{2D}(x - x_0)^2, \quad x_0^2 = \frac{2Dc_0}{m} \qquad (6.2.23)$$

is a solution of (6.2.20) to (6.2.22). As soon as the ripening time τ of the cheese is over, its surface $x = 0$ will be sealed. At this moment, the location of the diffusion front can be designated by $x_0(\tau)$. Then the mathematical problem to be solved reads:

$$\frac{\partial c}{\partial t} = D\frac{\partial^2 c}{\partial x^2} - m, \ 0 \le x \le x_0(\tau),$$

$$\frac{\partial c}{\partial x} = 0 \quad \text{for} \quad x = 0, \quad t \ge \tau,$$

$$c(x_0, t) = c = \frac{\partial c}{\partial x} = 0, \quad x = x_0(\tau), \quad t \geq \tau,$$

$$c(x, t) = c_1 = \frac{m}{2D}(x - x_0)^2, \quad 0 \leq x \leq x_0, \quad t = \tau. \tag{6.2.24}$$

Various numerical methods have been used [3.15] to solve this implicit problem. It is called an *implicit problem*, because $x_0(t)$ is absent. A transformation $c = \partial u/\partial t$ may create an explicit problem or the boundary condition (6.2.24) can be replaced by $\partial c/\partial x = f(t)$ for $0 \leq x \leq 1$ giving [3.15]

$$c(x, t) = \frac{2}{\pi^2} \sum_{p=1/2, 3/2} \frac{\exp\left(-p^2\pi^2 t\right)}{p^2} \cos p\pi x + \frac{1}{2}\left(x^2 - 1\right). \tag{6.2.25}$$

Ablation is the step-by-step removal of matter (ice, fuel for *thermonuclear fusion*, metals) mainly by direct transition of the solid into the gaseous state. This is of interest for *laser-induced fusion*. In this process, the ablation front penetrates into the material and one again has a moving boundary (interface between two thermodynamic phases). The penetration process can be described by $x_0(t) = s(t)$. Let l be the thickness of the layer in which the thermal ablation energy $q(t) > 0$ is supplied, then the local temperature $T(x, t)$ obeys:

$$\frac{\partial T}{\partial t} = \frac{\partial^2 T}{\partial x^2} \quad 0 < x < s(t),$$

$$T(x_0, t) = g(t), \quad \frac{\partial T}{\partial x} = \lambda\frac{dx_0}{dt} + q(t), \quad x_0 = s(t), \tag{6.2.26}$$

and

$$T = \varphi(x) < 0 \quad \text{for} \quad 0 < x < l, \quad t = 0,$$

$$T = f(t) < 0 \quad \text{for} \quad x = 0, \quad t > 0, \quad s(0) = l. \tag{6.2.27}$$

First, $\varphi(x)$ and $f(t)$ are unknown. Solving for $T(x, t)$ for a given path $s(t)$ is called the *inverse* STEFAN *problem*. A transformation on a co-moving coordinate system $\xi = x/s(t)$ is useful, see the remarks on progressing waves in the next section.

Free boundaries occur on surfaces of a free *jet flow*, where the atmospheric pressure determines the boundary. A very simple example of this type can be described by conformal mapping. The function

$$z = -\zeta + e^{-\zeta}, \quad x = -u + e^{-u}\cos v, \quad y = -v - e^{-u}\sin v \tag{6.2.28}$$

describes the *pouring out* of a liquid of an infinite half space tank (BORDA *outlet*). For the steady-state building of the jet, the nonlinear boundary condition is needed:

$$\left(\frac{\partial u}{\partial x}\right)^2 + \left(\frac{\partial u}{\partial y}\right)^2 = const = C. \tag{6.2.29}$$

$u(x, y)$ describes the streaming velocity and $v(x, y) = const$ describes the *streamlines*. It is not possible to directly solve (6.2.28) for $u(x, y), v(x, y)$. We make an *hodograph transformation* as in section 3.4. We replace the dependent variables $u(x, y), v(x, y)$ by the independent variables $x(u, v), y(u, v)$ becoming then the new dependent variables. Thus:

$$du = u_x dx + u_y dy, \quad dv = v_x dx + v_y dy,$$
$$dx = x_u du + x_v dv, \quad dy = y_u du + y_v dv, \tag{6.2.30}$$

so that

$$dx = (v_y du - u_y dv)/D, \quad dy = (-v_x du + u_x dv)/D, \tag{6.2.31}$$

where

$$D = \begin{vmatrix} u_x & u_y \\ v_x & v_y \end{vmatrix}. \tag{6.2.32}$$

Comparing the expressions for dx and dy in (6.2.30) as well as in (6.2.31) results in

$$x_u = v_y/D, \quad x_v = -u_y/D$$

and

$$x_u^2 + y_u^2 = \frac{1}{u_x^2 + v_y^2} = \frac{1}{C^2} = 1. \tag{6.2.33}$$

These equations now express the free nonlinear boundary condition on the free surface of the jet. But how do we find a mathematical expression for the free surface? We make a setup for the surface $v = const$ using an unknown function $F(u)$

$$x_u = F(u), \quad y_u = \sqrt{1 - F^2(u)}.$$

After some trial and error and inserting into (6.2.28) one obtains

$$F(u) = 1 - e^u, \quad y_u = \pm\sqrt{1 - (1 - e^u)^2}. \tag{6.2.34}$$

Integration yields

$$x(u) = u - e^u,$$
$$y(u) = const \pm \sqrt{2e^u - e^{2u}} + 2\arcsin\sqrt{e^u/2}. \tag{6.2.35}$$

An investigation of the solution demonstrates that the flow velocity makes a discontinuous jump on the free surface: from v(surface) to $v = 0$ in air. Thus the potential $u(x, y)$ is discontinuous too (LEVI-CIVITA *potential*). Problems of this kind occur in plasma physics (*hose instability*). If the pressure in fluid jets drops below the saturation vapour pressure of the fluid, *cavitation* sets in: the fluid evaporates and gas bubbles are formed. If they implode due to changes of the surrounding pressure, the solid material nearby may be damaged or even destroyed. (This may occur for turbine blades or ship's screws.)

Seepage flow through a retaining dam (if an earth-fill dam) is an important engineering problem connected with hydro-power generation and *flood disaster* protection. The surface of seepage water flow presents free boundary that is contained within the dam. Seepage of a fluid through a porous material in the x direction is described by the DARCY *law*

$$\vec{v} = -\mu \nabla(p + \rho g y) = -\nabla \varphi, \quad \Delta \varphi = 0.$$

Here μ is the porosity of the dam material and $\rho g y$ the weight of the water. Let $x = 0$ be the waterside of the dam and d its thickness, then the boundary conditions on the solids read

$$\varphi(x = 0, y) = H \text{ for } 0 \le y \le H,$$
$$\varphi(x = d, y) = h \text{ for } 0 \le y \le h,$$
$$\varphi(x = d, y) = y \text{ for } y > h. \tag{6.2.36}$$

Here H is the height of the water level and h is the height of the dam ($h > H$). Furthermore, one can assume that there is no seepage at $x = d, y = h$:

$$\frac{\partial \varphi}{\partial y} = 0 \quad \text{for} \quad 0 \le x \le d, \quad y = 0,$$

$$\frac{\partial \varphi}{\partial x} = 0 \quad \text{for} \quad x = d, \quad y > h. \tag{6.2.37}$$

Within the dam the free seepage water surface $y^*(x)$ is described by

$$\varphi = y, \quad \partial \varphi / \partial n = 0 \tag{6.2.38}$$

or by the BERNOULLI equation. One is now interested in the shape of the free boundary:

$$y^* = f(x), \quad f(0) = H, \quad f(d) \ge h,$$
$$\frac{d}{dx} f(x) \Big|_{x=0} = 0, \quad \frac{d}{dx} f(x) \Big|_{x=d} = \infty. \tag{6.2.39}$$

The problem described by the last equations (6.2.36) to (6.2.39) can be solved using a BAIOCCHI *transformation*

$$w(x, y) = \int_y^H (\varphi(x, \eta) - \eta) \, d\eta \tag{6.2.40}$$

leading to a partial differential equation for w. This high mathematics is, however, of very small practical meaning. In the engineering world, a dam is neither homgeneous nor isotropic, so that the LAPLACE equation has to be modified

$$\frac{\partial}{\partial x}\left[k_{11}(x,y)\frac{\partial\varphi}{\partial x} + k_{12}(x,y)\frac{\partial\varphi}{\partial y}\right]$$
$$+ \frac{\partial}{\partial y}\left[k_{21}(x,y)\frac{\partial\varphi}{\partial x} + k_{22}(x,y)\frac{\partial\varphi}{\partial y}\right] = 0. \tag{6.2.41}$$

For an isotropic homogeneous dam $k_{11} = k_{22} = 1$, $k_{12} = k_{21} = 0 = constants$ would be valid, for an isotropic inhomogeneous dam one would have functions $k_{12}(x,y) = k_{21}(x,y) = 0$, $k_{11}(x,y) = k_{22}(x,y)$. Engineering values of μ and the k_{ij} may vary 1:10 in nature!

In recent years there has been great interest in moving elastic boundaries, as in the *blood streaming* in an *artery* [6.2].

Problems

1. Using the command **FindRoot** find α from (6.2.14). You can assume:

   $\rho_{ice} = 0.917$ kg/dm^3, $\quad\rho_{water} = 0.998$ kg/dm^2,
   $c_{ice} = 2.1$ kJ/kg K, $\quad c_{water} = 4.182$ kJ/kg K,
   $\lambda_{ice} = 2.2$ W/m K, $\quad\lambda_{water} = 0.598$ W/m K,
   $a_{ice} = 0.0112$ cm^2/sec, $a_{water} = 0.0017$ cm^2/sec,
   $T_s = 0°$C, $T_0 = 5°$C [6.5].
   Remember: Joule J $=$ m^2 kg s^{-2}, Watt W $=$ Js$^{-1} =$ m^2 kg s^{-3}.
   (International Union of Pure and Applied Physics).
   Vary T_s, T_0.

2. Plot (6.2.5), (6.2.15) and (6.2.16) taking numerical values from problem 1. Use $T_0 = 0°$C, -5°C, -10°C and $0 \leq x \leq 5$. Do you observe a salient point on the x axis? Does it depend on $L\rho$? (Yes).

3. Plot (6.2.28) and separately (6.2.35). For the methods see section 4.2.

6.3 Waves of large amplitudes. Solitons

Phenomena with large amplitudes are mainly described by nonlinear equations and cannot be described by approximately linearized differential equations. Small amplitude pressure waves in gases can be described by the D'ALEMBERT *equation* (4.3.1), but high pressure waves cannot. Let us consider the *conti-nuity equation* for the one-dimensional time dependent compressible flow of a gas without sources. According to (3.3.22) one has

$$\rho_t = u\rho_x + \rho u_x = 0. \tag{6.3.1}$$

The EULER equation (3.3.23) assumes the form

$$\rho u_t + \rho u u_x + p_x = 0. \tag{6.3.2}$$

The assumption of adiabatic behavior (3.3.24) and the definition of the *velocity of sound* (3.3.25) allows us to rewrite (6.3.2) in the form (3.3.26) or

$$\rho u_t + \rho u u_x + a^2 \rho_x = 0. \tag{6.3.3}$$

Thus, the calculations (3.3.39) to (3.3.45) yield the one-dimensional *potential flow equation* (3.3.46) in the form

$$(a^2 - \varphi_x^2)\varphi_{xx} - 2\varphi_x\varphi_{xt} - \varphi_{tt} = 0. \tag{6.3.4}$$

Again, $u(x,t) = -\partial\varphi(x,t)/\partial x$ and the velocity of sound a is given by (3.4.38) in the form

$$a^2 = a_0^2 - \frac{(\kappa-1)}{2}\varphi_x^2 - (\kappa-1)\varphi_t. \tag{6.3.5}$$

This is a consequence of (3.3.24) and (3.3.44) or can be read off from (3.3.47). If $\varphi_x^2\varphi_t$ and φ_{xt} are small, one obtains a constant sonic speed $a \approx a_0$ and (3.3.46) becomes the linear wave equation $\varphi_{xx} = \varphi_{tt}/a_0^2$.

To solve the nonlinear partial equation(6.3.4) we follow the lines of thinking in section 3.4 and apply a LEGENDRE *transformation*. In full analogy to (3.4.53) we make the setup (where now x and t are dependent variables)

$$\varphi_x = -u, \ \varphi_t = -q, \ \varphi_{xx} = -u_x, \ \varphi_{tt} = -q_t,$$
$$d\varphi = \varphi_x d_x + \varphi_t dt = -udx - qdy,$$
$$\Psi(u,q) = ux + qt + \varphi(x,t),$$
$$d\Psi = \Psi_u du + \Psi_q dq = xdu + tdq,$$
$$\Psi_u = x, \ \Psi_q = t, \ \Psi_{uu} = x_u, \ \Psi_{uq} = t_q,$$
$$\varphi_{xx} = -\Psi_{qq}/D, \ \varphi_{xt} = \Psi_{qv}/D, \ \varphi_{tt} = -\Psi_{uu}/D, \tag{6.3.6}$$

where

$$D = \begin{vmatrix} \Psi_{uu} & \Psi_{uq} \\ \Psi_{uq} & \Psi_{qq} \end{vmatrix}. \tag{6.3.7}$$

We then obtain the exactly linearized potential equation (6.3.4) in the form

$$\Psi_{qq}\left(a^2 - u^2\right) + 2u\Psi_{uq} - \Psi_{uu} = 0. \tag{6.3.8}$$

This offers now a great advantage. Deriving for (6.3.4) the characteristics one obtains

$$w = \frac{\mathrm{d}x}{\mathrm{d}t} = u \pm a(u), \qquad (6.3.9)$$

which cannot be used, since the wave velocity w depends on the unknown solution $u(x, t)$. But the characteristics for (6.3.8)

$$\frac{\mathrm{d}q}{\mathrm{d}u} = -(u \pm a) \qquad (6.3.10)$$

are independent from q and can be used to solve (6.3.8) as we shall see in the next section.

Since the wave speed w depends on u, a wave propagates faster in relation to the speed of flow velocity u. This means that a pressure wave amplitude will steepen up during propagation. This continual steepening of the wavefront can no longer be described by single valued functions of location. One has to introduce a discontinuity into the flow to get over this difficulty. This discontinuity is called a *shock wave*, across which the flow variables change discontinuously. This jump is described by the HUGONIOT *equation* [3.30], [3.37]. The discontinuity vanishes if the basic fluid equations are modified to contain viscosity and heat conduction. Including these terms, BECHERT derived a nonlinear partial differential equation of fifth order, which could be solved by a similarity transformation [6.3]. This solution is no longer discontinuous but describes a continuous but very steep transition between the values of the flow parameters before and after the wave front.

It seems that PREISWERK in his thesis [6.4] was first to draw attention to the fact that the same gasdynamic equation that we just discussed describes a water flow with *hydraulic jumps* (water surges) in rivers when a water lock is suddenly opened up-stream. *Pressure jumps* occurring in *hydro-electric power stations* obey similar equations [6.5]. PREISWERK has shown that gasdynamic equations can be applied on water surface flow if $\kappa = c_p/c_\nu$ is put $= 2$ in these equations. Formally, the temperature T of the gas corresponds to the water depth.

We now have the mathematical tools to undertake the problem of *gravity waves* on the surface of a lake described by $\Delta\varphi = 0$ and equations (6.1.4), (6.1.5). An approximate linearized solution has been discussed in problem 1 of section 6.1. The nonlinearized solution starts with the BERNOULLI equation (6.1.4), which we write in the form

$$\varphi_t + \frac{1}{2}\left(\varphi_x^2 + \varphi_z^2\right) + g(h - h_0) = 0. \qquad (6.3.11)$$

Here h_0 is the constant depth of the unperturbed free water surface and $h(x, z, t)$ describes the disturbed lake surface. Now $z = 0$ designates the ground of the lake. This designation is in contradiction to section 3.1. Since the normal velocity component $v_z = -\partial\varphi/\partial z$ must vanish at the solid ground, one has the boundary condition

$$\varphi_z = 0 \quad \text{for} \quad z = 0. \tag{6.3.12}$$

On the free surface, the boundary condition reads

$$\varphi_z = h_t + \varphi_x h_x \quad \text{for} \quad z = h(x, z, t). \tag{6.3.13}$$

The nonlinear boundary condition is given by (6.3.11). Some authors add a term describing wind pressure or capillary tension to describe the initial excitation of the waves. We now have two unknown functions $\varphi(x, z, t)$ and $h(x, z, t)$ to be determined from $\Delta\varphi = 0$ and the three boundary conditions (6.3.11) - (6.3.13). Several methods exist to solve this problem. Using the transformations [3.30]

$$\xi = \sqrt{\varepsilon}(x - c_0 t), \quad \tau = \varepsilon^{3/2} t, \quad \psi = \sqrt{\varepsilon}\,\varphi \tag{6.3.14}$$

and the abbreviation

$$c_0 = \sqrt{gh_0} \tag{6.3.15}$$

one obtains

$$\varepsilon\psi_{\xi\xi} + \psi_{zz} = 0, \quad \psi_z = 0 \quad \text{for} \quad z = 0, \tag{6.3.16}$$

$$\psi_z = \varepsilon^2 h_\tau + \varepsilon(\psi_\xi - c_0)h_\xi \quad \text{for} \quad z = h, \tag{6.3.17}$$

$$\varepsilon^2 \psi_\tau - \varepsilon c_0 \psi_\xi + \frac{1}{2}\left(\varepsilon\psi_\xi^2 + \psi_z^2\right) + \varepsilon g(h - h_0) = 0. \tag{6.3.18}$$

Using a perturbation setup

$$h = h_0 + \varepsilon h_1 + \varepsilon^2 h_2, \quad \psi = \varepsilon\psi_1 + \varepsilon^2\psi_2, \tag{6.3.19}$$

one gets

$$\psi_{1zz} = 0, \quad \psi_{nzz} + \psi_{n-1\xi\xi} = 0,$$
$$\psi_1 = \psi_1(\xi, \tau), \quad \psi_2 = -\frac{1}{2}z^2\psi_{1\xi\xi}, \quad \psi_3 = z^4\psi_{1\xi\xi\xi\xi}/24,$$
$$\psi_{2z} = -c_0 h_{1\xi\xi} \quad \text{for} \quad z = h_0,$$
$$\psi_{3z} + h_1\psi_{2zz} = h_{1\tau} - c_0 h_{2\xi} + \psi_{1\xi}h_{1\xi} \quad \text{for} \quad z = h_0,$$
$$-c_0\psi_{1\xi} + gh_1 = 0, \quad \psi_{1\tau} - c_0\psi_{2\xi} + \psi_{1\xi}^2/2 + gh_2 = 0. \tag{6.3.20}$$

Elimination results in the KORTEWEG-DE VRIES *equation*

$$\frac{\partial h_1}{\partial \tau} + \frac{3c_0}{2h_0}h_1\frac{\partial h_1}{\partial \xi} + \frac{c_0 h_0^2}{6}\frac{\partial^3 h_1}{\partial \xi^3} = 0. \tag{6.3.21}$$

Writing it in simplified form [6.15]

$$v_t + \alpha v v_x + v_{xxx} = 0 \tag{6.3.22}$$

and using the setup

$$v(x,t) = v(\eta), \quad \eta = x - c_0 t \tag{6.3.23}$$

one obtains

$$v_\eta(\alpha v - c_0) + v_{\eta\eta\eta} = 0. \tag{6.3.24}$$

After two integrations one gets

$$v_{\eta\eta} = c_0 v - \alpha v^2/2, \quad \frac{1}{2}v_\eta^2 = \frac{c_0}{2}v^2 - \frac{\alpha}{6}v^3, \tag{6.3.25}$$

so that

$$\int\limits_{v_{max}}^{v} \frac{dv}{\sqrt{c_0 v^2/2 - \alpha v^3/6}} = \sqrt{2}\,\eta. \tag{6.3.26}$$

This results in a *soliton* solution

$$v(x,t) = \frac{3c_0}{\alpha}\operatorname{sech}^2\left(\frac{\sqrt{c_0}}{2}(x - c_0 t)\right). \tag{6.3.27}$$

A soliton or *solitary wave* is a nonlinear wave propagating without change of shape and velocity. In these waves an exact balance occurs between the nonlinearity effects steepening the wave front due to the increasing wave speed and the effect of dispersion tending to spread the wave front. Soliton solutions have been discussed in *plasma physics, solid state physics* [6.6], in *supraconductors* [6.7], in the FERMI-ULAM problem in *nuclear power stations* [6.8], in ionospheric problems [6.9], in nonlinear mechanics (pendulum), in general relativity, in lasers [6.10] as well as in the elementary particle theory.

Modulation of the amplitude of a wave presents an important physical and technical problem with many applications. The effect of *modulation* is a variation of the amplitude and the phase of a wave. Let us first investigate these effects on a generalized wave equation [6.12]. We shall investigate nonlinearity, dispersion and dissipation. To do this, we consider the weakly nonlinear wave equation

$$\frac{1}{c^2}\Phi_{tt} - \Phi_{xx} + b\,\Phi + \varepsilon g\,\Phi_t = -V'(\Phi) + \varepsilon N(\Phi) + \varepsilon\,\Phi_t G(\Phi), \tag{6.3.28}$$

where $V' = dV/d\Phi$, $N = -\Phi^{2n-1}$ and $G(\Phi)$ are nonlinear functions and c, b, g are constants which may depend on the frequency ω. ε is a small parameter. We now define a *phase surface*:

$$\Theta(x,t) = const, \tag{6.3.29}$$

which has the property that all points (x,t) on it have the same value of the wave function $\Theta(x,t)$ We thus have

$$d\Theta = \Theta_x dx + \Theta_t dt = 0, \tag{6.3.30}$$

so that points moving with the speed

$$\frac{dx}{dt} = -\frac{\Theta_t}{\Theta_x}, \tag{6.3.31}$$

see a constant phase Θ. Defining wave number k and frequency by

$$\Theta_x = k, \quad -\Theta_t = \omega \quad \text{or} \quad \frac{\partial \Theta}{\partial t} + \omega = 0, \tag{6.3.32}$$

we find that (6.3.31) is the phase speed. In the three-dimensional case we have $\nabla \Theta = \vec{k}$ (*wave vector*)

$$\text{curl} \vec{k} = 0, \tag{6.3.33}$$

which indicates that wave crests are neither vanishing nor splitting off. The last two equations result in the *conservation of wave crests*

$$\frac{\partial \vec{k}}{\partial t} + \nabla \omega = 0. \tag{6.3.34}$$

A point moving with the *group velocity*

$$\left(\frac{dx}{dt}\right)_g = \frac{d\omega}{dk} \tag{6.3.35}$$

sees ω unchanged. The equation (6.3.28) can be classified as follows:

1. $N = 0$, $G = 0$, $V' = 0$: the equation is linear, k and ω are independent of x und t.

   1.1 $b = 0$, $g = 0$:   no dispersion $\omega(k)$, no dissipation,
   $\Phi = A \exp(ikx - i\omega t) \quad \omega = ck;$ \hfill (6.3.36)

   1.2 $b \neq 0$, $g = 0$:   frequency dispersion $\omega(k)$, no dissipation,
   $\Phi = A \exp(ikx - i\omega t), \quad \omega(k) = \pm c\sqrt{k^2 + b^2} :$ \hfill (6.3.37)

   1.3 $b = 0$, $g \neq 0$:   dissipation, ω complex
   $\Phi = A \exp(ikx - i\omega t), \quad \omega = \frac{1}{2} ic^2 \varepsilon g \pm c\sqrt{k^2 - c^2 \varepsilon^2 g^2/4};$ \hfill (6.3.38)

   1.4 $b \neq 0$, $g \neq 0$:   dispersion and dissipation,
   $\Phi = A \exp(ikx - i\omega t), \quad D(\omega, k) = \omega^2/c^2 - k^2 - b + i\omega\varepsilon g = 0.$ (6.3.39)

2. $N \neq 0$, $G \neq 0$, $V' \neq 0$, the wave equation is nonlinear, V' describes strong nonlinearity, εN weak nonlinearity and εG weak dissipation. For weak nonlinearity defined by

$$\langle \Theta_{xx} \rangle = \frac{1}{2\pi} \int_0^{2\pi} \Theta_{xx} d\Theta \approx 0, \quad \langle \Theta_{tt} \rangle \approx 0, \quad \Theta_x = k = const, \quad \Theta_t = -\omega$$

$$\tag{6.3.40}$$

one has *frequency dispersion* $\omega(k)$ and *amplitude dispersion* $\omega(A, k)$.

Two conclusions can be drawn:

1. For a nonlinear, nondissipative wave equation (6.3.28) the amplitude A is constant and $\Theta(x, t) = k(x, t)x - \omega(x, t)t$ yields a dispersion relation.

2. For any nonlinear dissipative wave equation (6.3.28) the frequency is not modified by the dissipative terms in first order of ε.

Now we investigate a modulated wave. We assume a sinusoidal wave $a_0 \cos \Theta_0$, $\Theta_0 = k_0 x - \omega_0 t$ with the amplitude $a(x, t)$ and phase $\Theta(x, t)$ varying slowly in x and t:

$$\Theta(x, t) = k_0 x - \omega_0 t + \varphi(x, t), \quad \omega_0 = \omega_0(k_0, a_0^2). \tag{6.3.41}$$

According to (6.3.32) we redefine

$$w(x, t) = -\Theta_t = \omega_0 - \varphi_t, \quad k(x, t) = \Theta_x = k_0 + \varphi_x. \tag{6.3.42}$$

For weak modulation one may write [3.30]

$$\omega = \omega_0 + \frac{\partial \omega}{\partial a_0^2} \left(a^2 - a_0^2\right) + \frac{\partial \omega}{\partial k_0}(k - k_0) + \frac{\partial^2 \omega}{\partial k_0^2}(k - k_0)^2 + \dots . \tag{6.3.43}$$

If one makes the replacement

$$\omega - \omega_0 \text{ by } i\frac{\partial}{\partial t}, \quad k - k_0 \text{ by } i\frac{\partial}{\partial x} \tag{6.3.44}$$

one obtains the so-called *nonlinear* SCHROEDINGER *equation*

$$i\left[\frac{\partial a}{\partial t} + \frac{\partial \omega}{\partial k_0}\frac{\partial a}{\partial x}\right] + \frac{1}{2}\frac{\partial^2 \omega}{\partial k_0^2}\frac{\partial^2 a}{\partial x^2} - \frac{\partial \omega}{\partial a_0^2}|a|^2 a = 0. \tag{6.3.45}$$

In a frame of reference ξ, τ moving with the group velocity, equation (6.3.45) becomes [3.30]

$$i\frac{\partial a}{\partial \tau} + \frac{1}{2}\frac{\partial^2 a}{\partial \xi^2} + \alpha|a|^2 a = 0, \quad \alpha = -\frac{\partial \omega/\partial a_0^2}{\partial^2 \omega/\partial k_0^2}. \tag{6.3.46}$$

In plasma physics the nonlinear SCHROEDINGER equation describes electron waves. It also appears in nonlinear optics [6.13]. If one inserts

$$a(\xi, \tau) = U(\xi - c\tau)\exp(ik\xi - i\omega\tau), \quad |a|^2 = U^2 \tag{6.3.47}$$

into (6.3.46) one gets as the real part

$$U'' + U(2\omega - k^2) + 2\alpha|U|^2 U = 0. \tag{6.3.48}$$

Multiplication by U' and two integrations yield

$$\xi - c\tau = \int \frac{dU}{\sqrt{(2\omega - k^2)U^2 - \alpha U^4 + C_1}} + C_2, \qquad (6.3.49)$$

which represents an elliptic integral and then a JACOBI function (*cnoidal wave*). For $C_1 = 0$ one obtains an *envelope soliton*

$$U(\xi - c\tau) = const \cdot \text{sech}\left[(2\omega - k^2)(\xi - c\tau)\right], \qquad (6.3.50)$$

and the real part of the solution of (6.3.46) reads

$$a(\xi, \tau) = const \cdot \text{sech}\left[(2\omega - k^2)(\xi - c\tau)\right]\cos(k\xi - \omega\tau) \qquad (6.3.51)$$

since

$$\int \frac{dx}{x\sqrt{1 - x^2}} = -\text{arsech}\,x. \qquad (6.3.52)$$

The steepening of a wave front occurs not only for plane waves as discussed earlier, but also for spherical waves. Such waves may be inward running *compression waves* like in a *kidney stone destroyer* or in an *antitank rocket launcher* or in outward running *rarefication waves* (*explosion waves*) [6.14].

In spherical geometry the equations (6.3.1) and (6.3.2) read

$$\rho_t + u\rho_r + \rho\left(u_r + \frac{2u}{r}\right) = 0,$$

$$u_t + uu_r + \frac{1}{\rho}p_r = 0. \qquad (6.3.53)$$

But these waves are no longer adiabatic, so (3.3.24) is no longer valid and has to be replaced by the *polytropic equation*

$$\frac{d}{dt}\left(p\rho^{-n}\right) = \frac{\partial}{\partial t}\left(p\rho^{-n}\right) + u\frac{\partial}{\partial r}\left(p\rho^{-n}\right) = 0, \qquad (6.3.54)$$

given by

$$n = \frac{c - c_p}{c - c_V}. \qquad (6.3.55)$$

c, c_p, c_V are the specific heats for polytropic, isobaric and isochoric thermodynamic changes of state respectively. A *progressing wave* (*simple wave*) setup (based on similarity transformations) can be written down [3.37] to solve equations (6.3.53), (6.3.54)

$$u(r,t) = \alpha r t^{-1} U(\eta), \quad p(r,t) = \alpha^2 r^{\kappa+2} t^{-2} P(\eta),$$
$$\rho = r^\kappa \Omega(\eta), \qquad \eta = r^{-\lambda} t, \quad \lambda = \alpha^{-1}. \qquad (6.3.56)$$

Ordinary differential equations will be obtained and spherical shock waves may appear [6.3].

There are also *electromagnetic shock waves* [6.15], [6.16]. This is understandable since the *magnetic permeability* $\mu(H)$ may depend on the solution for the magnetic field H. Furthermore, in *ferroelectric material* or in plasmas, the *electric permittivity* ε may depend on the electric field E. It might be therefore of interest to investigate if the *electromagnetic pulse* (EMP) due to *nuclear explosions* could be explained by an electromagnetic shock wave in a plasma. A transverse electromagnetic wave $E_x = E(z,t)$ in a plasma satisfies a nonlinear wave equation

$$\frac{\partial^2 E(z,t)}{\partial z^2} = \tilde{\varepsilon}^{3/2}\frac{\partial^2 E(z,t)}{\partial t^2}, \tag{6.3.57}$$

where

$$\tilde{\varepsilon} = dD/dH. \tag{6.3.58}$$

The solution for the E-wave propagating with a phase speed $\tilde{\varepsilon}^{-3/2}$ may be

$$E(z,t) = \Phi\left(z - \tilde{\varepsilon}^{-3/2}t\right). \tag{6.3.59}$$

This wave will steepen if $\partial E/\partial z > 0$ or $\Phi'\tilde{\varepsilon}' > 0$. We thus investigate

$$\frac{\partial E}{\partial z} = \frac{\Phi'}{1 - \Phi'\frac{3}{2}t\tilde{\varepsilon}^{-2}\tilde{\varepsilon}'}, \quad \tilde{\varepsilon}' = -\frac{2e^2}{m^2}H\omega_P^2\left(\omega^2 - \frac{e^2H^2}{m^2}\right)^{-2} < 0. \tag{6.3.60}$$

If we designate $\Phi'(z,t=0) = \Phi_0$, then one obtains for the critical time t_c

$$t_c = 2\tilde{\varepsilon}^2/(3\Phi_0\tilde{\varepsilon}'). \tag{6.3.61}$$

As soon as this time is over, steepening will occur. EMP measurements gave $\Phi_0 < 0$, so that $\tilde{\varepsilon}' < 0$ becomes the critical condition. For a homogeneous isotropic nondissipative cold plasma this condition is satisfied since

$$\tilde{\varepsilon} = 1 - \frac{\omega_P^2}{\omega^2 - e^2H^2/m^2}. \tag{6.3.62}$$

Here ω_P is the *plasma frequency* and m the electron mass. One might thus speculate that the EMP is an electromagnetic shock wave [6.17].

The *inverse scattering method* [4.19] is an ingenious method to solve nonlinear partial differential equations. We will discuss the solution of the KORTEWEG-DE VRIES *equation* (6.3.22), which we write in the form

$$v_t - 6vv_x + v_{xxx} = 0. \tag{6.3.63}$$

The inverse scattering method does not directly solve this nonlinear partial differential equation, but instead, it solves two linear equations that have the same solution as (6.3.63). For this purpose, we consider the one-dimensional time-independent SCHROEDINGER *equation*

$$\psi_{xx} + [E - v(x,t)]\psi = 0 \tag{6.3.64}$$

for a fixed parameter t. The eigenvalues E may have discrete values $E_n = -k_n^2(t)$ as well as continuous values $E = k^2$, $E > 0$. To solve (6.3.64), two setups may be made:

$$\psi_n = c_n(t)\exp(-k_n x), \quad \psi = \exp(-ikx) + R(k,t)\exp(ikx). \qquad (6.3.65)$$

The bound states $E_n < 0$ are normalized by $\int_{-\infty}^{+\infty}\psi_n^2\,\mathrm{d}x = 1$ and the free states $E > 0$ correspond to waves inciding from $x = +\infty$ on the potential v. The part R of the particles presented by ψ will be reflected from the potential and the part $T(k,t)\exp(-ikx)$ penetrates into the potential. Due to particle conservation one has $|R|^2 + |T|^2 = 1$. In the next step, one solves (6.3.64) for v giving

$$v = (\psi_{xx} + E\psi)/\psi. \qquad (6.3.66)$$

Insertion into (6.3.63) results in

$$\psi^2\frac{\partial E}{\partial t} + \frac{\partial}{\partial x}\left(\psi\frac{\partial Q}{\partial x} - \frac{\partial\psi}{\partial x}\cdot\frac{\partial Q}{\partial x}\right) = 0, \qquad (6.3.67)$$

where

$$Q = \frac{\partial\psi}{\partial t} + \frac{\partial^3\psi}{\partial x^3} - 3(v+E)\frac{\partial\psi}{\partial x} \qquad (6.3.68)$$

is a formal abbreviation. We first consider only the ψ_n. Since $\psi_n \sim \exp(-k_n x) \to 0$ for $x \to 0$ and $\int_{-\infty}^{+\infty}\psi^2\,\mathrm{d}x \neq 0$ one obtains

$$\frac{\partial E_n}{\partial t} = 0, \quad k_n(t) = k_n(0). \qquad (6.3.69)$$

This indicates that the E_n do not depend on t and (6.3.67) is simplified. Two integrations yield

$$\frac{\partial\psi_n}{\partial t} + \frac{\partial^3\psi_n}{\partial x^3} - 3(v+E)\frac{\partial\psi_n}{\partial x} = F_n(t)\psi_n + D_n(t)\psi_n\int^x\psi_n^{-2}\,\mathrm{d}x. \qquad (6.3.70)$$

The integration constants $D_n(t)$ must vanish, since ψ_n^{-2} diverges for $x \to +\infty$. Then (6.3.70) is a linear partial differential equation containing the (still unknown) solution v of (6.3.63). Multiplication of (6.3.70) by ψ_n, integration from $-\infty$ to $+\infty$ and use of (6.3.66) result in

$$\frac{\mathrm{d}}{\mathrm{d}t}\int_{-\infty}^{+\infty}\frac{1}{2}\psi_n^2\,\mathrm{d}x = F_n(t)\int_{-\infty}^{+\infty}\psi_n^2\,\mathrm{d}x. \qquad (6.3.71)$$

Then (6.3.70) becomes

$$\frac{\partial\psi_n}{\partial t} + \frac{\partial^3\psi_n}{\partial x^3} - 3(v+E_n)\frac{\partial\psi_n}{\partial x} = 0. \qquad (6.3.72)$$

Since the potential v vanishes for $x \to \infty$, $E_n = -k_n^2(t)$, insertion of ψ_n yields

$$\frac{\mathrm{d}}{\mathrm{d}t} c_n(t) - 4k_n^3 c_n(t) = 0, \quad \text{thus} \quad c_n = c_n(0)\exp(4k_n^3 t). \tag{6.3.73}$$

If one repeats the calculation for $\psi = T(k,t)\exp(-ikx)$ one receives a differential equation for $T(k,t)$ from which $R(k,t) = R(k,0)\exp(8ik^3 t)$ follows. Quantum theory teaches that the knowledge of $c_n(t)$, $k_n(t)$ and $R(k,t)$ is sufficient to construct the scattering potential v, which is actually a solution of the KORTEWEG-DE VRIES equation. The construction of the potential v may be made using the GELFAND-LEVITAN *integral equation*. Let

$$B(x,t) = \frac{1}{2\pi} \int\limits_{-\infty}^{+\infty} R(k,t)\exp(ikx)\mathrm{d}k + \sum_{n=1}^{\infty} c_n^2(t)\exp(-k_n(t)x), \tag{6.3.74}$$

then the integral equation

$$K(x,y,t) + B(x+y,t) + \int\limits_{x}^{\infty} B(y+y',t)K(x,y',t)\mathrm{d}y' = 0 \tag{6.3.75}$$

determines K and the solution of the KORTEVEG-DE VRIES equation is given by

$$v(x,t) = -12\frac{\mathrm{d}K(x,y,t)}{\mathrm{d}x} \quad \text{for} \quad y = x. \tag{6.3.76}$$

The inverse scattering method can be used for the solution of many nonlinear partial differential equations [6.15].

Problems

1. The steepening up of the wave front of a large-amplitude wave in a *viscous* gas can be described [3.30] by

$$ax = \frac{u_2}{u_1 - u_2}\ln(u - u_2) - \frac{u_1}{u_1 - u_2}\ln(u_1 - u). \tag{6.3.77}$$

This solution describes a continuous variation between two asympyotic states $u(x \to -\infty) = u_1$ and $u(x \to \infty) = u_2$. Plot the function $u(x)$ for $u_2 + 0.0001 \le u \le u_1 - 0.0001$, $-6.5 \le x \le 8$.
Hint: use the command **ContourPlot**
Use $a = 1.$, $u_1 = 2.$, $u_2 = 1.0$.
Contours->{0},
PlotPoints->60.

2. Plot the soliton solution (6.3.27).
   Hint: plot $Sech^2(x)$ for $-3 \le x + 3$. Then $x \to x - c_0t$ in the argument.

3. Plot the envelope soliton (6.3.50).
   Hint:
   Plot[{Sech[x],-Sech[x],Sech[x]*Cos[4*x]},{x,-4.,4.}]

4. Solve (6.3.49) for $(2\omega - k^2) = 1$, $\alpha = 1$, $C_1 = 1$, $C_2 = 0$ by integration.

5. Derive the ordinary differential equations for $U(\eta), P(\eta), \Omega(\eta)$ resulting from (6.3.53), (6.3.54).

6. Derive the HUGONIOT *equation* (also called shock adiabatic curve). In a shock wave front the conservation of mass, momentum and energy, written for the normal components of the flow velocity connect the domains before and after the front [6.18]:

$$\rho_1 v_1 = \rho_2 v_2, \tag{6.3.78}$$

$$p_1 + \rho_1 v_1^2 = p_2 + \rho_2 v_2^2, \tag{6.3.79}$$

$$i_1 + \frac{v_1^2}{2} = i_2 + \frac{v_2^2}{2}, \tag{6.3.80}$$

where $i = c_pT$ is enthalpy, see next section. For given p_1 and $V_1 = 1/\rho_1$ the HUGONIOT curve $p_2(V_2)$ is given by

$$i_1 - i_2 + \frac{(V_1 + V_2)(p_2 - p_1)}{2} = 0. \tag{6.3.81}$$

Plot this curve and the adiabatic curve $pV^\kappa = const$, $\kappa = c_p/c_V$.

6.4   The rupture of an embankment-type water dam

A typical nonlinear engineering problem is the calculation of the possible rupture of an embankment dam. At the occasion of a rupture, a water surge shall propagate into the channel downstream of the dam. Such surge may generate heavy destructions along the channel or river. Let us assume that the channel has a width B and extends in the direction of the x axis. Let the water depth H in the storage lake be $H = 2.2$ m and the water depth h in the channel is assumed to be $h_0 = 1.2$ m. The water level after a dam rupture will be designated by $h(x,t)$. Thus the local water mass is given by $\rho_0 Bh(x) = q(x)\rho_0$, where ρ_0 is the (constant) water density. Let the storage lake have an extension of 5000 m in the x direction. The dam itself may be situated at $x = 5000$ m and rupture may occur suddenly at $t = 0$. Then we have the initial conditions for h and the stream velocity $u(x,t)$ at $t = 0$:

$$h(x,0) = H = 2.2 \text{ m}, \quad u(x,0) = 0, \quad 0 \le x \le 5000 \text{ m},$$
$$h(x,0) = h_0 = 1.2 \text{ m}, \quad u(x,0) = 0, \quad 5\,000 \le x \le \infty. \tag{6.4.1}$$

This indicates that at $t = 0$ and $x = 5000$ m a vertical water wall of a height $H - h_0 = 1$ m exists. At the other end of the lake ($x = 0$) no flow is present. The relevant equations describing the evolution in time of these nonlinear one-dimensional phenomena are:
the continuity equation

$$\frac{\partial}{\partial t}(\rho_0 q(x,t)) + \frac{\partial}{\partial x}(\rho_0 u(x,t) q(x,t)) = 0, \quad q_t + u_x q + q_x u = 0, \tag{6.4.2}$$

and the equation of motion

$$\rho_0 u_t + \rho_0 u u_x + p_x = 0. \tag{6.4.3}$$

The local hydrostatic pressure $p(x,t)$ per unit length is given by

$$p(x,t) = \rho_0 g q(x,t)/B. \tag{6.4.4}$$

Then we can now write for (6.4.3)

$$u_t + u u_x + \frac{g}{B} q_x = 0. \tag{6.4.5}$$

We thus have two nonlinear partial differential equations (6.4.2) and (6.4.5) for the two unknown functions $u(x,t)$ and $q(x,t)$. We use the method of characteristics developed in section 3.3 for such a system of two partial differential equations of first order. We compare our system of two partial equations with (3.3.20) and read off

$$a_{11} = 1, \quad a_{12} = 0, \quad a_{21} = 0, \quad a_{22} = 1,$$
$$b_{11} = u, \quad b_{12} = g/B, \quad b_{21} = q, \quad b_{22} = u. \tag{6.4.6}$$

Then (3.3.32) and (3.3.33) yield the propagation speed of *small* waves

$$\frac{dx}{dt} = u + \sqrt{\frac{gq}{B}} = u + \sqrt{gh} \quad \text{(downstream)},$$

$$\frac{dx}{dt} = u - \sqrt{\frac{gq}{B}} = u - \sqrt{gh} \quad \text{(upstream)} \tag{6.4.7}$$

and equations (3.3.30) and (3.3.31) result in

$$\pm \sqrt{\frac{gq}{B}} \, du + \frac{g}{B} \, dq = 0. \tag{6.4.8}$$

The two equations describe the modification of the *state variables* u, q along the characteristics (6.4.7). The problem is now that we cannot use or integrate the characteristics equation because they contain the still unknown solutions $u(x, t)$ and $q(x, t)$. We first make a transformation to a new variable λ [ms^{-1}]. We define

$$d\lambda = \sqrt{\frac{g}{B}}\frac{dq}{\sqrt{q}}, \quad \lambda(q) = \int_0^q \frac{dq}{\sqrt{q}}\sqrt{\frac{g}{B}} = \int_0^h \sqrt{\frac{g}{h}}\,dh = 2\sqrt{gh}. \tag{6.4.9}$$

Then we use the RIEMANN *invariants* defined by (3.4.30), (3.4.32). We use

$$r = u + \lambda \quad s = u - \lambda, \quad u = (r + s)/2, \quad \lambda = (r - s)/2, \tag{6.4.10}$$

$$du \pm d\lambda = 0, \quad u \pm \lambda = const = \begin{cases} r \\ s \end{cases}. \tag{6.4.11}$$

The r, s or u, q plane is called *state plane* by some authors. We now consider a mapping between the "linear" state plane r, s and the nonlinear physical plane described by x, t. Let us discuss the correspondence between the two planes. We allocate the point $P(r_1, s_1)$ of the state plane to the dam location point $\bar{P}$ (5000,0) of the physical plane. This expresses the fact that in the point $\bar{P}(x = 5000, t = 0)$ a local water wall of absolute height $h(5000, 0) = H = h_1$ or relative height 1 m above the normal water level in the channel exists with streaming velocity $u(5000, 0) = u_1 = 0$. According to (6.4.9) the height $h_1 = H = 2.2$ m corresponds to $r_1 = u_1 + \lambda_1 = \lambda_1, s_1 = -\lambda_1, \lambda_1 = 2\sqrt{gh_1}$. At the other end of the lake $x = 0$, one has $u(0, 0) = 0$ and $h(0, 0) = H = 2.2$ m. Thus the point $\bar{Q}(0, 0)$ corresponds to $Q(r_Q s_Q)$, where $u_Q = 0, \lambda_Q = 2\sqrt{gH}, r_Q = \lambda_Q, s_Q = -\lambda_Q, r_Q = -s_Q$. Inserting numbers for h, H and g (9.81 ms^{-2}) we receive for $\bar{P}$

$$\lambda_1 = 2\sqrt{2.2 \cdot 9.81} = 9.291, \; u_1 = 0, \; r_1 = 9.291, \; s_1 = -9.291, \tag{6.4.12}$$

all measured in [ms^{-1}]. On the other end of the lake we have for $\bar{Q}(0, 0)$

$$\lambda_Q = 9.291, \; u_Q = 0, \; r_Q = 9.291, \; s_Q = -9.291. \tag{6.4.13}$$

At the time $t = 0$ of the rupture of the dam the same physical states exist at $x = 5000$ and $x = 0$. But, at this time, the dam breaks down and elementary waves (composing later on a steepening surge downstream and a rarefaction wave upstream) start at $x = 5000$. Replacing in (6.4.7) the $dx \to \Delta x$, $dt \to \Delta t$ we can write for the wave speeds

$$\frac{\Delta x}{\Delta t} = u \pm \sqrt{gh} = u \pm \frac{\lambda}{2}. \tag{6.4.14}$$

Thus, the first elementary wave running to the left to $x = 0$ and upstream reduces the water level H in the lake. It has a wave speed $\Delta x/\Delta t = 0 - \lambda_1/2 = -4.646$[ms^{-1}] and $s_1 = -9.291 = const$, $r_1 = +9.291$. The wave running to the right (downstream) increases the water level $h(x, t)$ in the channel from

$h_0 = h_1$ to h_2 and has a wave speed $\Delta x/\Delta t = 0 + \lambda_1/2 = +4.646[\text{ms}^{-1}]$ and $s_1 = -9.291$, $r_1 = +9.291 = const$. Waves running to the left from P to Q transfer their s-value to Q, since $s = const$ is valid for waves running to the left: $s_Q = s_1$. Waves running to the right, downstream from P to $x \to \infty$ transfer their r-value, so that the whole domain right-hand of the dam ($x > 5000$) always has the same r-value. At the time $t = 0 + \Delta t$ the next two elementary waves start. Both waves now run into domains where the states had been modified by the first two waves. The upstream wave enters an area where the water level had been reduced from H to $H - h_1$ and might be reflected at the lake end $x = 0$. It will no longer return with 4.646 [ms^{-1}], because water level and driving pressure had been lowered. The second elementary downstream wave starting at $t = \Delta t$ will be faster than the first one because in the channel the water level had been increased by the first downstream wave and the water had started to stream to $x \to \infty$. In order to be able to calculate the $\Delta u, \Delta \lambda$, etc., we need to have some knowledge about the final state, $t \to \infty, x \to \infty$. For an infinitely long channel we define the point $R(r_N, s_N)$ in the state plane. Apparently the final conditions will read for R: $q(x, t \to \infty) = h_0 = 1.2$ m, $\lambda_N = 2 \cdot \sqrt{1.2 \cdot 9.81} = 6.862$, and for $\bar{R}(\infty, \infty)$ $u_N = 0$, $r_N = 6.862$, $s_N = -6.862$. The whole phenomenon of the break-down of the dam occupies a square in the s, r plane. The four corners are given by $R(s_N = -6.862, r_N = +6.862)$, $(s_N = -6.862, r_1 = 9.291)$, $(s_1 = -9.291, r_N = 6.862)$ and $P = Q(s_1 = -9.291, r_1 = 9.291)$. Now it the accuracy and our will that have to decide how many steps $i = 1 \ldots N$ we will calculate. For this decision we consider the pressure difference from $p_1 = \rho_0 g q(x, 0) = \rho_0 g B h(x, 0) = \rho_0 B x_1^2/4 = \rho_0 B \lambda_1^2/4$ down to $p_N = Br_N^2/4$. This concerns the variation $\lambda_1 \to \lambda_N$, $r_1 \to r_N$ etc. If we choose $N = 10$ pressure steps, then each elementary wave carries $\Delta r = (9.921 - 6.862)/10 = 0.243 = |\Delta s|$. This corresponds to an accuracy of 2.6 % (0.243:9.291). Table 6.1 describes the situation in detail.

Table 6.1 Pressure steps (for a downstream wave)

Nr	in front of the wave				behind the wave		
	s	λ	u	$\Delta x/\Delta t$	s	λ	u
1	−9.291	+9.291	0	4.646	−9.048	+9.170	+0.122
2	−9.048	+9.170	+0.122	4.707	−8.805	+9.048	+0.243
3	−8.805	+9.048	+0.243	4.767	−8.562	+8.927	+0.365
4	−8.562	+8.927	+0.365	4.829	−8.319	+8.805	+0.486
5	−8.319	+8.805	+0.486	4.889	−8.076	+8.684	+0.608
6	−8.076	+8.684	+0.608	4.950	−7.833	+8.562	+0.729
7	−7.833	+8.562	+0.729	5.010	−7.590	+8.441	+0.851
8	−7.590	+8.441	+0.851	5.072	−7.347	+8.319	+0.936
9	−7.347	+8.319	+0.936	5.096	−7.104	+8.198	+1.094
10	−7.104	+8.198	+1.094	5.193	−6.862	+8.077	+1.215
11	−6.862	+8.007	+1.215	5.254	[−6.619	+7.955	+1.336]

The grid of points within the square in the r, s plane can be mapped into the x, t plane: for every point in the r, s plane the values $u(x, t), q(x, t)$ in the x, t plane are defined by the equations (6.4.7) to (6.4.11). Interpolation within the grid delivers any wanted $u(x, t), q(x, t)$ and thus the solution of equations (6.4.2), (6.4.5).

6.5   Gas flow with combustion

Combustion of petrol or gunpowder within a gas flow has many practical applications: turbogas exhauster, jet engines, ram jets, rocket propulsion and, finally, guns. Depending on the type of propellant or gunpowder, the combustion or explosion process has quite different characteristic features. Usually, *combustion* is defined as the burning of a fuel associated with the generation of heat. The spreading out of a combustion may excite a combustion wave. A *detonation* or explosion is a very rapid chemical reaction of an oxidizer and a fuel with large release of heat and pressure waves. A *deflagration* is the burning of explosives or fuel at a rate slower than a detonation. Chemical reactions and thermodynamics enter into the description of these processes. Combustion of gases and in gases is always connected with a gas-dynamic (compressible) flow. In some combustion and detonation processes it may be necessary to include new source terms like in (3.3.22) and (3.3.23) into the basic equations. A source term $g(x, t)$ describing the increase of the gas mass by combustion may be $g(x, t) = D$, where D describes the gas production [g cm^{-3}s^{-1}] due to combustion. Let $\tilde{u}(x, t)$ be the flow velocity of the generated gas, then the source term $f(x, t)$ in (3.3.23) may read $f(x, t) = D(u(x, t) - \tilde{u}(x, t))$ describing a *jolting acceleration* of the new gas masses. The energy theorem may connect the area in front and behind the combustion front:

$$\rho\frac{\tilde{u}^2}{2} + \tilde{i} = \rho\frac{u^2}{2} + i + D, \qquad (6.5.1)$$

where $i(\sim c_p T)$ designates the *enthalpy* defined by $U + p/\rho$, U is the thermodynamic internal energy $(\sim c_V T)$. Since we do not intend to start an exposition of thermodynamics, we stop the presentation at this point. We just want to show that the characteristics method discussed in section 6.4 can be applied on combustion phenomena too.

In the frame of a research contract, we had the opportunity to investigate the *intake stroke* and the *compression stroke* of a DIESEL *engine* of type JW 15 [6.19]. The comparison between the values $p(t)$ calculated by the characteristics method as described in section 6.4 and the measured values of the pressure showed satisfactory agreement.

Problems

1. Calculate the efficiency of a *ram jet* (LORIN *engine*). Assume that the heat generated by the fuel combustion is given by $Q = c_p(T' - T)$. Here $G = f\rho w$ is the mass flow ratio, f is the cross section of the tube, w flow velocity per unit mass, $T' - T$ is the temperature increase. Energy conservation results in

$$c_p T + \frac{w^2}{2} = c_p T_\infty + \frac{w_\infty^2}{2} \quad \text{and} \quad c_p T' + \frac{w^2}{2} = c_p T_e + \frac{w_e^2}{2}. \quad (6.5.2)$$

Here the subscripts ∞ and e designate the values at infinity and at the exhaust of the tube. Assume that compression due to the stagnation pressure of the flying ram jet and the consecutive expansion are free of losses, so that entropy is conserved. Then $T/T_\infty = T'/T_e$ and

$$Q/G = c_p(T_e - T_\infty) + \frac{w_e^2}{2} - \frac{w_\infty^2}{2} = c_p T(T_e - T_\infty)/T_\infty$$

$$= \frac{w_e^2/2 - w_\infty^2/2}{1 - T_\infty/T}. \quad (6.5.3)$$

The useful power is defined by the *thrust power*: $Gw_\infty(w_e - w_\infty) = -Fw_\infty$, where F is the propulsion force. Thus, the solution for the efficiency is

$$\eta = \frac{-Fw_\infty}{Q} = \left(1 - \frac{T_\infty}{T}\right) \frac{2w_\infty}{w_e + w_\infty}. \quad (6.5.4)$$

2. Whereas a ram jet produces thrust only at flight, a *jet engine* produces thrust even at rest. This is due to the supercharger of the engine. The turbine generates the thrust. If, however, one assumes that the compressor power equals the turbine power, then (6.5.2), (6.2.3) are again valid. But now the temperature T before the combustion does depend on the compressor power and the engine works even for $w_\infty = 0$. Which engine has the higher efficiency?

3. Calculate the maximum exhaust speed v_{max} of a *rocket*. Assume that the whole enthalpy of the exhaust gases is transformed into kinetic energy.
   Hints: assume adiabatic behavior $p/p_0 = (\rho/\rho_0)^\kappa = (T/T_0)^{\kappa/(\kappa-1)}$ for the change of state during the exhaust. $\kappa = c_p/c_V$ is the ratio of the specific heats. Use the BERNOULLI *equation* in the form

$$\frac{1}{2}(v_0^2 - v^2) + \int_p^{p_0} \frac{1}{\rho(p)} dp = 0. \quad (6.5.5)$$

Result:

$$v^2 = \frac{2\kappa}{\kappa-1}\frac{p_0}{\rho_0}\left[1-\left(\frac{p}{p_0}\right)^{(\kappa-1)/\kappa}\right] = v_{max}^2\left[1-\left(\frac{p}{p_0}\right)^{(\kappa-1)/\kappa}\right].$$

(6.5.6)

This is the SAINT VENANT-WANTZEL *formula*. Calculate v_{max} for air ($\kappa = 1.405$, $T_0 = 288$ K). Result: 757 [ms^{-1}]. A rocket motor will reach this exhaust speed in space ($p = 0$).

References

a) References on *Mathematica*

[1] Wolfram, S. et al, *The Mathematica Book*, Mathematica Version 4, Wolfram Media, Fourth Edition, Cambridge University Press, 1999.

[2] Wolfram, S. et al, *Mathematica 3.0 Standard Add-On Packages*, Cambridge University Press.

[3] Wolfram, S., *Mathematica. A System for Doing Mathematics by Computer*, Second Edition, Addison-Wesley, Publishing Company, 1991.

[4] Gray, T., Glynn, J., *Exploring Mathematics with Mathematica*, Addison Wesley, Redwood City, CA 94065, 1991.

[5] Press, W. et al, *Numerical Recipes*, Cambridge University Press, Cambridge, 1987 (FORTRAN).

[6] Wickham-Jones, T., *Mathematica Graphics*, Telos-Springer, Santa Clara, CA, 1994.

[7] Gaylord, R. et al, *Introduction to Programming with Mathematica*, Telos-Springer, Santa Clara, CA, 1993.

[8] Maeder, R., *Programming in Mathematica*, Addison-Wesley, Reading, Mass, 1990.

[9] Maeder, R., *The Mathematica Programmer*, Academic Press, New York, 1996.

[10] Ganzha, V., Vorozhtsov, E., *Numerical Solutions of Partial Differential Equations*, CRC Press, Boca Raton, FL, 1996 (finite difference methods).

[11] Vvedensky, D., *Partial Differential Equations with Mathematica*, Addison Wesley, Reading, Mass., 1992.

[12] Blachmann, *Mathematica, A Practical Approach*, Prentice Hall, Englewood Cliffs, NJ, 1992.

[13] Kofler, M. et al, *Mathematica*, Addison Wesley, Reading, Mass., 2002.

b) References

[1.1] Ackeret, J., *Helvet. Physica Acta* $\underline{19}$, 103 (1946).

[1.2] Morse, P., Feshbach,H., *Methods of Theoretical Physics*, McGraw-Hill, New York, 1953 (2 volumes).

[1.3] Cap, F., *Phys. Fluids* $\underline{28}$ (6), 1766 (1985).

[1.4] Moon, P., Spencer, D.E., *Field Theory Handbook*, Springer, Berlin, New York, 1971.

[1.5] Bôcher, M., *Ueber die Reihenentwicklungen der Potentialtheorie*, B. Teubner, Leipzig, 1894.

[1.6] Knight, R., The Potential of a Sphere Inside an Infinite Circular Cylinder, *Quart. J. Math.*, Oxford, $\underline{7}$, 127 (1936).

[1.7] Collatz, L., *The Numerical Treatment of Differential Equations*, Springer, New York, 1966.

[1.8] Dive, P., *Ondes ellipsoidales et relativité*, Gauthier-Villars, Paris, 1950.

[2.1] Abramowitz, M., Stegun, I., *Handbook of Mathematical Functions*, Dover Publication, New York.

[2.2] Cap, F., Groebner, W. et al, *Solution of Ordinary Differential Equations by Means of Lie Series*, NASA Report CR-552, Washington, DC, August 1966.

[2.3] Groebner, W. et al, *Lie Series for Celestial Mechanics, Accelerators, Satellite Stabilization and Optimization*, NASA Contractor Report NASA CR-1046, Washington, DC, May 1968 (includes complete bibliography); Groebner, W., Knapp, H., *Contributions to the Method of Lie Series*, Bibliographisches Institut, Nr 802, Mannheim, 1967.

[2.4] Wanner, G., Ein Beitrag zur numerischen Behandlung von Randwertaufgaben gewoehnlicher Differentialgleichungen nach der Lie-Reihen-Methode, *Monatshefte f. Mathematik* $\underline{69}$ (1965), 431-449 and Numerical solution of ordinary differential equations by Lie series, MRC Report 880, Mathematics Research Center, University of Wisconsin, Madison, 1968.

[2.5] Cap, F., *Handbook on Plasma Instabilities*, Academic Press, New York, 1976, 1978, 1982, 3 Vols.

[2.6] Kamke, E., *Differentialgleichungen*, Akademische Verlagsgesellschaft, Leipzig, 1942.

[2.7] Cap, F., Lazhinsky, H., On an Equation Related to Nonlinear Saturation of Convection Phenomena, *Nonlinear Vibration Problems*, 14 (1974), 519-528.

[2.8] Cap, F., Averaging Method for the Solution of Nonlinear Differential Equations with Periodic Non-Harmonic Solutions, *Int. J. Non-Linear Mechanics* 9 (1974), 441-450; Cap, F., Langevin equation of motion for electrons in a non-homogeneous plasma, *Nuclear Fusion* 12 (1972), 125-126.

[2.9] Spanier, J., Oldham, K., *An Atlas of Functions*, Springer, Berlin, 1987.

[2.10] Schmidt, G., Tondl, A., *Non-Linear Vibrations*, Akademie-Verlag, Berlin, 1986.

[2.11] Knapp, H., Ergebnisse einer Untersuchung ueber den Wert der LIE-Reihen-Theorie fuer numerische Rechnungen in der Himmelsmechanik, *ZAMM* 42 (1962), T25-T27; Groebner, W., Cap, F., Perturbation Theory of Celestial Mechanics Using Lie series, *Proceedings 11th International Astronautical Congress*, Stockholm, 1960.

[2.12] Cap, F., Parametric machine, US patent 4.622.510; Cap, F., Elektrostatischer Hochspannungsgenerator, *Oesterr. Zeitschrift fuer Elektrizitaetswirtschaft* 41, Nr 7 (1988), pp 197-198; Cap, F., Leistung und Wirkungsgrad eines elektrostatischen Zylindergenerators, *Elektrotechnik und Maschinenbau* 105, Nr 2 (1988), pp 78-86 (and 104, Nr 4 (1987), pp 128-133) as well as *Progress in Astronautics* Vol 110 (1987), pp 475, Amer. Inst. Aeronautics and Astronautics.

[2.13] McLachlan, N., *Theory and Application of Mathieu Functions*, Clarendon Press, Oxford, 1947.

[3.1] Cap, F., *Lehrbuch der Plasmaphysik und Magnetohydrodynamik*, Springer, New York, 1994.

[3.2] Boozer, A., Guiding center drift equations, *Phys. Fluids* 23 (1980), Nr 5, 904 and Nakajiima, N. et al, On Relation between Hamada and Boozer Magnetic Coordinate Systems, *Research Report NIFS-173*, National Institute for Fusion Research, Nagoya, Japan, September 1992.

[3.3] Hamada, G., Hydromagnetic equilibria and their proper coordinates, *Nucl. Fus.* 2 (1962), p 23.

[3.4] Holbrook, J., *Laplace Transforms for Electronic Engineers*, Pergamon Press, Oxford, 1966.

[3.5] Titchmarsh, E., *Introduction to the Theory of Fourier Integrals*, University Press, Oxford, 1937.

[3.6] Erdélyi, A. et al, *Tables of Integral Transforms*, McGraw-Hill, New York, 1952.

[3.7] Arfken, G., *Mathematical Methods for Physicists*, Academic Press, New York, 1970.

[3.8] Birkhoff, G., *Hydrodynamics*, Princeton University Press, 1960.

[3.9] Ames, W., *Nonlinear Partial Differential Equations in Engineering*, Academic Press, New York, 1968.

[3.10] Hansen, A., *Similarity Analyses*, Prentice Hall, 1964.

[3.11] Bridgeman, G., *Dimensional Analysis*, Yale University Press, 1931.

[3.12] Cap, F. et al, Anwendungen von Aehnlichkeitstransformationen auf magnetogasdynamische Kanalstroemungen, *Elektrotechnik und Maschinenbau* Vol 68, Nr 11 (1969), 512-516.

[3.13] Courant, R., Hilbert, D., *Methods of Mathematical Physics*, Interscience Publisher, New York, 1953 (2 vols).

[3.14] Mathews, J., Walker, R., *Mathematical Methods of Physics*, W. Benjamin, New York, 1963.

[3.15] Crank, J., *Free and Moving Boundary Problems*, Clarendon Press, Oxford, 1988.

[3.16] Tolman, R., *The Principles of Statistical Mechanics*, Oxford University Press, 1959.

[3.17] Riedi, P., *Thermal Physics*, MacMillan Press, London, 1976.

[3.18] Thewlis, J. et al, *Encyclopaedic Dictionary of Physics*, Pergamon Press, London, 1962, vol 7, p 584.

[3.19] Thewlis, J., Cap, F., ibid., Vol 9, Multilingual Glossary, p 370.

[3.20] See [3.1], page 70.

[3.21] Cap, F., *Physik und Technik der Atomreaktoren* (Physics and Technology of Nuclear Reactors), Springer, Wien, 1957.

[3.22] Chapman, S., Cowling, T., *The Mathematical Theory of Non-Uniform Gases*, Cambridge University Press, Cambridge, 1964.

[3.23] Krall, N., Trivelpiece, A., *Principles of Plasma Physics*, McGraw-Hill, New York, 1973.

[3.24] Davison, B., Sykes, J., *Neutron Transport Theory*, Clarendon Press, Oxford, 1957.

[3.25] *Proceed. Sympos. Fast Reactor Physics*, Aix en Provence, 24-28 Sept. 1979, Internat. Atomic Energy Agency, Wien, 1980, 2 vols.

[3.26] Weinberg, A., Wigner, P., *The Physical Theory of Neutron Chain Reactors*, University of Chicago Press, 1958.

[3.27] Kuttner, P., *Beitraege zur Transporttheorie zu sphaerischen Reaktoren*, Ph.D. Thesis, Innsbruck, 1957.

[3.28] Spiegel, M., *Laplace Transforms*, Schaum Publishing, New York, 1965.

[3.29] Saely, R., *Development of New Methods for the Solution of Differential Equations by the Method of LIE Series*, W. Groebner et al, *Contract JA 37-68-C-1199*, European Research Office, July 1969.

[3.30] Shivamoggi, B., *Theoretical fluid dynamics*, Nijhoff, Dordrecht, 1985.

[3.31] Crocco, L., Eine neue Stromfunktion fuer die Erforschung der Bewegung der Gase in Rotation, *Z. angew. Math. Mech.* 17 (1937), 1.

[3.32] Vazsony, A., On rotational gas flow, *Q. Appl. Math.* 3 (1945), 29.

[3.33] Cap, F., The MHD Theorems by Kaplan and Crocco and their Consequences for MHD flow, *Sitzber. Oest. Akad. Wiss.* II/04 (2001), 25-31.

[3.34] Frankl, F., Karpovich., E., *Gas Dynamics of Thin Bodies*, Interscience, New York, 1953.

[3.35] Cap, F., Characteristics and Constants of Motion Method for Collisional Kinetic Equations, *Rev. Roumaine des Sciences Technique, Ser Mécanique appliquée* 17, Nr 3 (1972), p 485; Goddard Space Flight Center, Greenbelt, *Report X-640-71-123*, April 1971.

[3.36] Forsythe, G., Wasow, W., *Finite Difference Methods for Partial Differential Equations*, Wiley, New York, 1960, p 141; Cap, F., Ueber zwei Verfahren zur Loesung eindimensionaler instationaerer gasdynamischer Probleme, *Acta Physica Austriaca* 1 (1947), 89; Mathem. Review 9 (1949), p 216.

[3.37] Courant, R., Friedrichs, K., *Supersonic flow and shock waves*, Interscience, New York 1948.

[3.38] Carrier, G., On the nonlinear vibration problem of the elastic string, *Quart. Appl. Math.* 3, 157 (1954).

[4.1] Faber, G., Beweis, dass unter allen homogenen Membranen von gleicher Flaeche und gleicher Spannung die kreisfoermige den tiefsten Grundton gibt, *Sitzber. math. phys. Kl, Bayer. Akad. d. Wiss.* (1923), p 169-172.

[4.2] Kac., M., Can you hear the shape of a drum? *Amer. Math. Monthly* 73, Nr 2 (1966), p 1-23.

[4.3] Gordon, C.S., When you can't hear the shape of a manifold, *The Mathematical Intelligencer*, 11, Nr 3 (1989), p 39-47.

[4.4] Cap, F., *Nouvelle méthode de résolution de l'equation de Helmholtz pour une symmétrie cylindrique*, E. Beth Memorial colloquium, D. Reidel, Dordrecht, 1967; see also *Nukleonik* 6 (1964), 141-147 and *Nucl. Science and Engineering*, 26 (1966), 517-521.

[4.5] Sommerfeld, A., *Partial Differential Equations in Physics*, Academic Press, New York, 1949.

[4.6] Cap, F., Eigenfrequencies of membranes of arbitrary boundary and with varying surface mass density, *Appl. Math. and Computation* 124 (2001), 319-329.

[4.7] Landau, L., Lifschitz, E., *Elastizitaetstheorie*, Akademie Verlag, Berlin 1966 or Pergamon Press, Oxford (in English).

[4.8] Lebedev, N. et al, *Problems in Mathematical Physics*, Pergamon Press, Oxford, 1966.

[4.9] Timoshenko, S., *Theory of Plates and Shells*, McGraw-Hill, New York, 1959.

[4.10] Leissa, A., Vibration of Plates, National Aeronautics and Space Administration, *NASA Report SP-160*, U.S.Gov.Printing Office, Washington, DC.

[4.11] Taylor, G., Analysis of the swimming of microscopic organisms, *Proc. Roy. Soc. A* 209 (1951), 447-460.

[4.12] Surla, K., A Uniformly Convergent Spline Difference Scheme for a Singular Perturbation Problem, *ZAMM* 66, Nr 5 (1986), T 328-329.

[4.13] Groebner, W., *Lie Reihen und ihre Anwendung*, Deutscher Verlag der Wissenschaften, Berlin, 1960.

[4.14] Busch, G., Schade, H., *Lectures in Solid State Physics*, Pergamon, Oxford, 1976.

[4.15] Cap, F., Roever, W., Lichtwege in inhomogenen absorbierenden isotropen Medien (Light paths in inhomogeneous absorbing isotropic media), *Acta Physica Austriaca* 8, Nr 4 (1954), 346-355; Brandstatter, *Waves, Rays and Radiation in Plasma Media*, McGraw-Hill, New York 1963, p 36.

[4.16] Oden, J., Reddy, J., *Variational Methods in Theoretical Mechanics*, Springer, Berlin, 1983.

[4.17] Wallnoefer, M., *Anwendung der Methode von Ritz auf Delta u = 0 in einem zweifach zusammenhaengenden Gebiet*, M.S. diploma, 1964, Mathematical Institute University Innsbruck.

[4.18] Brebbia, C., Teller, J., Wrobel, L., *Boundary Element Techniques*, Springer, Berlin, 1984.

[4.19] Agranovich, Z., Martchenko, V., *The Inverse Problem of Scattering Theory*, Gordon and Breach, New York 1974.

[4.20] Kastlunger, K., in Groebner et al, *Development of New Methods for the Solution of Differential Equations by the Method of Lie Series*, Final

Report July 1969, European Research Office of the U.S. Government, contract JA-37-68-C-1199.

[4.21] Marsal, D., *Die numerische Loesung partieller Differentialgleichungen*, Bibliographisches Institut, Mannheim, 1976.

[4.22] Gerald, C., *Applied Numerical Analysis*, Addison-Wesley, Reading, Mass., 1986.

[4.23] Acton, F., *Numerical Methods That Work*, Harper & Row, New York, 1970.

[4.24] Noble, B., *Numerical Methods*, Oliver and Boyd, London, 1964.

[4.25] Gruber, R., *Finite Elements in Physics*, North-Holland, Amsterdam, 1987.

[4.26] Breitschuh, U., Jurisch, R., *Die Finite Element Methode*, Akademie Verlag, Berlin, 1993.

[4.27] Schatz, A. et al, *Mathematical Theory of Finite and Boundary Element Methods*, Birkhaeuser, Basel, 1990.

[4.28] Hartmann, F., *Methode der Randelemente*, Springer, Berlin, 1987.

[4.29] Langtangen, H., *Computational Partial Differential Equations*, "Diffpack", Springer, Berlin, 1999.

[4.30] Brenner, S., *The Mathematical Theory of Finite Element Methods*, Springer, Berlin 1996.

[4.31] Cap, F., Collocation Method to Solve Boundary Value Problems for Arbitrary Boundaries, *Proc.13th IMACS World Congress in Computation and Applied Mathematics*, Dublin, 22-26 July 1991, Vol 1, p 292.

[4.32] Cap, F., Collocation Method to Solve Analytically Nonseparable Boundary Problems, Computational Physics Conference, Amsterdam, 10 - 13 Sept 1990, *Europhys. Confer.* 14, p 50.

[4.33] Cap, F., Nichttriviale homogene Randwertaufgabe der Laplace Gleichung, *Z. f. Angew. Math. Mech. (ZAMM)* 73, Nr 10 (1993), p 284.

[4.34] Menke, K., Loesung des Dirichlet-Problems bei Jordangebieten mit analytischem Rand durch Interpolation, *Monatsh. f. Math.* 80 (1975), 297-306.

[4.35] Menke, K., Bestimmung von Naeherungen fuer die konforme Abbildung mit Hilfe von stationaeren Punktsystemen, *Numer. Math.* 22 (1974), 111-117.

[4.36] Gutknecht, M., Numerical conformal mapping methods based on function conjugation, *J. Comp. Appl. Math.* 14 (1986), 31-77.

[4.37] Grisvard, P., *Elliptic Problems in non Smooth Domains*, Pitman, London, 1985.

[4.38] Eisenstat, S., On the Rate of Convergence of the Bergmann Vekua Method for the Numerical Solution of Elliptic Boundary Value Problems, *SIAM Numer. Anal.* 11, Nr 3 (1974), 654-680. For the Vekua method see [5.12].

[4.39] Reichel, L., Boundary Collocation in Féjer Points for Computing Eigenvalues and Eigenfunctions of the Laplacian, in *Approximation Theory*, Ed. Chui, Academic Press, Boston, 1986, and *J. Comp. Appl. Math.* 11 (1984), 175, 15 (1986), p 59.

[4.40] Cap, F., A New Collocation Method for Boundary Value Problems, p 118-119, in Modelling Collective Phenomena in Complex Systems, *Europhysics Abstracts, Europ. Phys. Soc.* (Granada Conference 2-5 Sept 1998), Paris, 1998, Vol 22F.

[4.41] Courant, R., Ein allgemeiner Satz zur Theorie der Eigenfunktionen selbstadjungierter Differentialausdruecke, *Nachr. Ges. Goettingen, Math. Phys.* Kl, 1923, 13 July.

[4.42] Cap, F., A New Numerical Method to Calculate the Vibrations of Clamped Kirchhoff-Plates of Arbitrary Form, *Computing* 61, (1998), 181-188.

[4.43] Mathews, J., *Numerical Methods*, Prentice Hall, Englewood Cliff, NY, 1987.

[4.44] Forsythe, G. et al, *Computer Methods for Mathematical Computations*, ibid., 1977.

[4.45] King. J., *Introduction to Numerical Computation*, McGraw-Hill, New York, 1984.

[5.1] Cap, F., Axisymmetric toroidal MHD equailibria of arbitrary cross section as an eigenvalue problem, *Sov. J. Plasma Phys.* 10(2), Mar-Apr 1984, 255.

[5.2] Cap, F., Toroidal Resonators and Waveguides of Arbitrary Cross Section, *IEEE Trans. Microwave Theory, Tech. MTT* 29, Nr 10 (1981), 1053-1059.

[5.3] Keil, R., Numerical Calculation of Electromagnetic Toroidal Resonators, *Arch. El. Ueb.* 38, Nr 1 (1984), 30-36, ibid., 37, Nr 11 (1983), 359-365.

[5.4] Cap, F., Khalil, S., Eigenvalues of Relaxed Axisymmetric Toroidal Plasmas of Arbitrary Aspect Ratio and Arbitrary Cross Section, *Nuclear Fusion* 29, Nr 7 (1989), 1166-1170.

[5.5] Knight, J., On Potential Problems Involving Spheroids inside a Cylinder, *Quart. J. Math. Oxford* 7 (1936), 127-133.

[5.6] Cap, F., Eine neue analytische Methode zur Loesung von Randwert-problemen der Neutronendiffusion bei Raendern, die nicht mit Koordi-natenflaechen uebereinstimmen, *Atomkernenergie-Kerntechnik* 44, Nr 4 (1984), 273-275.

[5.7] Buchanan, G. et al, Vibration of circular annular membranes with variable density, *J. Sound and Vibrations* 226, Nr 2 (1999), 379-382.

[5.8] Gottlieb, H., Exact solutions for vibrations of some annular membranes with inhomogeneous densities, *J. Sound and Vibrations* 233 Nr 1 (2000), 165-170.

[5.9] Conway, H., Leissa, A., A Method for Investigating Certain Eigenvalue Problems of the Buckling and Vibration of Plates, *J. Appl. Mech.* 27 (1960), 557-558.

[5.10] Conway, H., Dubil, J., Vibration Frequencies of Truncated-Cone and Wedge Beams, ibid. 32 (1965), 932-934

[5.11] Conway, H., Furnham, K., The Flexural Vibrations of Triangular, Rhombic and Parallelogram Plates and Some Analogies, *Int. J. Mech. Sci* 7 (1965), 811-816.

[5.12] Vekua, I., *New Methods for Solving Elliptic Equations*, John Wiley, New York, 1967.

[5.13] Fox, L. et al, Approximations and bounds for eigenvalues of elliptic operators, *SIAM J. Numer. Anal.* 4 Nr 1 (1967).

[5.14] Donnely, J., Eigenvalues of membranes with reentrant corners, *SIAM J. Numer. Anal.* 6 Nr 1(1969), 47.

[5.15] Kondratev, V. et al, Boundary value problems for partial differential equations in non-smooth domains, *Uspekhi Mat. Nauk* 38, Nr 2 (1983), 3-76.

[5.16] Kuttler, J. et al, Eigenvalues of the Laplacian in two dimensions, *SIAM Review* 26, Nr 2 (1984), 163-193.

[5.17] Reichel, L., On the computation of eigenvalues of the Laplacian by the boundary collocation method in *Approximation Theory* Vol 5, Academic Press, New York, 1986, (p 539-543).

[5.18] Still, G., Computable Bounds for Eigenvalues and Eigenfunctions of Elliptic Differential Operators, *Numer. Math.* 54, 201-223 (1988) (defect minimization method).

[5.19] Dauge, M., *Elliptic Boundary Value Problems on Corner Domains*, Springer, Berlin, 1989.

[5.20] Rvachev, V., Sheiko, T., R functions in boundary value problems in mechanics, *Appl. Mech. Reviews*, ASME 48, Nr 4 (1995), 151-187.

[5.21] Still, G., *Defektminimisierungsmethoden zur Loesung elliptischer Rand-
        und Eigenwertaufgaben*, Habilitationsschrift Trier, 1989.

[5.22] Cap, F., Groebner, W., New Method for the solution of the Deuteron
        Problem, and its application to a Regular Potential, *Il Nuovo Cimento*
        X, $\underline{1}$ (1955), 1211-1222.

[6.1]   Meirmanov, A., *The Stefan Problem*, de Gruyter, Berlin, 1992

[6.2]   Tang, D., Yang, J., A Free Moving Boundary Iteration Method for Un-
        steady Viscous Flow in Stenotic Elastic Tubes, *SIAM J., Scient. Com-
        puting* $\underline{21}$, Nr 4 (2000), 1370-1386.

[6.3]   Bechert, K., Ueber die Differentialgleichungen der Wellenausbreitung in
        Gasen, *Annalen d. Phys.* (5) $\underline{39}$, Nr 5 (1941), 357-372.

[6.4]   Preiswerk, E., *Anwendung gasdynamischer Methoden auf Wasserstroe-
        mungen mit freier Oberflaeche*, ETH, Thesis Nr 1010, Zuerich, Verlag
        Leemann, 1938.

[6.5]   Bollrich, G., *Technische Hydrodynamik*, Verlag fuer Bauwesen, Berlin,
        1996.

[6.6]   Tamana, T. et al, *Jap. J. Appl. Phys.* $\underline{14}$ (1975), 367.

[6.7]   Kumar, P. et al, *Phys.Rev.* $\underline{14B}$, (1976), 118.

[6.8]   Fermi, E., Ulam, S. et al, *Los Alamos Report LA-E1* (1940) and *Phys.
        Rev. Lett.* $\underline{37}$ (1976), 69.

[6.9]   Petviashvilli, S., *Sov. J. Plasma Phys.* $\underline{2}$ (1976), 246.

[6.10] Lamb, G., *Rev. Mod. Phys.* $\underline{43}$ (1971), 99.

[6.11] Gervais, J. et al, *Physics Reports* $\underline{23}$, C (1976), 240.

[6.12] Cap, F., Amplitude Dispersion and Stability of Nonlinear Weakly Dis-
        sipative Waves, *J. Math. Phys.* $\underline{13}$, Nr 8 (1972), 1126-1130.

[6.13] Akhmanov, R. et al, *Problems on Non-linear Optics*, Gordon and
        Breach, New York, 1972.

[6.14] Cap, F., Explosions in an Ionosphere with Finite Electric Conductivity
        in the Presence of a Uniform Magnetic Field, *Z. f. Flugwissenschaften* $\underline{18}$,
        Nr 2/3 (1970), 98-100, and *Proc. Sec. Intern. Colloquium on Gasdynam-
        ics of Explosions*, Novosibirsk, USSR, 24-29 August 1969 and *Astronaut.
        Acta* $\underline{15}$ Nr 5/6 (1970), 642.

[6.15] Karpman, V. et al, *Nonlinear Waves in Dispersive Media*, Pergamon
        Press, Oxford, 1975.

[6.16] Katayev, I., *Electromagnetic Shock Waves*, Iliffe, London, 1966.

[6.17] Cap, F., Ist der EMP eine elektromagnetische Stosswelle? *Elektrotechnik u. Maschinenbau (EuM)* 101, Nr 7 (1984), 332-336.

[6.18] Landau, L. et al, *Hydrodynamics*, Pergamon Press, Oxford.

[6.19] Ladurner, O., *Gasdynamische Durchrechnung der Gasstroemung im Zylinder der Verbrennungskraftmaschine*, Ph.D. Thesis, Innsbruck, Austria, 1954.

[6.20] Finsterwalder, Die Theorie der Gletscherschwankungen, *Z. Gletscherk.* 2 (1907), 81-193; for modern publications see the next references.

[6.21] Hutter. K., *Theoretical Glaciology*, Reidel Book Comp., Dordrecht-Boston, 1983.

[6.22] Goedert, G. and Hutter, K., Material update procedure for planar flow of ice with evolving anisotropy, *Annales of Glaciology* 30 (2000), 107-114.

[6.23] Hutter. K., Zryd, A. and Roethlisberger, H., On the numerical solution of Stefan problems in temperate ice, *Glaciology* 36 (1990), 41-48.

Appendix

Mathematica **commands used in this book.**

(Numbers indicate the pages where the command has been used.)

1. General commands

;	semicolon	3,10	allows the use of several commands on the same line
/a	divide	3	division by a
*	multiply	3	multiplication, also uses 1 space
+ -	add, minus	3	addition, subtraction
^	power	10	raise to a power
%	raw percent	10	gives the last result generated
%%			gives the result before the last
%%...%k times			gives the k-th previous result
=	set	3	assigns a to have value b
==	equal	3,30	x==y tests if x and y are equal
:=	define	30	assignment without actual value
/;	condition	248,268	the command before /; is only then executed, if the condition placed after /; is satisfied
/.	replacement	80,200	replace x by y: x/.−>y
<<	get	79,92,162	<<package load package
echo	$Packages	157,162	ask the computer which packages are available
AppendTo[$Path		157	typing this into the *Mathematica* window loads a package
command1//command2			execute command2 after the result of command1 has been obtained
N[Pi]		236	give the numerical value of (π), make a numerical operation
(*comment*)		11,50	commentary text within a code
Clear[y,...]		30,42	clears values and definitions
Messages		243	Off[General::spell], Off[Symbol::tag] switches messages off
?, ??		161	gives information about expression (full information)
Information[Symbol]		-	gives information about a symbol

Definition[Symbol]	-	gives definition
Join	237	concatenates lists together
Table (4.8.14)	58,79,236	builds lists, matrices, arrays
List	-	defines a list of objects
[[ ]]	236	pick out, fill element of Table
First	62,249	picks out the first element of an expression
Name giving	42,164	result of numerical solution of differential equation (2.2.6) gets name for later plotting or plot gets name
Print	67,245,248	print on screen "text", and numbers
Attributes[ ]		expresses properties and characteristics of a symbol [ ]
Options[ ]		gives list of default options of a symbol
Prolog	267	option for graphics
Timing	244	measures the time the operation needed

2. Input and Output

InputForm	69,82,133,243	expressions, results are given in a form
InputForm[%]	30	suitable for next input
OutputForm		the reaction, result, given by *Mathematica*
StandardForm	69	suitable for input, equation (2.6.9)
TraditionalForm	-	usual mathematical notation
MatrixForm	239	represents a matrix in the usual form in an array (lines and columns) (4.8.24)
FortranForm		gives an expression in FORTRAN notation
TeXForm	-	expression pretended suitable to be exported into a LaTeX-file (congratulations if you succeed)
Display	-	writes graphics and plots
FullForm	-	gives an expression in the interior form

3. Algebra

N[expression,n]	236	gives numerical value of expression with n-digit precision

Re[z] - gives real part of number z, works only for numbers

Im[z] - gives imaginary part

Accuracy[x] - gives the number of digits after the decimal point in the number x

Precision[x] - gives the number of digits in the number x

Expand to separate 99,102,135 helps to separate partial differential
part.d.equ. equations (3.1.79)

Expand 99,103,134,172 multiply out products and powers

Factor writes an expression as a product

Collect 199,200 groups together powers

Simplify 10 simplifies expression by algebraic transformations

FullSimplify tries more transformations

Cancel 161 cancels common factors between numerator and denominator, does not work for common functions

Apart - separate into terms with simple denominators

Together - put all terms over a common denominator

Rationalize 277 converts into exact rational numbers

Solve 119 solve equations with respect to the given variables

Sum 225 evaluates the sum

Chop 240 replaces numbers smaller than 10^{-10} by exact zero

Det[M] 30 calculates the determinant of a square matrix M

Matrix 30,242,239 to be filled later, a matrix should be defined $M=\{\{a_{11}, a_{12}\}\{a_{21}, a_{22}\}\}$

Array - builds a m×n matrix

LinearSolve 237 solves linear equations (4.8.16)

Roots[f[x]==0,k] - looks for the k-th root of a polynomial equation f(x)=0

Tensorproduct ** 237 Noncommutative multiplication or two spaces: (4.8.17), (4.8.28)

LUSolve 239 solves lu=b. (4.8.22)

LUFactor 239 lower and upper triangularization of a matrix (4.8.21)

Eliminate - eliminates variables between a set of simultaneous equations (see Solve)

4. Calculus

Table[Evaluate[y[x]/.%] {x,0.1,1.5,0.5}]	58,62	calculate and print on screen a list of $y(x), 0.1 \leq x \leq 1.5$, with a step 0.5, where $y(x)$ is the result of the last calculation
NIntegrate	243	numerical calculation of definite integral over x (and y)
Redefine a solution	62,63	if T is the name of the numerical solution of a differential equation and if $y(x)$ is the solution function to be evaluated for plotting a new function u[x_]=y[x]/. First [T] may be defined
FindRoot[f[x]==0,x,1.]	57,236	you may prefer to plot $f(x)$ to find a root (4.8.15)
FindRoot[f[x]==0, {x,xstart,xmin,xmax.}]		looks into xmin $\leq$ x $\leq$ xmax
FindRoot[f[x]==0, {x,{x_0,x_1}]		x_0,x_1 two initial points instead of a possibly missing derivative $f'(x)$
Series[f[x]{x,0,7}]	66	in equation (2.5.54), expands $f(x)$ into a power series about x=0 up to order 7
FourierTrigSeries [f[u],u,3]	79	expands $f(u)$ up to order 3 into a Fourier series with respect to u
Normal	79	generates a series or a polynomial without the remainder of order 6
FourierTransform	79	<<Calculus`FourierTransform` loads the package

5. Functions

x,x_,u[t_] (underline)	3,30	x is the formal variable x_ is the actual variable
f[x_]	3	creates a function that may be evaluated at any arbitrary value of x
u[x], u[x_]:=	30	compare these two expressions
InverseFunction[f]	162	expresses the inverse function of $f(x)$
Log	3	Log[x] designates the natural logarithm $\ln(x)$
Cosh, Sinh	3	cosh, hyperbolic cosine or sine
Sign [x]	-	gives -1 or $+1$, depends on whether $x<0$ or $x>0$. If $x = 0$, it gives 0.
Sec	11	secans function
Sech	306	hyperbolic secans

Sqrt	3	square root
Exp[3*x]	11,26	exponential function, $\exp(3x)$
AiryAi,AiryBi	188	$Ai(x),Bi(x)$ Airy functions
ArcTan	10	$\arctan(x)$
/;ArcTan	268	conditional use of arctan
N[ArcTan[x,y]]/;y>=0;	269	for correct definition
2*Pi+N[ArcTan[x,y]];<0; [x,y]];<0;	269	of the multivalued function
AppellF1	161	hypergeometric function of two variables (4.2.17)
BesselJ[n,x]	38,57	Bessel function of order n, $J_n(x)$
BesselY[n,x]	38	Neumann function of order n, $Y_n(x)$, $N_n(x)$
Bessel fractional order	265, 279	(5.2.10), (5.3.1)
BesselI[n,z]	42,187	modified Bessel function $I_n(z)$
BesselK[n,z]	42,187	and $K_n(z)$
ChebyshevT[n,x]	47	Chebyshev polynomial $T_n(x)$
ChebyshevU[n,x]	47	Chebyshev function $U_n(x)$
CosIntegral	43	cosine integral function $Ci(x)$ $$Ci(x) = -\int\limits_0^x \frac{\cos t}{t}\mathrm{d}t$$
SinIntegral	43	sine integral function $Si(x)$ $$Si(x) = \int\limits_0^x \frac{\sin t}{t}\mathrm{d}t$$
DiracDelta	115	defined on page 115
EllipticF	82,159	elliptic integral of the first kind
EllipticK	82	complete elliptic integral of the first kind $K(m)$
HermiteH[n,x]	43,49	Hermite polynomial $H_n(x;\alpha)$
HypergeometricU	50	confluent hypergeometric function $U(a,b,z)$
Hypergeometric2F1	43,113	$_2F_1(a;b;c;z)$
Hypergeometric1F1	43	$_1F_1(a;b;z)$ Kummer confluent function
JacobiAmplitude	82	$am(u,m)$, the inverse function of the elliptic integral of first kind
JacobiSN,CN	82	$sn(u,m)$, $cn(u,m)$ Jacobi elliptic functions
LaguerreL[n,x]	48,50	Laguerre polynomial $L_n(x)$
LaguerreL[n,a,x]	48	Laguerre polynomial $L_n^a(x)$
LegendreP[l,m,x]	46	associated Legendre polynomial $P_l^m(x)$
LegendreQ[n,m,z]	46	associated Legendre function $Q_n^m(z)$

6. Vector analysis

7. Plotting and Graphics

Plot Options:

		ratio height to width
Frame−>True	237,267	or−>False
PlotPoints−>100	83	number of points
PlotRange−>{0., 1.8}	124,276	restricts plot range
PlotStyle	-	style of lines (thickness, dashing ...)
ClipFill−> (for Plot3D only)	218	specifies whether to fill (or not to fill) areas of a plot outside the bounding box
PointSize[1/40]	276	the radius of the circle points should be 1/40 of the total width of the graph
AbsolutePointSize[6]	243,267	the filled circles representing a point should have a radius 6 points (1 point=1/72 inch)
PlotRegion (only 3D)		specifies what region a plot should fill
Prolog−>	243	gives options to a plot
Mesh (3D)	83	−>True or False to draw an x,y mesh
ViewPoint−>{1.3, −2.4, 2.}	278	default view, Figure 5.13
ViewPoint−>{0., −2.0, 0.}	278	sideways, Figure 5.14
ViewPoint−>{0., 0., 2.}	270,278	top view, Figure 5.12, Figure 5.8
PlotJoined (ListPlot)	267	−>True or False, whether the points should be joined by a line
Shading (3D)	83	−>True or False, to generate shading surface
ContourShading	83, 158	−>True or False, domain to be shaded or not
Contours−>10	-	number of contour lines
Contours−>Range [-5.,5.,0.45]	83	see page 83
ContourSmoothing−>10		−>4 is default
ContourLines−>True	-	or−>False
Contours−>{0}	167,277	draw only the contour line with value 0

Index

Printed in the United States
by Baker & Taylor Publisher Services